Lecture Notes in Mathematics

Edited by A. Dold and B. Eckmann

1317

J. Lindenstrauss V. D. Milman (Eds.)

Geometric Aspects of Functional Analysis

Israel Seminar (GAFA) 1986–87

Springer-Verlag
Berlin Heidelberg New York London Paris Tokyo

Editors

Joram Lindenstrauss
Hebrew University of Jerusalem
Givat Ram, 91 904 Jerusalem, Israel

Vitali D. Milman
School of Mathematical Sciences
The Raymond and Beverly Sackler Faculty of Exact Sciences
Tel Aviv University
Ramat Aviv, 69 978 Tel Aviv, Israel

Mathematics Subject Classification (1980): 46 B 20, 52 A 20, 52 A 40, 47 A 15, 28 D 99

ISBN 3-540-19353-7 Springer-Verlag Berlin Heidelberg New York
ISBN 0-387-19353-7 Springer-Verlag New York Berlin Heidelberg

© Springer-Verlag Berlin Heidelberg 1988
Printed in Germany

Printing and binding: Druckhaus Beltz, Hemsbach/Bergstr.
2146/3140-543210

FOREWORD

This is the third published volume of the proceedings of the Israel Seminar on Geometric Aspects of Functional Analysis. The first volume (1983-84) was published privately by Tel Aviv University and the second volume (1985-86) is volume 1267 of the Springer Lecture Notes in Mathematics. This volume covers the 1986-87 session of the seminar. As in previous years the seminar was partially supported by the Israel Mathematical Union.

The large majority of the papers in this volume are original research papers. As should be clear from the contents of this volume and the titles of the actual lectures delivered at the seminar, there was last year a strong emphasis on classical finite-dimensional convexity theory and its connection with Banach space theory. In recent years, it has become evident that the notions and results of the local theory of Banach spaces are useful in solving classical questions in convexity theory. We hope that the present volume will help in clarifying this point.

The papers in this volume are arranged in accordance with the order of their presentation at the seminar. The last four papers are based on talks delivered at conferences in the summer of 1987.

In preparing this volume we received invaluable help in editing and typing from Mrs. M. Hercberg. We are very grateful to her.

Joram Lindenstrauss, Vitali Milman

GFA

1986–1987

GAFA 1986-1987

List of Seminar Talks

October 31, **G. Schechtman**(Weizmann Institute), "Balancing vectors in the max-norm" (after J. Spencer).

November 16, **L. Tzafriri** (Hebrew University), "On invertibility of large matrices in Euclidean space" (joint work with J. Bourgain).

November 23, **L. Tzafriri** (Hebrew University), "Invertibility of large submatrices" (joint work with J. Bourgain).

December 5, 1. **N. Alon** (Tel Aviv University), "Splitting measures".
2. **C.J. Read** (Cambridge University, England), "Solution to the invariant subspace problem on a class of Banach spaces".

December 19, **Y. Yomdin** (Beersheva University), "Approximational versus topological complexity of functions on infinite dimensional spaces".

January 2, **B. Bollobás** (Cambridge University, England), "Random graphs".

January 9, 1. **N. Alon** (Tel Aviv University), "The number of d-polytopes".
2. **J. Bourgain** (IHES, Paris), "Non-linear convolutions".

January 18, 1. **J. Bourgain** (IHES), "Approximation of zonoids by zonotopes" (joint work with J. Lindenstrauss and V. Milman)
2. **S. Reisner** (Haifa University), "On two theorems of Lozanovskii concerning intermediate Banach lattices".

February 1, 1. **J. Bourgain** (IHES), "Ruzsa's problem on recurrent sets".
2. **Y. Peres** (Tel Aviv University), "Applications of Banach limits to recurrence and equidistribution".
3. **Y. Gordon** (Technion, Haifa), "A simple new proof of Milman's inequality".

March 27, 1. **V. Milman** (Tel Aviv University), "A new approach to isomorphic symmetrization".
2. **R. Schneider** (University of Freiburg, Germany), "On the Alexandrov-Fenchel inequality involving zonoids".

April 10, **E. Bachmat** (Hebrew University), "Dense packing of $\mathbb{R}^n$ with balls" (after Kabatyanskii and Levenstein).

May 1, 1. **G. Schechtman** (Weizmann Institute), "Almost isometrical natural embeddings in L_p" (after A. Arias).
2. **M. Meyer** (Université Paris VI, France), "Sections of the unit ball of ℓ_p^n", (joint work with A. Pajor).

May 5, 1. **M. Zippin** (Hebrew University), "Banach spaces with a separable dual".
2. **A. Pajor** (Université Paris VII, France), "Centroid bodies".

The following talks were presented at the GAFA-Seminar in the framework of the D.P. Milman Memorial Conference on Geometric Aspects of Functional Analysis, organized by Tel Aviv University, May 28 - May 31.

May 28, 1. **D. Amir** (Tel Aviv University), Opening address.
2. **V. Milman** (Tel Aviv University), "A very old joint result with D.P. Milman".

3. **M. Gromov** (IHES, France), "Topological spectra of normed spaces".
4. **J. Lindenstrauss** (Hebrew University), "Some useful facts about Banach spaces".
5. **W.B. Johnson** (Texas A & M University), "Homogeneous Banach spaces".
6. **L. Tzafriri** (Hebrew University), "Invertibility of operators on ℓ_p^n and other finite dimensional spaces".

May 29, 1. **G. Pisier** (Université Paris VI, Texas A & M University) "On the p-th variations of martingales and related Banach spaces".
2. **B. Maurey** (Université Paris VII), "ℓ^p structures in B-spaces and related unsolved questions".
3. **M. Marcus** (City University of NY), "On the distribution of the maximum of Gaussian processes".
4. **J. Bourgain** (IHES, University of Illinois), "Some new pointwise ergodic theorems".
5. **M. Gromov** (IHES), "Atiyah's results on convex bodies in R^n".

May 31, 1. **K. Ball** (Cambridge University, England), "Norms generated by logarithmically concave functions".
2. **I. Gohberg** (Tel Aviv University), "Band bases and applications".
3.**P. Milman** (University of Toronto), Transforming an analytic function to normal crossings by blowing up (a variant of resolution of singularities)".
4. **N. Tomczak-Jaegermann** (University of Alberta, Edmonton), "Entropies of convex bodies and operators".
5. **M. Talagrand** (C.N.R.S., Ohio State University), "Mixed volumes of convex bodies and Gaussian processes".

June 11, **J. Bourgain** (IHES), "Finite-dimensional homogeneous Banach spaces".

Table of Contents

THE INVARIANT SUBSPACE PROBLEM ON A CLASS OF NONREFLEXIVE BANACH SPACES, 1

C.J. Read

Department of Pure Mathematics
and Mathematical Statistics
Cambridge University
Cambridge, England

Abstract

In [1] we gave a fairly short proof that there is an operator on the space ℓ_1, without nontrivial invariant subspaces, and we conjectured that the same might be true of any space $\ell_1 \oplus W$ where W is a separable Banach space. This conjecture turns out to be true, and by proving it here we give the first example of a reasonably large class of Banach spaces for which the solution to the invariant subspace problem is known. This continues the sequence of counter-examples which began on an unknown Banach space (Enflo [2], Read [4], Beauzamy [3], (simplification of [2])), proceeded to the space ℓ_1 (Read [5,1]) and here continues with the case of any separable Banach space containing ℓ_1 as a complemented subspace. No counter-example is known to the author for a Banach space which does not contain ℓ_1.

1. Introduction

Here and elsewhere in this paper, we follow as closely as possible the style of the proof in [1]. The proof here is a modified version of [1], changed as little as we could in order to incorporate the Banach $\ell_1 \oplus W$, rather than ℓ_1. Throughout this paper, the underlying field may be either $\mathbb{R}$ or $\mathbb{C}$.

1.1. Let $(X, \|\cdot\|)$ denote the Banach space $\ell_1 \oplus W$, where W is a given separable Banach space. We quote the following result of R. Ovsepian and A. Pelcynski ([6]).

Let us choose, once and for all, such a biorthogonal system $(x_i), (x_i^*)$ for the given Banach space W. We may assume that $\|x_i\| = 1$ for all i (1.1.1), and if we take the liberty of renorming our space W with the equivalent norm

$$\|x\|' = \inf\left\{20\|y\| + \sum_1^\infty |x_i^*(z)| : y + z = x\right\} ,$$

we may assume that $\|x_i^*\| = 1$ also (1.1.2). Let $(g_i)_{i=0}^\infty$ be the unit vector basis of ℓ_1, and let F be the dense subspace of X spanned by $\{x_i\} \cup \{g_i\}$. The norm of any linear map $S : W \cap F \to Z$ is then bounded by

$$\|S\| \leq \sum_{i=1}^\infty \|Sx_i\| , \tag{1.1.3}$$

and indeed the right hand side is an upper bound for the nuclear norm of S.

We may also assume that the norm on $X = \ell_1 \oplus W$ is taken in the sense of ℓ_1, that is

$$\|x + w\|_X = \|x\|_{\ell_1} + \|w\|_W \tag{1.1.4}$$

for every $x \in \ell_1$, $w \in W$. Then the norm of an operator $S : F \to Z$ is bounded by

$$\|S\| \leq \max\left(\sup_{i \geq 0} \|Sg_i\| , \sum_{i=1}^\infty \|Sx_i\|\right) . \tag{1.1.5}$$

Definition 1.2. Let $\underset{\sim}{d} = (d_i)_{i=1}^\infty$ denote a strictly increasing sequence of positive integers. This sequence will be involved in most of the rest of our definitions, and it will be required to increase "sufficiently rapidly" in the following sense.

If $P(\underset{\sim}{d})$ is a proposition depending on the sequence $\underset{\sim}{d}$, we say (just as we did in [4, §1]) that $P(\underset{\sim}{d})$ is true "provided $\underset{\sim}{d}$ increases sufficiently rapidly", if there is a constant $c > 0$ and functions $f_i : \mathbb{N}^i \to \mathbb{N}$ $(i = 1, 2, 3, \ldots)$ such that the following holds.

If $d_1 > c_1$ and for all $i > 1$, we have

$$d_i > f_{i-1}(d_1, d_2, \ldots, d_{i-1})$$

then $P(\underset{\sim}{d})$ is true.

So if $P(\underset{\sim}{d})$ is true "provided $\underset{\sim}{d}$ increases sufficiently rapidly", there is a sequence $\underset{\sim}{d}$ such that $P(\underset{\sim}{d})$ is true; and if $P_1, P_2, \ldots, P_k$ are a finite collection of propositions, each of which is true provided $\underset{\sim}{d}$ increases sufficiently rapidly, then the proposition $\bigwedge_{i=1}^k P_i(\underset{\sim}{d})$ is true "provided $\underset{\sim}{d}$ increases sufficiently rapidly".

Given such a sequence $\underset{\sim}{d}$, let us write $a_i = d_{2i-1}$ and $b_i = d_{2i}$ for each $i \in \mathbb{N}$.

1.3. For the sake of convenience, we shall use the notation p.$\underline{d}$. to mean 'provided $\underline{d}$ increases sufficiently rapidly'.

1.4. We define $a_0 = 1$, $v_0 = 0$, $v_n = n(a_n + b_n)$ $(n \in \mathbb{N})$.

1.5. Let $|p|$ denote the sum of the absolute values of the coefficients of the polynomial p.

1.6. Given the space X with norm as in (1.1.4), with unit vector basis $(g_i)_{i=0}^{\infty}$ for ℓ_1, and a biorthogonal system $(x_i), (x_i^*)$ for W satisfying (1.1.0-2), we define a sequence $(f_i)_{i=0}^{\infty} \subset X$ such that $\{f_i\} = \{g_i\} \cup \{x_i\}$, in terms of the sequence $\underline{d}$. We define

$$f_i = \begin{cases} x_n & \text{if } i = a_n - 1 \text{ for some } n \in \mathbb{N}\ , \\ g_i & \text{if } 0 \le i < a_1 - 1, \\ g_{i-n} & \text{if } a_n \le i < a_{n+1} - 1 \text{ for some } n \in \mathbb{N}\ . \end{cases}$$

1.7. If $S \subset \{g_i : i \ge 0\}$ we write Π_S for the norm 1 projection $X \to \overline{\text{lin}}(S)$ such that $\Pi_S(w) = 0$ if $w \in W$, and

$$\Pi_S(g_i) = \begin{cases} g_i\ , & \text{if } \ g_i \in S \\ 0\ , & \text{if } \ g_i \notin S \end{cases}$$

2. Defining some operators on X, given the sequence $\underline{d}$

Let the sequence $\underline{d}$ be given. Certainly we have

$$0 = v_0 < a_1 + b_1 = v_1 < 2(a_2 + b_2) = v_2 < \dots\ ,$$

so the sets $\{0\}$, $(v_{n-1}, v_n]$ $(n = 1, 2, 3, \dots)$ form a partition of $\mathbb{Z}^+$. P.$\underline{d}$, we have $v_{n-1} < a_n$ for each $n \in \mathbb{N}$, so each interval $(v_{n-1}, v_n]$ can be partitioned into the disjoint union of the sets

$$(v_{n-1}, na_n]$$

and

$$(na_n, v_n]$$

Moreover, $(v_{n-1}, na_n]$ is the disjoint union of the sets

$$(v_{n-1}, a_n)\ \ ,\ \ [ra_n, ra_n + v_{n-r}] \qquad (r = 1, \dots, n)\ ,$$

and

$$(ra_n + v_{n-r}, (r+1)a_n) \qquad (r = 1, \dots, n-1)\ .$$

Furthermore, $(na_n, v_n]$ is the disjoint union of the sets

$$(na_n + rb_n, (r+1)(a_n + b_n)) \qquad (r = 0, \ldots, n-1)$$

and

$$[r(a_n + b_n), na_n + rb_n] \qquad (r = 1, \ldots, n) .$$

So $\mathbb{Z}^+$ is (p.$\mathcal{d}$) the disjoint union of all the intervals $\{0\}$, (v_{n-1}, a_n) $(n \in \mathbb{N})$, $[ra_n, ra_n + v_{n-r}]$ $(n \in \mathbb{N},\ r = 1, \ldots, n)$, $(ra_n + v_{n-r}, (r+1)a_n)$ $(n \geq 2,\ r = 1, \ldots, n-1)$, $(na_n + rb_n, (1+r)(a_n + b_n))$ $(n \in \mathbb{N},\ r = 0 \ldots n-1)$, and $[r(a_n + b_n), na_n + rb_n]$ $(n \in \mathbb{N},\ r = 1 \ldots n)$.

Bearing this in mind, we make the following definition.

Definition 2.1. Let the sequence $\mathcal{d} \subset \mathbb{N}$ be given. We shall show that, p.$\mathcal{d}$., there is a unique sequence $(e_i)_{i=0}^{\infty} \subset F$ with the following properties:

(2.1.0)
$$f_0 = e_0 .$$

(2.1.1) If integers r, n, i satisfy $0 < r \leq n$, $i \in [o, v_{n-r}] + ra_n$, we have

$$f_i = a_{n-r} \cdot (e_i - e_{i-ra_n}) .$$

(2.1.2) if integers r, n, i satisfy $1 \leq r < n$, $i \in (ra_n + v_{n-r}, (r+1)a_n)$, (respectively, $1 \leq n$, $i \in (v_{n-1}, a_n)$) then

$$f_i = 2^{(h-i)/\sqrt{a_n}} \cdot e_i$$

where $h = (r + \frac{1}{2})a_n$ (respectively, $h = \frac{1}{2}a_n$).

(2.1.3) If integers r, n, i satisfy $1 \leq r \leq n$, $i \in [r(a_n + b_n), na_n + rb_n]$, then

$$f_i = e_i - b_n \cdot e_{i-b_n} .$$

(2.1.4) If integers r, n, i satisfy $0 \leq r < n$, $i \in (na_n + rb_n, (r+1)(a_n + b_n))$ then

$$f_i = 2^{(h-i)/\sqrt{b_n}} \cdot e_i$$

where $h = (r + \frac{1}{2})b_n$.

Note 2.2. P.$\mathcal{d}$., Definition 2.1 gives $f_i = \sum_{j=0}^{i} \lambda_{ij} e_j$ uniquely for each $i \geq 0$, and since λ_{ii} is never zero, this linear relationship is invertible. So the e_i exist and are unique, and indeed

$$\operatorname{lin}\{e_i : 0 \leq i \leq n\} = \operatorname{lin}\{f_i : 0 \leq i \leq n\} = F_n \qquad \text{say} ,$$

for all $n \geq 0$. If $x = \sum_{i=0}^{N} \lambda_i e_i \in F$, we write $|x| = \sum_{i=0}^{N} |\lambda_i|$.

Definition 2.3. For each $n \geq 0$, let $I_n : (F_n, \|\cdot\|) \to (F_n, |\cdot|)$ be the identity map. We note that if $n = ma_m$ (respectively, $n = v_m$) for some $m \in I\!N$, then the norm $\|\cdot\|$ on F_n depends only on the underlying space W, the value of m, and the elements $\{a_i\}_{i=1}^m$, $\{b_i\}_{i=1}^{m-1}$ (respectively, $\{a_i\}_{i=1}^m$, $\{b_i\}_{i=1}^m$) of the sequence $\underset{\sim}{d}$. So for a given space X, we may choose functions $M_1 : I\!N^2 \to I\!N$ and $M_2 : I\!N^2 \to I\!N$ such that for all $m \in I\!N$, and for all $\underset{\sim}{d}$ such that definition 2.1 is meaningful, we have

$$\|I_{ma_m}\| \vee \|I_{ma_m}^{-1}\| \leq M_1(m, a_m) \tag{2.3.1}$$

and

$$\|I_{v_m}\| \vee \|I_{v_m}^{-1}\| \leq M_2(m, b_m) \,. \tag{2.3.2}$$

Definition 2.4. Let Q_m $(m \geq 1)$ denote the projection $F \to F_{ma_m}$ such that

$$Q_m(f_j) = \begin{cases} f_j \;, & 0 \leq j \leq ma_m \\ f_{j-ra_n+(r-n+m)a_m} \;, & j \in [0, v_{n-r}] + ra_n \;, \quad 0 < n-m < r \leq n \,, \\ 0 \;, & \text{otherwise} \,. \end{cases}$$

Definition 2.5. Let Q_m^0 $(m \geq 1)$ denote the projection $F \to F_{ma_m}$ such that

$$Q_m^0(f_j) = \begin{cases} f_j \;, & 0 \leq j \leq ma_m \\ -a_{n-r} \cdot e_{j-ra_n} \;, & j \in [0, v_{n-r}] + ra_n \;, \quad 0 < n-m < r \leq n \,. \\ 0 \;, & \text{otherwise} \,. \end{cases}$$

Definition 2.6. Let $P_{n,m}$ $(m > n \geq 1)$ be the operator $\tau_{n,m} \circ Q_m$, where $\tau_{n,m} : F_{ma_m} \to F_{ma_m}$,

$$\tau_{n,m}(e_j) = \begin{cases} e_j \;, & 0 \leq j < (m-n)a_m \\ 0 \;, & (m-n)a_m \leq j \leq ma_m \,. \end{cases}$$

Definition 2.7. Let $T : F \to F$ be the linear map such that $Te_i = e_{i+1}$ $(i \geq 0)$. Also, let I denote the identity map.

3. Continuity of Q_m and other projections

Lemma 3.1. *$\|Q_m\| \leq m$ for all m.*

Proof: By (1.1.5),

$$\|Q_m\| \leq \max\Big(\sup_{i \geq 0} \|Q_m g_i\| \,,\, \sum_{i=1}^{\infty} \|Q_m x_i\|\Big) \tag{3.1.1}$$

Now Definition 2.4 gives $Q_m f_i = f_j$ $(j \leq i)$ or zero for every i; so $Q_m f_i$ is a vector of norm at most one. Since $\{f_i\} \supset \{g_i\}$, we conclude that

$$\|Q_m\| \leq \max\left(1\ ,\ \sum_{i=1}^{\infty} \|Q_m x_i\|\right) . \tag{3.1.2}$$

But $Q_m x_i = Q_m f_{a_i - 1} = f_{a_i - 1}$ $(1 \leq i \leq m)$ or zero $(i > m)$(3.1.3) (by Definition 2.4). Hence,

$$\|Q^0_m\| \leq m .$$

Lemma 3.2. *$\|Q^0_m\| \leq a_m$ for all m, p.d..*

Proof: Again, (1.1.5) gives

$$\|Q^0_m\| \leq \max\left(\sup_{i \geq 0} \|Q^0_m g_i\|\ ,\ \sum_{i=1}^{\infty} \|Q^0_m x_i\|\right) .$$

Definition 2.5, like Definition 2.4 gives $Q^0_m(x_i) = x_i$ $(i \leq m)$ or zero $(i > m)$ (3.2.1), so

$$\|Q^0_m\| \leq \max\left(\sup_{i \geq 0} \|Q^0_m g_i\|\ ,\ m\right) . \tag{3.2.2}$$

Now $\{g_i\} \subset \{f_i\}$, and $Q^0_m(f_i)$ is f_i or zero unless $i \in [o, v_{n-r}] + ra_n$, $0 < n - m < r \leq n$, when

$$\begin{aligned} \|Q^0_m f_i\| &= a_{n-r} \cdot \|e_{i-ra_n}\| \\ &\leq a_{m-1} \cdot \|e_{i-ra_n}\| \\ &\leq a_{m-1} \cdot \|I^{-1}_{v_{m-1}}\| \cdot |e_{i-ra_n}| \end{aligned}$$

since $i - ra_n \leq v_{m-1}$;

$$\leq a_{m-1} \cdot M_2(m-1, b_{m-1}) \quad \text{by} \quad (2.3.2) .$$

Hence,

$$\begin{aligned} \|Q^0_m\| &\leq \max\left(a_{m-1} \cdot M_2(m-1, b_{m-1}), m\right) \\ &\leq a_m \qquad \text{p.d} . \end{aligned}$$

Lemma 3.3. *P.d, we have $\|P_{n,m}\|, \|\tau_{n,m}\| \leq \max(a_{n+1}, m)$, and $\|Q_n P_{n,m}\| \leq a_{n+1}$ for all $n < m$.*

Proof: Using (1.1.5), we observe that (2.6) gives

$$\begin{aligned} P_{n,m}(x_i) &= \tau_{n,m} Q_m(x_i) \\ &= \begin{cases} \tau_{n,m}(x_i) & ,\ i \leq m \\ 0 & ,\ i > m \end{cases} \\ &= \begin{cases} x_i & ,\ i \leq m \\ 0 & ,\ i > m \end{cases} \end{aligned} \tag{3.3.1}$$

hence,

$$\|P_{n,m}\| \le \max(\sup \|P_{n,m} g_i\|, m)$$
$$\le \max(\sup \|P_{n,m} f_i\|, m) \; .$$

Now $P_{n,m} f_i = \tau_{n,m} Q_m f_i$ and by (2.4), $Q_m f_i$ is either zero of f_j for some $0 \le j \le m a_m$. Thus

$$\sup \|P_{n,m} f_i\| \le \max_{0 \le j \le m a_m} \|\tau_{n,m} f_j\| \; . \tag{3.3.2}$$

But (2.6) and (2.1) give for all $0 \le j \le m a_m$,

$$\tau_{n,m} f_j = \begin{cases} -a_{m-r} \cdot e_{j-r a_m} \; , \; j \in [0, v_{m-r}] + r a_m \; , \; m-n \le r \le m \; , \\ f_j \quad \text{or zero , otherwise .} \end{cases}$$

Hence

$$\begin{aligned} \max_j \|\tau_{n,m} f_j\| &\le \max(1, \sup_{\substack{j \in [0, v_n] + r a_m \\ r \in [m-n, m]}} a_{m-r} \cdot \|e_{j-r a_m}\|) \\ &\le \max(1, a_n \cdot \sup_{i \in [0, v_n]} \|e_i\|) \\ &\le a_n \cdot M_2(n, b_n) \quad \text{by (2.3.2)} \; ; \end{aligned} \tag{3.3.3}$$

then

$$\begin{aligned} \|P_{n,m}\| &\le \max\left(a_n \cdot M_2(n, b_n), m\right) \\ &\le \max(a_{n+1}, m) \qquad \text{p.}\not{d}\text{. .} \end{aligned}$$

We note also that since Q_m acts as the identity on the domain of $\tau_{n,m}$, we have $\|\tau_{n,m}\| \le \|\tau_{n,m} Q_m\| = \|P_{n,m}\|$.

Using (1.1.5) to consider the operator $Q_n P_{n,m}$ we consider first the sum $\sum_{i=1}^{\infty} \|Q_n P_{n,m} x_i\|$. By (3.3.1),

$$\begin{aligned} Q_n P_{n,m} x_i &= \begin{cases} Q_n x_i & , \; i \le m \\ 0 & , \; i > m \; . \end{cases} \\ &= \begin{cases} x_i & , \; i \le n \\ 0 & , \; i > n \; , \quad \text{by (3.1.3) .} \end{cases} \end{aligned}$$

Thus $\sum_i \|Q_n P_{n,m} x_i\| = n$. Also, by (3.3.2), (3.3.3), (3.1), we have $\sup_i \|Q_n P_{n,m} f_i\| \le \|Q_n\| \cdot \max_i \|P_{n,m} f_i\| \le n a_n M_2(n, b_n)$. Therefore, (1.1.5) gives

$$\begin{aligned} \|Q_n P_{n,m}\| &\le \max\left(\sup_i \|Q_n P_{n,m} g_i\| \; , \; \sum_i \|Q_n P_{n,m} x_i\|\right) \\ &\le \max\left(n a_n M_2(n, b_n), n\right) \\ &\le a_{n+1} \qquad \text{p.}\not{d}\text{. .} \end{aligned}$$

4. Continuity of T

Lemma 4. *Let $\eta > 0$ be given. The following are true p.d.;*

(1) $\|T\| \le 1+\eta$, *indeed* $\|T|_{\ell_1}\| \le 1+\eta$ *while* $T|_W$ *has nuclear norm at most* η.

(2) $\|T^{a_n+b_n}(I - Q_n^0)\| < 1+\eta$ *for all* $n \in \mathbb{N}$.

4.1 Proof of Lemma 4(1).

By (1.1.5) it is sufficient to show that $\|T|_{\ell_1}\| < 1+\eta$, while $\sum_{i=1}^{\infty} \|Tx_i\| < \eta$ (4.1.1). (4.1.1) is easy to prove; by (2.1), (1.6.1) we have

$$x_i = f_{a_i-1} = 2^{(1-\frac{1}{2}a_i)/\sqrt{a_i}} \cdot e_{a_i-1} ;$$

$$Tx_i = 2^{(1-\frac{1}{2})/\sqrt{a_i}} \cdot e_{a_i} = 2^{(1-\frac{1}{2}a_i)/\sqrt{a_i}} \cdot (f_0 + a_{i-1}^{-1} \cdot f_{a_i})$$

(by (2.1));

$$\|Tx_i\| \le 4 \cdot 2^{-\frac{1}{2}\sqrt{a_i}}$$

since $\|f_j\| = 1$ for all j; then $\sum_{i=1}^{\infty} \|Tx_i\| \le \sum_{i=1}^{\infty} 2^{-\frac{1}{2}\sqrt{a_i}} \le \eta$ p.d. (4.1.2)

We now examine the behavior of T on ℓ_1. Let $R, S \subset \mathbb{Z}^+$ be as follows. $R = \bigcup_{i=0}^{4} R_i$ and $S = \bigcup_{i=1}^{4} S_i$ where

$$R_0 = \{o\} ,$$

$$R_1 = \{ra_n + v_{n-r} : 0 < r \le n\} ;$$

$$R_2 = \{ra_n - 1 : 2 \le r \le n\} ;$$

$$R_3 = \{na_n + rb_n : 1 \le r \le n\};$$

$$R_4 = \{r(a_n + b_n) - 1 : 1 \le r \le n\} ;$$

$$S_1 = \bigcup_{\substack{0<r\le n \\ n\ge 1}} ([0, v_{n-r}) + ra_n) ;$$

$$S_2 = \bigcup_{n=1}^{\infty} S_{2,n} \quad \text{where} \quad S_{2,n} = \Big(\bigcup_{r=1}^{n-1} (ra_n + v_{n-r}, (r+1)a_n - 1)\Big) \cup (v_{n-1}, a_n - 1)$$

$$S_3 = \bigcup_{\substack{1\le r\le n \\ n\ge 1}} \big[r(a_n + b_n), na_n + rb_n\big] ;$$

$$S_4 = \bigcup_{n=1}^{\infty} S_{4,n} \quad \text{where} \quad S_{4,n} = \bigcup_{r=0}^{n-1} (na_n + rb_n, (r+1)(a_n + b_n) - 1) .$$

Note that, p.$\not{d}$., $R \cup S = \mathbb{Z}^+ \backslash \{a_n - 1 : n \geq 1\}$ so for all i, $g_i = f_j$ for some $j \in R \cup S$. Thus

$$\sup_i \|Tg_i\| = \sup_{j \in R \cup S} \|Tf_j\| . \tag{4.1.3}$$

We now investigate the right hand side of (4.1.3), considering first those $j \in S$.

If $i \in S$ then by Definitions 2.1, 2.7, we have

$$Tf_i = \begin{cases} f_{i+1} & , \quad i \in S_1 \cup S_3 \\ 2^{1/\sqrt{a_n}} \cdot f_{i+1} & , \quad i \in S_{2,n} \\ 2^{1/\sqrt{b_n}} \cdot f_{i+1} & , \quad i \in S_{r,n} . \end{cases}$$

So $\sup\limits_S \|Tf_i\| = 2^{1/\sqrt{a_1}} < 1 + \eta$ p.$\not{d}$..

Definitions 2.7 and 2.1 give the following slightly more elaborate action for T on those f_i with $i \in R$ (the 'bad cases'). If $i \in R$, then

$$Tf_i = \begin{cases} 2^{(1-\frac{1}{2}a_1)/\sqrt{a_1}} \cdot f_i \ , \ i = 0 \in R_0 \\ a_{n-r} \cdot (\varepsilon_1 f_{i+1} - \varepsilon_2 f_{v_{n-r}+1}) \ , \ i = ra_n + v_{n-r} \in R_1 \ , \text{ where} \\ \qquad \varepsilon_2 = 2^{(1+v_{n-r}-\frac{1}{2}a_{n-r+1})/\sqrt{a_{n-r+1}}} \ , \text{ and} \\ \qquad \varepsilon_1 = \begin{cases} 2^{(1+v_{n-r}-\frac{1}{2})/\sqrt{a_n}} & , r < n \ , \\ 2^{(1+na_n-\frac{1}{2}b_n)/\sqrt{b_n}} & , r = n \ . \end{cases} \\ 2^{(1-\frac{1}{2}a_n)/\sqrt{a_n}} \cdot \left(f_0 + \dfrac{1}{a_{n-r}} f_{i+1} \right) \ , \ i = ra_n - 1 \in R_2 \ , \ 2 \leq r \leq n \\ \varepsilon_1 f_{i+1} - b_n \varepsilon_2 f_{i+1-b_n} \ , \ i = na_n + rb_n \in R_3 \ , \text{ where} \\ \qquad \varepsilon_2 = 2^{(na_n+1-\frac{1}{2}b_n)/\sqrt{b_n}} \ , \\ \qquad \varepsilon_1 = \begin{cases} \varepsilon_2 \ , \ r < n \ , \\ 2^{(v_n+1-\frac{1}{2}a_{n+1})/\sqrt{a_{n+1}}} \ , \ r = n \ . \end{cases} \\ 2^{(1-(r+1)a_n-\frac{1}{2}b_n)/\sqrt{b_n}} \cdot \left(\sum\limits_{j=0}^{r-1} b_n^j f_{i-jb_n} + b_n^r \cdot \left(f_0 + \dfrac{1}{a_{n-r}} f_{ra_n} \right) \right) \ , \\ \qquad i = r(a_n + b_n) - 1 \in R_4 \ . \end{cases} \tag{4.1.4}$$

It is not hard to see that, p.$\not{d}$., these values have in common that they are all of norm less than η, for any fixed $\eta > 0$. So $\sup\limits_R \|Tf_i\| < \eta$ p.$\not{d}$.. But then (4.1.3) gives $\sup\limits_i \|Tg_i\| \leq 1 + \eta$ (4.1.5), and then (1.1.5) gives

$$\begin{aligned} \|T\| &\leq \max\left(\sup_i \|Tg_i\| \ , \ \sum_i \|Tx_i\|\right) \\ &\leq \max(1 + \eta, \eta) = 1 + \eta \ . \end{aligned}$$

by (4.1.2), (4.1.5). This concludes the proof of Lemma 4(1).

4.2 Proof of Lemma 4(2).

From now on we will assume that $\underset{\sim}{d}$ increases sufficiently rapidly that $\|T\| \leq 2$. Let us show that p.$\underset{\sim}{d}$.,

$$\sum_{i=1}^{\infty} \|T^{a_n+b_n} \circ (I - Q_n^0)x_i\| < \eta \quad \text{for all} \quad n \in \mathbb{N} \tag{4.2.1}$$

and

$$\sup_{i \in R \cup S} \|T^{a_n+b_n} \circ (I - Q_n^0)f_i\| \leq 1 + \eta \quad \text{for all} \quad n \in \mathbb{N}\ . \tag{4.2.2}$$

This last supremum is equal to $\sup_i \|T^{a_n+b_n} \circ (I - Q_n^0)g_i\|$ by the argument of §4.1, so (1.1.5), (4.2.1) and (4.2.2) together give $\|T^{a_n+b_n} \circ (I - Q_n^0)\| \leq \max(\eta, 1+\eta) = 1+\eta$ for each n. We first establish (4.2.1). By (3.2.1),

$$(I - Q_n^0)x_i = \begin{cases} x_i & , \quad i > n \\ 0 & , \quad i \leq n \end{cases}$$

so

$$\sum_{i=1}^{\infty} \|T^{a_n+b_n} \circ (I - Q_n^0)x_i\| = \sum_{i=n+1}^{\infty} \|T^{a_n+b_n}x_i\| \leq \sum_{i=n+1}^{\infty} 2^{a_n+b_n-1}\|Tx_i\|$$

since $\|T\| \leq 2$ by hypothesis. The inequality which leads to (4.1.2) gives $\|Tx_i\| \leq 4.2^{-\sqrt{a_i}}$, hence

$$\sum_{i=1}^{\infty} \|T^{a_n+b_n} \circ (I - Q_n^0)x_i\| \leq 2^{a_n+b_n+1} \sum_{i=n+1}^{\infty} 2^{-\sqrt{a_i}} < 2$$

for all n, p.$\underset{\sim}{d}$..

Now we prove (4.2.2). We consider the effect of the operator $T^{a_n+b_n} \circ (I - Q_n^0)$ on each f_i, considering several different cases.

Case 0. If $i \in [0, na_n]$ then $Q_n^0 f_i = f_i$; hence $(I - Q_n^0)f_i = 0$.

Case 1. If for some $m > n$ we have $i \in [0, v_{m-r}] + ra_m$, $0 < r \leq m$, let us consider separately the cases $m - n < r$ and $m - n \geq r$. If $m - n < r$ then by (2.5), $Q_n^0(f_i) = -a_{m-r} \cdot e_{i-ra_m}$; so by (2.1), $(I - Q_n^0)f_i = a_{m-r} \cdot e_i$. Thus

$$T^{a_n+b_n} \circ (I - Q_n^0)f_i = a_{m-r} \cdot e_{i+a_n+b_n}\ .$$

Since $m - r < n$ we will have $i + a_n + b_n > ra_m + v_{m-r}$, hence

$$\|e_{i+a_n+b_n}\| = \|T^{a_n+b_n-(v_{m-r}+ra_m+1-i)}e_{ra_m} + v_{m-r} + 1\|$$

$$\leq 2^{a_n+b_n}\|e_{1+ra_m+v_{m-r}}\|$$

$$= \begin{cases} 2^{a_n+b_n} \cdot 2^{(1+v_{m-r}-\frac{1}{2}a_m)/\sqrt{a_m}} & , \quad r < m \\ 2^{a_n+b_n} \cdot 2^{(1+ma_m-\frac{1}{2}b_m)/\sqrt{b_m}} & , \quad r = m , \end{cases}$$

(by (2.1));

$$\leq 2^{-\frac{1}{4}\sqrt{a_m}}$$

for all $0 < m - n < r$, p.d.. Then

$$\|T^{a_n+b_n} \circ (I - Q_n^0)f_i\| \leq a_{m-r} \cdot 2^{-\frac{1}{4}\sqrt{a_m}} < 1 \qquad \text{p.d.}$$

If $m - n \geq r$ then $Q_n^0 f_i = 0$ by (2.5); so

$$T^{a_n+b_n} \circ (I - Q_n^0)f_i = T^{a_n+b_n} f_i = a_{m-r} \cdot (e_{i+a_n+b_n} - e_{i-ra_m+a_n+b_n}) \,.$$

If $i+a_n+b_n \leq ra_m+v_{m-r}$, then the right hand side is just $f_{i+a_n+b_n}$. If $i+a_n+b_n > ra_m+v_{m-r}$ then

$$\|e_{i+a_n+b_n} - e_{i-ra_m+a_n+b_n}\| = \|T^{i+a_n+b_n-ra_m-v_{m-r}-1}(e_{ra_m+v_{m-r}+1} - e_{v_{m-r}+1})\|$$

$$\leq 2^{a_n+b_n}(\|e_{ra_m+v_{m-r}+1}\| + \|e_{v_{m-r}+1}\|)$$

$$= 2^{a_n+b_n}(2^{(1+v_{m-r}-\frac{1}{2}a_m)/\sqrt{a_m}} + 2^{(1+v_{m-r}-\frac{1}{2}a_{m-r+1})/\sqrt{a_{m-r+1}}})$$

(by (2.1.2))

$$\leq 2^{a_n+b_n-\frac{1}{4}\sqrt{a_{m-r+1}}} \qquad \text{p.d.}\,.$$

Then

$$\|T^{a+b_n} \circ (I - Q_n^0)f_i\| \leq a_{m-r} \cdot 2^{a_n+b_n-\frac{1}{4}\sqrt{a_{m-r+1}}}$$

$$< 1 \qquad \text{p.d.}\,,$$

since $m - r + 1 > n$. This concludes Case 1; in all subcases we have found that $\|T^{a_n+b_n} \circ (I - Q_n^0)f_i\| \leq 1$.

Case 2. If for some $m > n$ we have $i \in (ra_m + v_{m-r}, (r+1)a_m)$ (respectively, $(v_{m-1}, a_m - 1)$) where $1 \leq r < m$, then $Q_n^0(f_i) = 0$ and by (2.1.2),

$$f_i = 2^{(h-i)/\sqrt{a_m}} \cdot e_i$$

where $h = (r + \frac{1}{2})a_m$ (respectively, $h = \frac{1}{2}a_m$). If $i + a_n + b_n < (r+1)a_m$ (respectively, $i + a_n + b_n < a_m$) then (2.1.2) gives

$$T^{a_n+b_n} f_i = 2^{(h-i)/\sqrt{a_m}} \cdot e_{i+a_n+b_n}$$

$$= 2^{(a_n+b_n)/\sqrt{a_m}} \cdot f_{i+a_n+b_n} \,,$$

so

$$\|T^{a_n+b_n} \circ (I - Q_n^0) f_i\| = \|T^{a_n+b_n} f_i\| = 2^{(a_n+b_n)/\sqrt{a_m}}$$
$$\leq 1 + \eta$$

for all $n < m$ p.$\not{d}$.. If $i + a_n + b_n \geq (r+1)a_m$ (respectively, $\geq a_m$), then

$$T^{a_n+b_n} f_i = 2^{(h-i)/\sqrt{a_m}} \cdot e_{i+a_n+b_n}$$
$$= 2^{(h-i)/\sqrt{a_m}} \cdot T^{i+a_n+b_n-(r+1)a_m} \cdot e_{(r+1)a_m}$$
$$\text{(respectively,} \quad 2^{(h-i)/\sqrt{a_m}} \cdot T^{i+a_n+b_n-a_m} \cdot e_{a_m})$$

But for all $1 \leq r \leq m$,

$$e_{ra_m} = e_0 + a_{m-r}^{-1} \cdot f_{ra_m} ;$$
$$\|e_{ra_m}\| \leq 2 ; \tag{4.2.3}$$

thus

$$\|T^{a_n+b_n} f_i\| \leq 2^{(h-i)/\sqrt{a_m}} \cdot 2^{a_n+b_n} \cdot \max_{1 \leq r \leq m} \|e_{ra_m}\|$$
$$\leq 2^{(h-i)/\sqrt{a_m}} \cdot 2^{a_n+b_n+1} .$$

But $i + a_n + b_n \geq (r+1)a_m$ (resp., $\geq a_m$); so $i - h \geq \frac{1}{2}a_m - a_n - b_n$. Therefore

$$\|T^{a_n+b_n} \circ (I - Q_n^0) f_i\| = \|T^{a_n+b_n} f_i\| \leq 2^{a_n+b_n+1} \cdot 2^{(a_n+b_n-\frac{1}{2}a_m)/\sqrt{a_m}}$$
$$\leq 1$$

for all $n < m$ p.$\not{d}$. This concludes Case 2. In all subcases we have found that $\|T^{a_n+b_n} \circ (I - Q_n^0) f_i\| \leq 1 + \eta$.

Case 3. If for some $m \geq n$ we have $i \in [r(a_m + b_m), ma_m + rb_m]$ then $Q_n^0 f_i = 0$ so we consider $T^{a_n+b_n} f_i = T^{a_n+b_n}(e_i - b_m \cdot e_{i-b_m}) = e_{i+a_n+b_n} - b_m \cdot e_{i+a_n+b_n-b_m}$. If $m > n$ and $i + a_n + b_n \leq ma_m + rb_m$, then by (2.1.3), this is just $f_{i+a_n+b_n}$. This is also true if $m = n$ and $i + a_n \leq na_n + rb_n$. If $m > n$ and $i + a_n + b_n > ma_m + rb_m$, then

$$\|e_{i+a_n+b_n} - b_m \cdot e_{i+a_n+b_n-b_m}\|$$
$$= \|T^{i+a_n+b_n-(ma_m+rb_m+1)}(e_{ma_m+rb_m+1} - b_m \cdot e_{ma_m+(r-1)b_m+1})\|$$
$$\leq 2^{a_n+b_n}(1 + b_m) \cdot \max_{0 \leq r \leq m} \|e_{ma_m+rb_m+1}\| .$$

By (2.1.4), (2.1.2) we have for all $0 \leq r \leq m+1$,

$$e_{ma_m+rb_m+1} = \varepsilon(r,m) f_{ma_m+rb_m+1}$$

where

$$\varepsilon(r,m) = \begin{cases} 2^{(ma_m - \frac{1}{2}b_m + 1)/\sqrt{b_m}} & , \quad r < m \\ 2^{\left(1+ma_m+rb_m-\frac{1}{2}a_{m+1}\right)/\sqrt{a_{m+1}}} & , \quad r = m, m+1 \, . \end{cases} \tag{$*$}$$

So we have

$$\|e_{i+a_n+b_n} - b_m \cdot e_{i+a_n+b_n-b_m}\| \le 2^{a_n+b_n} \cdot (1+b_m) \cdot \varepsilon(r,m) \le 1$$

for all $n < m$, $0 \le r \le m$ p.d.. If $m = n$ and $i + a_n > na_n + rb_n$ then

$$T^{a_n+b_n} f_i = e_{i+a_n+b_n} - b_n \cdot e_{i+a_n}$$
$$= T^{i+a_n-(na_n+rb_n+1)}(e_{na_n+(r+1)b_n+1} - b_n \cdot e_{na_n+rb_n+1})$$

$$\|T^{a_n+b_n} f_i\| \le 2^{i+a_n-(na_n+rb_n+1)} \cdot (1+b_n) \cdot \max_{1 \le r \le n+1} \varepsilon(r,n)$$
$$\le 2^{a_n} \cdot (1+b_n) \cdot \max_{1 \le r \le n+1} \varepsilon(r,n)$$

(since $i \le na_n + rb_n$)

$$\le 1$$

for all n p.d., in view of ($*$). This concludes Case 3, and, once again, we have

$$\|T^{a_n+b_n} \circ (I - Q_n^0) f_i\| = \|T^{a_n+b_n} f_i\| \le 1$$

in all cases.

Case 4. If $i \in \left(ma_m + (r-1)b_m, r(a_m + b_m)\right)$ for some $m \ge n$, $1 \le r \le m$ we note first that $Q_n^0 f_i = 0$; so we are considering $T^{a_n+b_n} f_i$. If $m = n$ and $r < n$ then (2.1.4) gives very conveniently

$$T^{a_n+b_n} f_i = 2^{a_n/\sqrt{b_n}} f_{i+a_n+b_n}$$

for all such i; so

$$\|T^{a_n+b_n} f_i\| = 2^{a_n/\sqrt{b_n}} < 1 + \eta \qquad \text{p.d.}\,.$$

If $m = n$ and $r = n$ then $i + a_n + b_n > v_n$ hence

$$\|T^{a_n+b_n} f_i\| = \|2^{(h-i)/\sqrt{b_n}} \cdot e_{i+a_n+b_n}\|$$

where $h = \left(n - \frac{1}{2}\right) b_n$,

$$\le 2^{a_n+b_n} \cdot 2^{(h-i)/\sqrt{b_n}} \cdot \|e_{1+v_n}\|$$
$$= 2^{a_n+b_n} \cdot 2^{(h-i)/\sqrt{b_n}} \cdot 2^{(1+v_n-\frac{1}{2}a_{n+1})/\sqrt{a_{n+1}}}$$
$$< 1 \quad \text{for all} \quad n, i > 0 \quad \text{p.d.}\,.$$

If $m > n$ and $i + a_n + b_n$ is still in the interval $\big(ma_m + (r-1)b_m, r(a_m + b_m)\big)$ then (2.1.4) gives

$$\begin{aligned}\|T^{a_n+b_n} f_i\| &= \|2^{(a_n+b_n)/\sqrt{b_m}} f_{i+a_n+b_n}\| \\ &= 2^{(a_n+b_n)/\sqrt{b_m}} < 1 + \eta\end{aligned}$$

for all $n < m$ p.$\mathcal{d}$. If on the other hand $i + a_n + b_n \geq r(a_m + b_m)$ then familiar arguments give

$$\|T^{a_n+b_n} f_i\| = 2^{(h-i)/\sqrt{b_m}} \cdot \|e_{i+a_n+b_n}\|$$

(where $h = (r - \frac{1}{2})b_m$)

$$\begin{aligned}&\leq 2^{a_n+b_n} \cdot 2^{(h-i)/\sqrt{b_m}} \cdot \|e_{r(a_m+b_m)}\| \\ &= 2^{a_n+b_n} \cdot 2^{(h-i)/\sqrt{b_m}} \cdot \|b_m^r e_{ra_m} + \sum_{j=1}^{r} b_m^{r-j} f_{jb_m+ra_m}\| \\ &\leq 2^{a_n+b_n+(h-i)/\sqrt{b_m}} \cdot \big(2b_m^r + \sum_{j=1}^{r} b_m^{r-j}\big)\end{aligned}$$

for we noted (4.2.3) that $\|e_{ra_m}\| \leq 2$. Now $i + a_n + b_n \geq r(a_m + b_m) \geq h + \frac{1}{2}b_m$ so $i - h \geq \frac{1}{4}b_m$ p.$\mathcal{d}$.. Then

$$\|T^{a_n+b_n} f_i\| \leq 2^{a_n+b_n-\frac{1}{4}\sqrt{b_m}} \cdot (m+2)b_m^m < 1$$

for all $n < m$ p.$\mathcal{d}$.. This concludes Cases 0 to 4, which exhaust all possibilities and establish that p.$\mathcal{d}$.,

$$\|T^{a_n+b_n} \circ (I - Q_n^0) f_i\| \leq 1 + \eta$$

for all $n \in \mathbb{N}$, $i \in R \cup S$. Hence, (4.2.2) is established, and therefore, we have proved Lemma 4.2.

5. Showing that T has no non-trivial closed invariant subspaces

For any $n < m$, let $K_{n,m} \subset F_{ma_m}$ be the set

$$\left\{x \in F_{ma_m} : \|x\| \leq a_m \quad , \quad \|\tau_{n,m}(x)\| \geq \frac{1}{a_m}\right\} .$$

Now $K_{n,m}$ is a compact subset of F_{ma_m}, depending only on the choice of n, m, $(a_i)_{i=1}^n$, $(b_i)_{i=1}^{m-1}$. Let T_m denote the "truncated" version of T on F_{ma_m};

$$T_m : F_{ma_m} \to F_{ma_m} : e_i \longrightarrow \begin{cases} e_{i+1} & , \quad i < ma_m \\ 0 & , \quad i = ma_m \, . \end{cases}$$

Given $x \in K_{n,m}$, write $x = \sum_{i=\alpha}^{ma_m} \lambda_i e_i$ where $\lambda_\alpha \neq 0$. Then

$$\mathrm{lin}\{T_m^r x : a_m \leq r \leq ma_m\} = \mathrm{lin}\{e_{\alpha+a_m}, e_{\alpha+a_m+1}, \ldots, e_{ma_m}\} .$$

In particular, since $r_{n,m}(x) \neq 0$, we have $\alpha < (m-n)a_m$, so that certainly

$$e_{(m-n+1)a_m} \in \operatorname{lin}\{T_m^r x : a_m \leq r \leq ma_m\} .$$

Since $K_{n,m}$ is compact, there are a finite number $p_1, p_2, \ldots, p_K$ of polynomials $p_j(t) = \sum_{i=a_m}^{ma_m} \lambda_{ij} t^i$ such that for all $x \in K_{n,m}$ there is a j for which

$$\|p_j(T_m)_x - e_{(m-n+1)a_m}\| < 1/a_m .$$

Writing $N = \max |p_j|$, note that we can write $N \leq N_1(m, a_m)$ for a suitable function $N_1 : \mathbb{N}^2 \to \mathbb{N}$ (since things depend only on the choices of n, m, $a_1 \ldots a_m$, $b_1 \ldots b_{m-1}$). Since $\|e_{(m-n+1)a_m} - e_0\| = a_{n-1}^{-1}$ (in view of (2.1.1)), we have the following lemma.

Lemma 5.1. *There is a function $N_1 : \mathbb{N}^2 \to \mathbb{N}$ with the following property. Provided $\underline{d}$ increases sufficiently rapidly, for all $n < m$ and $x \in K_{n,m}$ there is a polynomial p with $|p| \leq N_1(m, a_m)$, $p(t) = \sum_{a_m}^{ma_m} \lambda_i t^i$ and*

$$\|p(T_m)x - e_0\| \leq \frac{1}{a_m} + \frac{1}{a_{n-1}} .$$

With the notation of Lemma 5.1, let us take such a polynomial p, and write $q(t) = t^{b_m} p(t)/b_m$. Let us consider the vector $q(T)x$. For all $i \in [a_m + b_m, ma_m + b_m]$, we have $f_i = e_i - b_m \cdot e_{i-b_m}$ so if we write

$$p(T_m)x = \sum_{j=a_m}^{ma_m} \lambda_j e_j$$

then

$$\left\| \frac{T^{b_m}}{b_m} p(T_m)x - p(T_m)x \right\| = \left\| \sum_{j=a_m}^{ma_m} \lambda_j f_{j+b_m}/b_m \right\| \tag{5.1.1}$$

$$= \frac{1}{b_m} \sum_{j=a_m}^{ma_m} |\lambda_j| = \frac{1}{b_m} |p(T_m)x| \leq \frac{1}{b_m} \cdot |p| \cdot |x|$$

$$\leq \frac{1}{b_m} \cdot N_1(m, a_m) \cdot M_1(m, a_m) \cdot a_m$$

(since $x \in K_{n,m} \subset F_{ma_m}$, $\|x\| \leq a_m$)

$$\leq \frac{1}{a_m} \qquad \text{p.}\underline{d} .$$

Furthermore,

$$T^{b_m}(p(T) - p(T_m))x \in T^{b_m} \operatorname{lin}\{e_j : ma_m < j \leq 2ma_m\}$$

$$= \operatorname{lin}\{e_j : b_m + ma_m < j \leq b_m + 2ma_m\} ,$$

and provided that $b_m + 2ma_m < 2(a_m + b_m)$ we have by (2.1.4) that, for all $j \in (b_m + ma_m, b_m + 2ma_m]$,

$$\|e_j\| = 2^{(j-\frac{3}{2}b_m)/\sqrt{b_m}} < 2^{-\frac{1}{3}\sqrt{b_m}} \qquad \text{p.}\mathscr{d}. \, .$$

So

$$\begin{aligned}\left\|\frac{T^{b_m}}{b_m}\cdot(p(T)-p(T_m))x\right\| &\le 2^{-\frac{1}{3}\sqrt{b_m}}\cdot\left|\frac{T^{b_m}}{b_m}(p(T)-p(T_m))x\right| \\ &\le 2^{-\frac{1}{3}\sqrt{b_m}}\cdot\frac{1}{b_m}\cdot|p|\cdot|x| \le 2^{-\frac{1}{3}\sqrt{b_m}}\cdot\frac{1}{b_m}\cdot N_1(m,a_m)\cdot M_1(m,a_m)\cdot\|x\| \\ &< \frac{1}{a_m} \qquad \text{p.}\mathscr{d}. \end{aligned} \tag{5.1.2}$$

Combining Lemma 5.1, (5.1.1) and (5.1.2) we have the following.

Lemma 5.2. *With the notation of Lemma 5.1, the polynomial* $q(T) = (T^{b_m}/b_m)\cdot p(T)$ *satisfies* $t^{a_m+b_m}|q(t)$, $\deg q \le b_m + ma_m$, $|q| \le N_1(m,a_m)/b_m$ *and* $\|q(T)x - e_0\| \le \frac{1}{a_{n-1}} + \frac{3}{a_m}$, *for* $x \in K_{n,m}$.

Proof: For

$$\begin{aligned}\|q(T)x - e_0\| &\le \|p(T_m)x - e_0\| + \left\|\frac{T^{b_m}}{b_m}p(T_m)x - p(T_m)x\right\| + \left\|\frac{T^{b_m}}{b_m}(p(T)-p(T_m))x\right\| \\ &\le \left(\frac{1}{a_{n-1}} + \frac{1}{a_m}\right) + \frac{1}{a_m} + \frac{1}{a_m}\end{aligned}$$

(by Lemma 5.1, (5.1.1) and (5.1.2)),

$$= \frac{1}{a_{n-1}} + \frac{3}{a_m}\, .$$

Lastly we observe that the polynomials q in Lemma 5.2 satisfy

$$\begin{aligned}\|q(T)\circ(I-Q_m^0)\| &= \left\|\left(\frac{q}{t^{a_m+b_m}}\right)(T)\circ T^{a_m+b_m}\circ(I-Q_m^0)\right\| \\ &\le |q|\cdot(\|T\|\vee 1)^{\deg q - a_m - b_m}\cdot\|T^{a_m+b_m}\circ(I-Q_m^0)\| \\ &\le |q|\cdot 2^{(m-1)a_m+1}\end{aligned}$$

(since both operator norms are less than 2 p.$\mathscr{d}$. by Lemma 4). Hence

$$\|q(T)\circ(I-Q_m^0)\| \le \left(N_1(m,a_m)\cdot 2^{(m-1)a_m+1}\right)/b_m\, . \tag{5.2.1}$$

We are now in a position to prove the main result.

Theorem 5.3. *Provided $\underline{d}$ increases sufficiently rapidly, the continuous extension of T to X has no nontrivial closed invariant subpace.*

Proof: We shall just write T for its extension to all of X, since, by Lemma 4, we now know that such an extension exists. Let $x \in X$, $\|x\| = 1$ and $N > 0$. It is sufficient to show that for all such x and N there is a polynomial q such that

$$\|q(T)x - e_0\| < 2/a_{N-1} ,$$

(because e_0 is certainly cyclic for T). We first assert that there is an $m > n > N$ such that $\|\tau_{n,m} Q_m^0 x\| \geq \frac{1}{a_m}$. Writing $x = y + w$ ($y \in \ell_1$, $w \in W$) we note that by (3.1.3),

$$Q_n(x) = Q_n(y) + \sum_{i=1}^{n} x_i^*(w) x_i .$$

Now $\max_i \|Q_n f_i\| = 1$ since $Q_n f_i$ is either zero or some f_j; hence, $\|Q_n|_{\ell_1}\| = 1$. But $F \cap \ell_1$ is dense in ℓ_1, and for all $y_0 \in F$, $Q_n(y_0) = y_0$ for all but finitely many n. Hence, $Q_n(y) \xrightarrow{n\to\infty} y$. It follows that

$$\begin{aligned} \liminf_{n\to\infty} \|Q_n x\| &= \|y\| + \liminf_{n\to\infty} \Big\|\sum_{1}^{n} x_i^*(w) x_i\Big\| \\ &\geq \|y\| + \sup_i |x_i^*(w)| \\ &= |||x||| \quad , \qquad \text{say} , \\ &> 0 \end{aligned}$$

Let us choose an $n > N$ so large that

$$\|Q_n x\| \geq \tfrac{1}{2}|||x||| \geq \frac{2}{a_n} .$$

We know that $\|Q_n P_{n,k}\| \leq a_{n+1}$ for all $k > n$, yet for all $z \in F$, $Q_n P_{n,k} z = Q_n z$ for all but finitely many k, hence

$$Q_n P_{n,k} x \xrightarrow{k\to\infty} Q_n x .$$

Choose k so large that

$$\|Q_n P_{n,k} x\| \geq \frac{1}{a_n} .$$

Then

$$\|P_{n,k} x\| \geq \frac{1}{a_n \cdot \|Q_n\|} \geq \frac{1}{n a_n} .$$

Now if $\|\tau_{n,k} \cdot Q_n^0 x\| > \frac{1}{2na_n} > \frac{1}{a_k}$ p.$\underline{d}$., then our assertion is proved. If not then since

$$\|P_{n,k} x\| = \|\tau_{n,k} \circ Q_k x\| \geq \frac{1}{na} ,$$

we have

$$\|\tau_{n,k} \circ (Q_k^0 - Q_k)x\| > \frac{1}{2na_n} . \tag{5.3.1}$$

From (2.1.1) and Definitions 2.4,5 we obtain for all $j \geq 0$,

$$(Q_k - Q_k^0)f_j = \begin{cases} a_{m-r} \cdot e_{j-ra_m+(r-m+k)a_k} & , j \in [0, v_{m-r}] + ra_m ,\\ & \quad 0 < m-k < r \leq m \\ 0 & , \text{otherwise} . \end{cases}$$

Hence

$$\tau_{n,k} \circ (Q_k - Q_k^0)f_j = \begin{cases} a_{m-r} \cdot e_{j-ra_m+(r-m+k)a_k} & , j \in [0, v_{m-r}] + ra_m ,\\ & \quad 0 < m-k < r < m-n \\ 0 & , \text{otherwise} . \end{cases}$$

It follows that

$$\tau_{n,k} \circ (Q_k - Q_k^0) = \tau_{n,k} \circ (Q_k - Q_k^0) \circ \Pi_S$$

where

$$S = \{f_j : j \in \bigcup_{\substack{m>k \\ r\in(m-k,m-n)}} ([0, v_{m-r}] + ra_m)\} \subset \{g_i\} \; ; \text{ so}$$

$$S = \bigcup_{m>k} S_m$$

where

$$S_m = \bigcup_{r\in(m-k,m-n)} ([0, v_{m-r}] + ra_m) .$$

So if

$$\begin{aligned} \|\tau_{n,k}(Q_k - Q_k^0)x\| &= \|\tau_{n,k}(Q_k - Q_k^0)\Pi_S x\| \\ &> \frac{1}{2na_n} , \end{aligned}$$

then

$$\begin{aligned} \|\Pi_S x\| &> (2na_n \cdot \|\tau_{n,k}\| \cdot \|Q_k - Q_k^0\|)^{-1} \\ &> (2na_n \cdot \max(k, a_{n+1}) \cdot (k + a_k))^{-1} \end{aligned}$$

(by Lemmas 3.3,3.1,3.2). This implies that for some $m > k$,

$$\|\Pi_{S_m} x\| > 2^{k-m} \cdot (2na_n \cdot \max(k, a_{n+1}) \cdot (k + a_k))^{-1} .$$

However, it is not hard to check from the definition that if $j \in S_m$ then $\tau_{n,m}Q_m^0 f_j = f_j$. Moreover, for all j, $Q_m^0 f_j$ is either f_j or an element in $F_{v_{m-1}}$. Hence, if $j \notin S_m$, then $\Pi_{S_m} \circ (\tau_{n,m}Q_m^0 f_j) = 0$. It follows that

$$\Pi_{S_m} = \Pi_{S_m} \circ \tau_{n,m} \circ Q_m^0 .$$

Therefore, if

$$\|\Pi_{S_m} x\| > 2^{k-m} \cdot \left(2na_n \cdot \max(k, a_{n+1}) \cdot (k + a_k)\right)^{-1}$$

then, since $\|\Pi_{S_m}\| = 1$, we have

$$\begin{aligned} \|\tau_{n,m} \circ Q_m^0 x\| &\geq 2^{k-m} \cdot \left(2na_a \cdot \max(k, a_{n+1}) \cdot (k + a_k)\right)^{-1} \\ &\geq \frac{1}{a_m} \qquad \text{for all } n < k < m \text{ p.}\not{d}\ . \end{aligned}$$

So there is indeed an $m > n > N$ such that

$$\|\tau_{n,m} \circ Q_m^0 x\| > \frac{1}{a_m} .$$

Write $y = Q_m^0 x$; we have that $\|y\| \leq \|Q_m^0\| \leq a_m$ but $\|\tau_{n,m} y\| > 1/a_m$, and so $y \in K_{n,m}$. By Lemma 5.2 and (5.2.1) there is a polynomial q such that

$$\|q(T)y - e_0\| < \frac{1}{a_{n-1}} + \frac{3}{a_m}$$

and

$$\|q(T)(I - Q_m^0)\| < \left(N_1(m, a_m) \cdot 2^{(m-1)a_m + 1}\right)/b_m$$

Therefore

$$\begin{aligned} \|q(T)x - e_0\| &\leq \|q(T)y - e_0\| + \|q(T)(x - y)\| \\ &= \|q(T)y - e_0\| + \|q(T)(I - Q_m^0)x\| \\ &\leq \frac{1}{a_{n-1}} + \frac{3}{a_m} + \frac{N_1(m, a_m) \cdot 2^{(m-1)a_m + 1}}{b_m} \\ &< \frac{2}{a_{n-1}} \qquad \text{p.}\not{d}.\ . \end{aligned}$$

This is the required inequality; hence every vector x in the unit ball of X is cyclic for T, and T has no nontrivial closed invariant subspace.

6. Further Remarks

This section does not contain proofs, because the proofs of the assertions are almost identical to those of §6 of [1].

6.1. We have already established that $T|_W$ is nuclear. Investigating T further, one may also establish that $T|_{\ell_1}$ is a perturbation of a weighted shift operator by a nuclear operator.

6.2. The spectrum of T is equal to its approximate point spectrum, which is the unit disk.

6.3. If we choose $\underset{\sim}{d}$ so that $\underset{\sim}{d}$ increases sufficiently rapidly and $b_i \equiv -1 \bmod i!$ for all i, then T^k has no nontrivial closed invariant subspaces for any $k > 0$.

6.4. If we choose $\underset{\sim}{d}$ so that $n!|a_n$, $n!|b_n$ for all n, then every positive power of T except T itself will have an invariant subspace; in fact, T^n willl have as an invariant subspace the subspace

$$\overline{\text{lin}}\{e_i : n|i\} .$$

7. References

(1) C.J. Read, A short proof concerning the invariant subspace problem, J. Lond. Math. Soc., (2) 33 (1986).

(2) P. Enflo, Acta Math, to appear.

(3) B. Beauzamy, Operators without invariant subspaces, simplification of a result of P. Enflo, J. Integral Equations and Operator Theory (1985).

(4) C.J. Read, A solution to the invariant subspace problem, Bull. London. Math. Soc., 16 (1984) 337-401.

(5) C.J. Read, A solution to the invariant subspace problem on the space ℓ_1, Bull. Lond. Math. Soc., 17 (1985), 305-317.

(6) R. Ovsepian and A. Pelcynski, The existence in every separable Banach space of a fundamental and total bounded biorthogonal sequence and related constructions of unifromly bounded orthonormal systems in L^2. Studia Math, 54 (1975) 149-159.

APPROXIMATIONAL COMPLEXITY OF FUNCTIONS

Y. Yomdin

Ben Gurion University of the Negev
Beer-Sheva, Israel

Introduction

Topological complexity of real functions can be measured in various ways. For example, we can count the number of connected components of the level sets $f = \text{const}$; we can study the distribution of critical points or critical values of f, etc.

It is well-known that some classes of functions are "simple" from a topological point of view. The number of connected components of any level set of a polynomial, as well as the number of its critical values, is bounded by a constant, depending only on the degree of this polynomial. The set of critical values of any sufficiently smooth function has measure zero (Morse-Sard theorem [15]), etc.

Now assume that f is "simple" and g is close to f (in some reasonable metric). Unfortunately, in general we can say nothing about g; the number of the critical points and values of g, connected components of the level sets of $g = \text{const}$, etc., may jump to infinity. The traditional topological information is not stable with respect to perturbations. (See examples below.) Moreover, the results of the type of the Morse-Sard theorem simply disappear in considerations with finite accuracy. The exact zero does not exist for a computer and hence the computer does not understand the definition of a critical point as well as the property of the set to be of measure zero.

The requirement for complexity notions we consider to be stable with respect to perturbations of functions is the starting point of our approach. This requirement leads ultimately to the following procedure: we look at our function f through some device with a finite resolution ε. We denote the complexity we can distinguish by $\sigma(f, \varepsilon)$. Varying the resolution ε, we get the complexity function $\sigma(f, \cdot)$.

Thus, in the case of the Morse-Sard theorem, we replace critical points of f by "ε-near critical", where the norm of the gradient of f is at most ε. The Lebesgue measure is replaced by the metric entropy – the minimal number of balls of radius ε, we need to cover the considered set (see section 2 below).

Note that a priori it is not clear whether any nontrivial bound on $\sigma(f,\varepsilon)$ exists, even in the simplest case of polynomials. The topological "simplicity" is a highly unstable effect, as the example below shows. Nondegeneracy assumptions, which usually provide some stability, are irrelevant in our situation: we know nothing about the polynomials, arising in an approximation process, except their degrees and rough bounds on coefficients.

Thus, the fact that the perturbation stable, "quantitative" bounds on topological complexity of polynomials do exist, seems to be important in many relations. In this paper we give some of its consequences in smooth analysis, and smooth dynamics via the approximation process, described above. Returning to this general process we note that the following statement becomes self-evident:

If two functions f and g are ε-close to one-another in an appropriate norm, then $\sigma(f,\varepsilon)$ and $\sigma(g,\varepsilon)$ coincide: in resolution ε we see no difference between f and g.

Now, assume that some class P is given, such that the complexity $\sigma(p,\cdot)$ is known for any p in P.

If a function f is in the closure of P and if we know the rate of its approximation by functions in P, we immediately get, by the above statement, an upper bound for $\sigma(f,\cdot)$.

Thus in any concrete application of this approach two main difficulties should be settled:

a. Specifying the complexity property for which the "ε-quantification" as above makes sense.

b. Finding a sufficiently wide class P of model functions p, for which nontrivial bounds on $\sigma(p,\cdot)$ can be given.

Taking P-polynomials and the approximated functions those of finite smoothness, one gets essential new information on topological structure of smooth functions. Some of the results in this direction obtained in [18]-[28], are presented in section 1.

In section 2 we extend P to contain piecewise-polynomial functions. Consequently, the class of approximated functions expands far beyond the smooth ones. (For a polynomial approximation, by the theorems of Jackson and Bernstein, the rate of approximations determines, and is determined, by the degree of differentiability.) Most of the results of section 1 can be extended respectively.

An important conclusion is that even in such classical results as the Sard theorem, not the degree of differentiability, but namely the approximational complexity, is the adequate notion.

On the other hand, our approach works for functions better than C^k, e.g., for analytic ones. While the "qualitative" topological complexity properties usually are self-evident for analytic functions, the quantitative ones remain nontrivial and even more interesting. Thus the investigation of this direction of the complexity scale (this scale contains analyticity as an important threshold but goes beyond it) seems to be very interesting. Abundance of very concrete examples and computations, and strong relations with classical analysis: approximation theory, theory of analytic and quasianalytic functions, etc., seem to indicate an adequacy of the approximational complexity techniques in this situation. In sections 3, we give some of the simplest examples of this kind.

The general approach above works equally well for functions on infinite dimensional spaces. However, its specific realization becomes much more delicate, first of all, since the potential "model" functions (e.g., the polynomials depending only on a finite number of variables) are not dense. In section 3 we give some examples, where we compute complexity and, consequently, establish the properties like Sard theorem, for rather wide classes of functions on infinite dimensional spaces. These examples seem to be promising especially because the high differentiability in this context is completely irrelevant (see, e.g., [7]).

The approximational complexity tools seem to be applicable in many problems of smooth dynamical systems (e.g., continuity properties of the entropy, generic properties of systems, etc.). The investigation in this direction may clarify the dynamic meaning of smoothness or analyticity of the iterated mapping. In section 2 we give some results in this direction.

In conclusion let us mention, that by the essence of our approach all the results we get for polynomials and smooth functions remain meaningful in computations with a finite accuracy. In particular, the results of section 1 on polynomial functions seem to be useful in constructing algorithms for numerical processing of the standard operations in a semialgebraic category.

1. Sample of Results for Polynomials and Functions of Finite Smoothness

In this section we state our main "quantitative" results for polynomials. Then we state their main consequences, via approximation, for C^k-functions. Thus the approximating class P here consists of polynomials, approximated functions are those of finite smoothness. However, in this section, we do not follow the general scheme above. In particular, we do not define the complexity function $\sigma(f, \varepsilon)$. Let us start with some classical results on polynomials.

a. Bounds on topological complexity.

Let $f(x_1, \ldots, x_n)$ be a polynomial of degree d. The following result is well-known.

Proposition 1.1. *The number of connected components of the set $Y = \{f = c\}$ in $\mathbf{R}^n$, is bounded in terms of d and n only (see, e.g., [1], [28]). For $n = 1$ this means simply that the number of the roots of f is at most d. For $n = 2$, Harnack's theorem says that the number of ovals of a regular real curve S of degree d in $\mathbf{RP}^2$ is at most $1/2(d-1)(d-2)+1$.*

(Proof: consider the complexification S_c of this curve; S separates S_c into two parts, which are mapped one into another by a conjugation. Then if the number of the ovals of S is q, it is clear that any $q-1$ are nonhomologic to zero. Hence $q - 1 \leq$ genus $(S_c) = 1/2(d-1)(d-2)$.)

This property can be destroyed by an arbitrary small perturbation of our polynomial in C^∞-topology (or even by addition of an arbitrary small polynomial of higher degree).

Example. $f(x,y) = y^2$. The perturbation f_t is given by $f_t(x,y) = y^2 + t\sin\omega x$. Then for $t \neq 0$ the line $\{f_t = 0\}$ has, on $\mathbf{R}^2$, infinitely many ovals. Taking ω sufficiently large we can obtain arbitrarily many ovals inside the unit disk in $\mathbf{R}^2$.

Of course, the reason for such an instability in our example is that the line $y = 0$ is singular for the function y^2; however, approximating smooth functions by polynomials we cannot expect a priori any nonsingularity of level sets. Indeed, any closed set can be the set of zeros of a C^∞-function.

Thus the way to exploit the above property in approximation process is to combine it with the information on distribution of the values a smooth function with respect to their regularity. (See Theorems 1.7 and 1.9 below.)

In turn, this distribution can be studied on a basis of the second property of polynomials which we consider in this section.

Proposition 1.2. *Let $f : \mathbf{R}^n \to \mathbf{R}$ be a polynomial of degree d. Then the number of the critical values of f (i.e., the values of f at points where* grad $f = 0$*) does not exceed $(d-1)^n$.*

Proof: Critical points of f are determined by the system of n equations

$$\frac{\partial f}{\partial x_1} = 0, \dots, \frac{\partial f}{\partial x_n} = 0$$

each of degree $d-1$. If the solutions are nondegenerate, their number is bounded by $(d-1)^n$ by Bezout's theorem. Then the same is true also for critical values.

(Warning: The number of real degenerate solutions of a polynomial system may be higher than the bound given by the Bezout theorem.

Example. Consider the following system:

$$f_1(x_1,\ldots,x_{n-1}) = 0,\ldots, \qquad f_{n-1}(x_1,\ldots,x_{n-1}) = 0\ ,$$

each polynomial of degree d, and let this system have d^{n-1} nondegenerate solutions.

Now consider the system in $\mathbf{R}^n$.

$$g = f_1^2 + \ldots + f_{n-1}^2 = 0\ , \qquad x_n = 0\ ,\ x_n = 0,\ldots,\ x_n = 0,\ ((n-1) - \text{times}).$$

Clearly, this system has d^{n-1} isolated (degenerate) solutions. But the product of degrees is $2d < d^{n-1}$ for d sufficiently large and $n \geq 3$.)

However, the bound of Proposition 1.2 is valid also in a degenerated situation: The number of connected components of the critical set of a complexification f_c is always bounded by $(d-1)^n$, and the value of f_c on each component is constant.

This property is also non-stable. For example, the only critical value of $f(x,y) = y^2$ is 0. But for any $t \neq 0$ the function $f_t = y^2 + te^x \sin \omega x$ has infinitely many critical values. Allowing finite smoothness, we can easily get infinitely many critical values on the unit disk: $f_t = y^2 + tx^k \sin(1/x)$.

b. Bounds on Geometry.

Let $f(x_1,\ldots,x_n)$ be a polynomial of degree d, $Y = \{f = 0\}$. Let B_r^n denote the ball of radius r centered at $0 \in \mathbf{R}^n$.

Proposition 1.3. $\mathrm{Vol}_{n-1}(Y \cap B_r^n) \leq C(n) \cdot d \cdot r^{n-1}$, *where the constant C depends only on n.*

This result follows easily from the integral-geometric formula for the volume: almost any straight line intersects Y on at most d points. By similar reasoning, the following proposition is true.

Proposition 1.4. *Let $f = (f_1,\ldots,f_n) : \mathbf{R}^m \to \mathbf{R}^n$ be a polynomial mapping of degree d ($\max(\deg f_i) = d$), $m < n$. Then*

$$\mathrm{Vol}_m(f(\mathbf{R}^m) \cap B_r^n) \leq C(n,m,d) \cdot r^m\ .$$

c. Resolution of Singularities.

We just mention the following fundamental result in algebraic geometry: the existence of a resolution for any singularity. Roughly, this means that any polynomial by a generically one-to-one polynomial change of variables can be brought to the following simple form: $y_1^{m_1} \dots y_n^{m_n}$ (locally, in an appropriate coordinate system at any point of the source, and up to a nonzero factor). The resolution procedure is also highly unstable with respect to perturbations.

d. Computational Complexity.

We do not give precise definitions and results here. However, it is clear, that in any reasonable setting, the number of operations one needs to evaluate a polynomial $f(x_1, \dots, x_n)$ of degree d at any given point x, is of order at most d^n.

There is evidence (see e.g., the results on "fewnomials" ([1],[6],[12,13]), and on the additive complexity ([12,13]) that computational complexity of a polynomial bounds in some sense, its topological complexity. The connections here seem to be very deep, but not much is known in this direction.

As we mentioned above, some of these results allow "quantitative versions" which are stable with respect to perturbations. We start with the "quantification" of the Proposition 1.2.

Let $f : \mathbf{R}^n \to \mathbf{R}$ be a C^1-function. Let, for $\gamma \geq 0$, $\Sigma(f,\gamma) = \{x / \|\operatorname{grad} f(x)\| \leq \gamma\}$.

Let $B_r^n \subseteq \mathbf{R}^n$ be some ball of radius r. We denote $\Sigma(f,\gamma) \cap B_r^n$ by $\Sigma(f,\gamma,r)$ and $f(\Sigma(f,\gamma,r)) \subseteq \mathbf{R}$ by $\Delta(f,\gamma,r)$. $\Sigma(f,\gamma,r)$ and $\Delta(f,\gamma,r)$ are the sets of γ-critical points and γ-critical values of f on B_r^n, respectively. For $\gamma = 0$ we get the usual critical points and values.

Theorem 1.5. *Let $f : \mathbf{R}^n \to \mathbf{R}$ be a polynomial of degree d. Then, for any $\gamma \geq 0$, the set $\Delta(f,\gamma,r)$ can be covered by $N(n,d)$ intervals of length γr. The constant $N(n,d)$ here depends only on n and d. The ball B_r^n in the definition of $\Delta(f,\gamma,r)$ can be replaced by the r-cube $Q_r^n = [0,r]^n \subseteq \mathbf{R}^n$.*

Proof of this theorem is given in [18,28]. Computations in [28] give $N(n,d) = (2d)^n$.

First of all, for $\gamma = 0$, we obtain that the set of critical values of f contains at most $N(n,d)$ points, which is the result of Proposition 1.2 (with the constant $(d-1)^n$ slightly better than the value of N given above).

Now, the property given by Theorem 1.5 is compatible with approximations. Let $g : B_r^n \to \mathbf{R}$ be a k times differentiable function, and let p be the Taylor polynomial of degree $k-1$ of g

at the center of B_r^n. We have

$$\max_{x \in B_r^n} |g(x) - P(x)| \le R_k(g)$$

$$\max_{x \in B_r^n} \|dg(x) - dP(x)\| \le \frac{k}{r} R_k(g) ,$$

where $R_k(g) = \frac{1}{k!} \max \|d^k g\| \cdot r^k$ is the remainder term in the Taylor formula.

Hence, the critical points of g are at most γ_0 critical for P, where $\gamma_0 = (k/r)R_k(g)$, i.e., $\Sigma(g,0,r) \subseteq \Sigma(p,\gamma_0,r)$. Hence $\Delta(g,0,r) = g(\Sigma(g,0,r)) \subseteq g(\Sigma(P,\gamma_0,r))$. Finally, since $|g - P| \le R_k(g)$, $g(\Sigma(P,\gamma_0,r))$ is contained in a $R_k(g)$-neighborhood of $P(\Sigma(P,\gamma_0,r)) = \Delta(P,\gamma_0,r)$.

Now, by Theorem 1.5, $\Delta(P,\gamma_0,r)$ can be covered by at most $N(n,k-1)$ intervals of length $\gamma_0 r = kR_k(g)$. Thus, the $R_k(g)$-neighborhood of $\Delta(P,\gamma_0,r)$, and hence $\Delta(g,0,r)$, can be covered by the same number of intervals of length $(k+2)R_k(g)$, or by $k+2$ times this number of $R_k(g)$-intervals. We proved the following result:

Theorem 1.6. *For g as above, the critical values of g can be covered by at most $N_1(n,k)$ intervals of length $R_k(g)$, where $N_1(n,k) = (k+2)N(n,k-1)$ depends only on n and k.*

This result can be considered as a "Taylor formula" for the property of polynomials given by Proposition 1.2.

Let us continue in this direction, considering the following question: for an arbitrary $\varepsilon > 0$, how many intervals of length ε do we need to cover $\Delta(g,0,r)$? To answer this question, we first find r' such that the remainder term of g on any ball of radius r' is at most ε:

$$\frac{1}{k!}\max \|d^k g\| \cdot (r')^k = R_k(g) \cdot (r'/r)^k = \varepsilon , \qquad \text{i.e., } r' = r \cdot [\varepsilon / R_k(g)]^{1/k} .$$

Now we cover B_r^n by sub-balls of radius r'. We need at most $C(n)(r/r')^n = C(n)\left[R_k(g)/\varepsilon\right]^{n/k}$ such balls.

Theorem 1.6 guarantees that critical values of g on each small ball can be covered by at most $N_1(n,k)$ ε-intervals, and to cover all the critical values of g we need, therefore, at most $N_1(n,k) \cdot C(n) \cdot \left[R_k(g)/\varepsilon\right]^{n/k}$ such intervals. We proved the following theorem:

Theorem 1.7 ([18]). *For g as above and for any $\varepsilon, 0 < \varepsilon \le R_k(g)$, the set of critical values of g can be covered by $N_2(n,k)\left[R_k(g)/\varepsilon\right]^{n/k}$ intervals of length ε. Similarly, the set $\Delta(f,\gamma,r)$, for any $\gamma \ge 0$, can be covered by $C(n,k)\left[\left(R_k(g)/\varepsilon\right)^{n/k} + \gamma(r/\varepsilon)\left(R_k(g)/\varepsilon\right)^{(n-1)/k}\right]$ intervals of length ε.*

Corollary 1.8. (Morse-Sard theorem [15]). *If $g \in C^k$ with $k > n$, then the measure of critical values of g is zero.*

Proof: For any ε the measure of critical values of g is bounded by ε times the number of ε-intervals covering our set. Hence

$$m(\Delta) \leq \lim_{\varepsilon \to 0} \varepsilon \cdot C \cdot \left(\frac{1}{\varepsilon}\right)^{\frac{n}{k}} = \lim_{\varepsilon \to 0} C\varepsilon^{1-\frac{n}{k}} = 0 \quad \text{for} \quad \frac{n}{k} < 1 .$$

Now Theorem 1.7 combined with Proposition 1.1, allows us to prove the following bound for the average topological complexity of the fibers of a smooth function.

Theorem 1.9 ([20]). *For g as above and $\xi \in \mathbf{R}$, let $B_i(f^{-1}(\xi))$ denote the i-th Betti number of the fiber $f^{-1}(\xi)$. In particular, B_0 is the number of connected components.*

Then for any positive exponent ν

$$\int_{-\infty}^{\infty} [B_i(f^{-1}(\xi))]^{\nu} d\xi \leq B_i^{\nu}(k,n)\left[1 + R_k(f)\right]^{\nu} \int_1^{\infty} \left(\frac{1}{t}\right)^{\frac{k-\nu}{n}} dt .$$

In particular, this integral is finite if the degree of differentiability k is greater than $n + \nu$.

We turn now to the bounds on the volumes of algebraic sets, given above. To give a version of Proposition 3.1, more stable with respect to perturbations, let us denote by $M(\varepsilon, A)$ the minimal number of ε-balls, covering the set $A \subseteq \mathbf{R}^n$.

Theorem 1.10 ([28]). *Let $f : \mathbf{R}^n \to R$ be a polynomial of degree d. Denote for $\eta \geq 0$ by Y_η the set $\{x \in \mathbf{R}^n \ , \ |f(x)| \leq \eta\}$. Then for any $\varepsilon > 0$, $\eta > 0$, and the ball $B_r^n \subseteq \mathbf{R}^n$,*

$$M(\varepsilon, Y_\eta \cap B_r^n) \leq C_1(n,d)\frac{\eta^{\frac{1}{d}}}{r}\left(\frac{r}{\varepsilon}\right)^n + C_2(n,d)\left[\left(\frac{r}{\varepsilon}\right)^{n-1} + 1\right] .$$

In particular, for $\eta = 0$, this inequality implies the bound of Proposition 1.3.

Quantitative geometry of polynomials seems to be a very rich field, far from being completely understood. Various results in this direction are given in [28], Chapter 6. Via the approximation process, each of these results implies its smooth counterpart. Here we mention only the following rather unexpected fact: it is well known, that any closed set $F \subseteq \mathbf{R}^n$ is the set of zeroes of some C^∞-smooth function. It turns out that if, for some k, the k-th derivative of g is not "too big", $g^{-1}(0)$ looks like an algebraic set.

Theorem 1.11 ([19]). *Let $g : B_r^n \to \mathbf{R}$ be a C^∞ function. If for some $k = 2, 3, \dots$,*

$$R_k(g) \leq \frac{1}{2^{k+1}k!} R_0(g) , \qquad \text{then}$$

a. *$Y(g) = g^{-1}(0)$ is contained in a countable union of compact smooth hypersurfaces in B_r^n.*

b. *There is an open subset U in the set of all the straight lines in $\mathbf{R}^n$, such that the lines of U cover all the space B^n and any such line intersects $Y(g)$ in not more than $k-1$ points.*

c. $\mathrm{Vol}_{n-1}\left(Y(g)\right) \leq C(n,k) r^{n-1}$.

The following result we state here is close in spirit to a "quantitative resolution of singularities". We try, by changes of coordinates, to simplify a given polynomial. But in our approach "to simplify" means to get small derivatives, up to a fixed order.

Theorem 1.10 ([3,22,23]). *Let $f : B_1^n \to \mathbf{R}$ be a polynomial of degree d, and let $k \geq 1$ be fixed. Then the set $S = \left\{x \in B_1^n \ , \ |f(x)| \leq 1\right\}$ can be covered by the images of the mappings $\psi_i : B_1^n \to \mathbf{R}^n$, $i = 1, \dots, \kappa(n,d,k)$, such that $\|d^s \psi_i(x)\| \leq 1$ and $\|d^s\left(f \circ \psi_i(x)\right)\| \leq 1$ for $x \in B_1^n$, $s = 1, \dots, k$, $i = 1, \dots, \kappa$. The constant κ depends only on n, d and k.*

This theorem is the main ingredient of the proof of the entropy conjecture for C^∞-smooth dynamical systems ([22,23], see also [9,11,16,17]).

Let us state the main result of [22]:

Theorem 1.11. *For $f : N \to N$, f C^k-smooth, and $h(f)$ – the topological entropy of f ([16]),*

$$v(f) \leq h(f) + \frac{m}{k} \mathrm{Lip}(f) ,$$

where $v(f)$ is the volume growth of f and $\mathrm{Lip}(f)$ – the Lipschitz constant of f, and $m = \dim N$.

The expression m/k appeared in many results above, concerning C^k-functions of m variables. In this and the next section we show that in fact this expression, at least in part of the results, can be replaced by some kind of "complexity".

Below we prove Theorem 1.10 (and the further results towards the volume growth estimates) in the one-dimensional case, where all the technical difficulties disappear, but some of the main ideas can still be illustrated.

In dimension one, Theorem 1.10 has the following form:

Proposition 1.12. *Let $P : \mathbf{R} \to \mathbf{R}$ be a polynomial of degree d. Then the set $S = \left\{x \mid |p(x)| \leq 1\right\}$ can be covered by images of $\kappa(d,k)$ mappings $\psi_i : [0,1] \to \mathbf{R}$, such that $|d^s \psi_i| \leq 1$ and for $P_i = P \circ \psi_i$, $|d^s P_i| \leq 1$, $s = 0, 1, \dots, k$, $i = 1, \dots, \kappa$ ($k \geq 1$ fixed in advance).*

Proof: S consists of at most d intervals. Cover each of them by linear mapping ψ_i of [0,1]. Then $P_i = P \circ \psi_i$ is a polynomial degree d on [0,1], bounded on [0,1] by 1. By Markov theorem (see [5,10]), $|d^s P_i|$ on [0,1] is bounded by $C_s(d) = 2^s d^2 (d-1)^2 \dots (d-s)^2$. Subdividing each interval [0,1] into $C_k(d)$ parts and reparametrizing linearly once more, we get the required inequalities. We also get $\kappa(d,k) = k \cdot C_k(d)$.

The result of Proposition 1.11 is "stable under approximation":

Proposition 1.13. *Let $f : [0,1] \to \mathbf{R}$ be a C^k-function, with $\max |f^{(k)}| \le k!$. Then the set $S = \{x \mid |f(x)| \le 1\}$ can be covered by images of at most $\kappa(k)$ mappings $\psi_i : [0,1] \to [0,1]$ such that $f_i = f \circ \psi_i$ satisfy $|d^s f_i| \le 1$, $s = 0, 1, \dots, k$.*

Proof: Let P be a Taylor polynomial of f of degree $k-1$. We have $|p - f| \le 1$ on [0,1]. Let S' be the set $\{x \in [0,1] / |p(x)| \le 2\}$. Then $S \subseteq S'$, and $|f|_{/S'} \le 3$.

S' consists of at most $k-1$ intervals. Reparametrize these intervals linearly by $\psi_i : [0,1] \to [0,1]$. Then for $f_i = f \circ \tilde{\psi}_i$ we have: $|f_i| \le 3$ on [0,1] and $|f_i^{(k)}| \le k!$ on [0,1]. But then the well-known bound for intermediate derivatives gives $|d^s f_i| \le C_s'(k)$, $s = 1, \dots, k$. Subdividing and reparametrizing once more we get the required inequalities.

The following is the "easy part" of the volume growth estimate, which shows, however, why the degree of differentiability enters into the denominator of the exponent. We apply to a given "simplex" σ a big linear mapping L.

Theorem 1.14. *Let $\sigma : [0,1] \to \mathbf{R}^n$ be a C^k-smooth mapping. Let $L : \mathbf{R}^n \to \mathbf{R}^n$ be a linear transformation. Then the length of the curve $L \circ \sigma([0,1]) \cap B_1^n$ does not exceed $C \|L\|^{\frac{1}{k}}$, where $C = C_1(k,n) \big(R_k(\sigma)\big)^{\frac{1}{k}}$ depends only on d, k, n and σ, but not on L.*

Proof: $\max \|d^k (L \circ \sigma)\| \le \|L\| \cdot \max \|d^k \sigma\| = M$. Subdividing [0,1] into $[(M/k!)^{1/k}] + 1$ parts and reparametrizing linearly by $\varphi_j : [0,1] \to [0,1]$, we get $\max \|d^k (L \circ \sigma \circ \varphi_j)\| \le k!$.

Now, Proposition 1.12 is clearly true also for mappings of [0,1] into $\mathbf{R}^n$ (with $\kappa(k,n)$ now depending also on n). Hence, for each $L \circ \sigma \circ \varphi_j : [0,1] \to \mathbf{R}^n$ we can reparametrize the part of [0,1] which is mapped into the unit ball B_1^n, by $\kappa'(k,n)$ mappings ψ_i, as above. But certainly the length of the image of $L \circ \sigma \circ \varphi_j \circ \psi_i$ is at most 1. Hence the total length of $L \circ \sigma([0,1])$ inside B_1^n does not exceed $2\kappa''(k,n) \cdot (M/k!)^{1/k}$.

2. Complexity of Real Functions via Piecewise Polynomial Approximation

In this section we give some examples of an implementation of a general scheme described above. The complexity properties, we study here, are suggested by the results of section 1: the structure of critical values, from one side (Theorems 1.5-1.7 above) and the volume growth under big dilations, from another (Theorem 1.13).

The approximating class P now consists of piecewise polynomial functions. Extend of the quantitative results of section 1 to this class is usually straightforward.

To simplify definitions we shall assume that all the approximated functions are C^1, and the approximation will be considered in C^1-norm (these assumptions can be relaxed).

So let $f : Q_r^n \to \mathbf{R}$ be a C^1-function where $Q_r^n = [0, r]^n$ is the cube of the size r in $\mathbf{R}^n$. We want to study the critical points and values of f "with finite resolution $\varepsilon > 0$". Clearly we cannot distinguish the exactly critical points of f from the ε-critical, where $\| \operatorname{grad} f \| \leq \varepsilon$. On the other hand also the position of critical values we can identify only up to ε. Therefore, the following definition becomes quite natural:

Definition 2.1. *For $f : Q_r^n \to \mathbf{R}$ - a C^1-function and $\varepsilon > 0$, $C(f, \varepsilon)$ is equal to $M\big(\varepsilon, \Delta(f, \varepsilon)\big)$ (where $\Delta(f, \varepsilon)$, as above, is the set $f\big(\{x \mid \| \operatorname{grad} f(x) \| \leq \varepsilon\}\big)$.*

Theorem 1.5 immediately implies the following:

Proposition 2.2. *For $f : Q_r^n \to \mathbf{R}$ a polynomial of degree d, $C(f, \varepsilon) \leq (2d)^n$.*

To simplify the presentation we shall consider the piecewise-polynomial functions h of a very special type: we subdivide Q_r^n into some parallelepipeds Q_i, and assume that on each Q_i, h is equal to some polynomial p_i of degree d_i. It is technically convenient not to assume continuity of h and not to define it on the faces of Q_i. For Q_r^n fixed, we denote the class of all the h as above by P.

All the notions used above to define $C(f, \varepsilon)$, have meaning also for the functions in P, if we consider only the interior points of Q_i. The following corollary is implied then by Proposition 2.2.

Corollary 2.3. *For $h \in P$, $C(h, \varepsilon) \leq \sum\limits_i (2d_i)^n \stackrel{\text{def}}{=} \frac{1}{4}\sigma(h)$.*

Now the norm $\|f\|$ we define as the $\sup\limits_{x \in Q_r^n} \big(\|f(x)\| + \| \operatorname{grad} f(x) \|\big)$. If the function is only piecewise-smooth, we exclude the boundaries of the pieces of smoothness.

Proposition 2.4. *If $\|f - g\| \leq \varepsilon$, then $C(f, \varepsilon) \leq 4C(g, 2\varepsilon)$ and $C(g, \varepsilon) \leq 4C(f, 2\varepsilon)$.*

Proof: $\Sigma(f,\varepsilon) \subseteq \Sigma(g,2\varepsilon)$, since $\|f-g\| \leq \varepsilon$. Hence, $\Delta(f,\varepsilon) = f(\Sigma(f,\varepsilon)) \subseteq f(\Sigma(g,2\varepsilon))$, which, in turn, is contained in the ε-neighborhood of $g(\Sigma(g,\varepsilon))$, by the same inequality $\|f-g\| \leq \varepsilon$. But if the intervals of the length 2ε cover some set A, the intervals of the length 4ε with the same centers, certainly cover the ε-neighborhood of A.

Corollary 2.5. *For* $f : Q^n_r \to \mathbf{R}$, $C(f,\varepsilon) \leq 4 \sup_{h\in P, \|f-h\|\leq\varepsilon} C(h,2\varepsilon)$.

To make this bound more computable, we give the following definition:

Definition 2.6. *For* $f : Q^n_r \to \mathbf{R}$, $\varepsilon > 0$, $\sigma(f,\varepsilon)$ *is defined as* $\sigma(f,\varepsilon) = \inf_{h\in P, \|f-h\|\leq\varepsilon} \sigma(h)$.

By Corollary 2.3 and Corollary 2.5 we get

Theorem 2.7. *For* $f : Q^n_r \to \mathbf{R}$, $\varepsilon > 0$, $C(f,\varepsilon) \leq \sigma(f,\varepsilon)$. *In particular,* $M(\varepsilon, \Delta(f)) \leq \sigma(f,\varepsilon)$, *where* $\Delta(f) = \Delta(f,0,r)$ *is the set of exactly critical values of* f.

Theorem 2.7 is one of the main results of this paper, relating approximational complexity of a function with the structure of its critical values. Below we evaluate the complexity of various functions, and discuss the conclusions of Theorem 2.7 in each case.

Along the same lines we can construct complexity functions $C^{i,\nu}_1(f,\varepsilon)$, responsible for an average topological complexity of the fibers of f. As above, $C^{i,\nu}_1(f,\varepsilon)$ are majorated by (more computable) $\sigma^{i,\nu}_1(f,\varepsilon)$, and the main result, generalizing Theorem 1.8 is the following:

Theorem 2.8. *For* $f : Q^n_r \to \mathbf{R}$, $\int_R \left[B_i(f^{-1}(\xi))\right]^\nu d\xi \leq \int_1^\infty \sigma^{i,\nu}_1\left(f, \frac{1}{t}\right) dt$.

We do not give here the detail constrution of $C^{i,\nu}_1(f,\varepsilon)$.

The last example we give concerns generalizations of Theorems 1.10, 1.11, 1.12, 1.13. Following the informal procedure above, we consider our functions, $f : Q^n \to \mathbf{R}$ "in resolution $\varepsilon > 0$". We are interested in the structure of the set $(f = 0)$, but we distinguish only the set $\|f\| \leq \varepsilon$. The "complexity unit" now is a reparametrization $\varphi : Q^n \to Q^n$, such that $\|d^s\varphi\| \leq 1$, $s = 1,\ldots,k$ (where k is fixed in advance) and such that $f \circ \varphi$ looks as a constant in ε-resolution (and in C^k norm):

$$\|d^s(f\circ\varphi)\| \leq \varepsilon\,, \quad s = 1,\ldots,k\,.$$

Maybe this explanation is not very convincing. Indeed, the kind of complexity involved in volume estimates under iterations, seems to be understood much less than that of individual functions. However the following formal definition leads to a required generalization of Theorems 1.13 and 1.14:

Definition 2.9. *For k fixed, $f : Q_1^n \to \mathbf{R}^m$ a C^k-mapping and $\varepsilon > 0$, $C_2^k(f,\varepsilon)$ is equal to the minimal number κ of C^k-mappings $\varphi_i : Q_1^n \to Q_1^n$, such that the images $\operatorname{Im}\varphi_i$, $i = 1,\dots,\kappa$, cover the set $\{\|f\| \le \varepsilon\}$, $\|d^s\varphi_i\| \le 1$, and $\|d^s(f \circ \varphi_i)\| \le \varepsilon$, $s = 1,\dots,k$, $i = 1,\dots,\kappa$.*

Theorem 1.10 above implies immediately the following:

Proposition 2.10. *Let $f : Q_1^n \to \mathbf{R}^m$ be a polynomial mapping of degree d. Then for any $\varepsilon > 0$, $C_2^k(f,\varepsilon) \le \kappa(n,m,d,k)$.*

The proof of Theorem 1.10 is technically rather complicated, and the bounds for $\kappa(n,m,d,k)$ we get are very high (see [2,22,23]). For $n = 1$, $m = 1$, the proof of Proposition 1.11 above gives $\kappa(1,1,d,k) \le 2^k d \cdot \left(d!/(d-k)!\right)^2 \le 2^k d^{2k+1}$.

We consider the same approximating class P, as above, and get the following:

Corollary 2.11. *For $h \in P$,*

$$C_2^k(h,\varepsilon) \le \sum_i \kappa(n,m,d_i,k) \stackrel{\text{def}}{=} s_2^k(h) \ .$$

In particular, for $n = 1$, $m = 1$, $s_2^k(h) \le 2^k \sum_i d_i^{2k+1}$.

Defining the norm $\|f\|^k$ as the

$$\sup_{x\in Q_1^n} \Big(\sum_{s=0}^{k} \|d^s f(x)\| \Big) \ ,$$

we get

Proposition 2.12. *If $\|f - g\|^k < \varepsilon$, then $C_2^k(f,\varepsilon) \le \alpha(n,m,k) C_2^k(g,2\varepsilon)$.*

Proof: Since $\max\|f - g\| \le \varepsilon$, $\{\|f\| \le \varepsilon\} \subset \{\|g\| \le 2\varepsilon\}$. Let $\varphi_i : Q_1^n \to Q_1^n$ be the mappings appearing in Definition 2.9 for $C_2^k(g,2\varepsilon)$, $i = 1,\dots,C_2^k(g,2\varepsilon)$. We have $d^s\big(f \circ \varphi_i(x)\big) = \sum_{j=1}^{s} d^j f(\varphi_i(x)) \circ P_j(d\varphi_i(x))$, $s = 1,\dots,k$, where $P_j(d\varphi_i)$ are the universal polynomials in partial derivatives of φ_i. The same expression can be written for $d^s(g \circ \varphi_i)$.

Hence $\|d^s(f \circ \varphi_i - g \circ \varphi_i)\| \le \sum_{j=1}^{s} \|d^j(f - g)\| \cdot \|P_j(d\varphi_i)\| \le \varepsilon\alpha(n,m,k)$, since $\|f - g\|^k \le \varepsilon$ and $\|d^s\varphi_i\| \le 1$, $s = 1,\dots,k$. By Definition 2.9, $\|d^s(g \circ \varphi_i)\| \le 2\varepsilon$, therefore, $\|d^s(f \circ \varphi_i)\| \le \big(2 + \alpha'(n,m,k)\big)\varepsilon$, $s = 1,\dots,k$. Subdividing each Q_1^n into the subcubes of the size $1/\big(2+\alpha'(n,m,k)\big)$ and reparametrizing linearly we get new φ_i', such that $\|f \circ \varphi_i'\|^k \le \varepsilon$. Their number is at most $\big(2 + \alpha'(n,m,k)\big)^n C_2^k(g,2\varepsilon) = \alpha(n,m,k) C_2^k(g,2\varepsilon)$.

Corollary 2.13. For $f : Q_1^n \to \mathbf{R}^m$

$$C(f,\varepsilon) \le \alpha(n,m,k) \inf_{h\in P, \|f-h\|^k \le \varepsilon} C(h, 2\varepsilon) \ .$$

The more computable bound is given in the following definition and theorem.

Definition 2.14. *For $f : Q_1^n \to \mathbf{R}^m$, f C^k, and $\varepsilon > 0$, $\sigma_2^k(f,\varepsilon)$ is defined as $\sigma_2^k(f,\varepsilon) = \inf_{h\in P, \|f-h\|^k \le \varepsilon} \alpha(n,m,k) \cdot s_2^k(h)$.*

By Corollaries 2.11 and 2.13 we get:

Theorem 2.15. *For f as above, $\varepsilon > 0$, $C_2^k(f,\varepsilon) \le \sigma_2^k(f,\varepsilon)$.*

The following result generalizes Theorem 1.13.

Theorem 2.16. *Let $f : Q_1^n \to \mathbf{R}^m$, $n \le m$, be a C^1-mapping. Then for any real $M > 0$, the n-volume of $M \cdot f(Q_1^n) \cap B_1^m$ does not exceed $C_2^1(f, 1/M) \le \sigma_2^1(f, 1/M)$.*

Proof: The set $\{x \in Q_1^n \ , \ M \cdot f(x) \in B_1^m\}$ coincides with the set $\{\|f\| \le 1/M\}$. By definition of $C_2^1(f,\varepsilon)$, it can be covered by the images of $C_2^1(f, 1/M)$ mappings $\varphi_i : Q_1^n \to Q_1^n$, such that $\|d(f\circ\varphi_i)\| \le 1/M$, or $\|d(Mf\circ\varphi_i\| \le 1$. Hence the n-volume of $Mf(Q_1^n)$ is at most $C_2^1(f, 1/M)$. Below we compute $\sigma_2^1(f,\varepsilon)$ for some functions of one variable.

3. Estimates for Complexity of Some Classes of Functions

We start with C^k-functions. Let $f : Q_r^n \to \mathbf{R}$ be such a function, and let $\varepsilon > 0$ be given. We subdivide Q_r^n into subcubes of the size r', such that the Taylor approximation on each subcube is ε-close to f in C^1-norm.

Since $\|df - dP\| \le \frac{1}{(k-1)!} \max \|d^k f\| (r')^{k-1} = \frac{1}{r} k \cdot R_k(f) \cdot \left(\frac{r'}{r}\right)^{k-1}$, we obtain that r' should be equal to $\left(r\varepsilon / kR_k(f)\right)^{1/(k-1)} \cdot r$. Defining h to be equal to the Taylor polynomial of f on each subcube $Q_{r'}^n$, we get $\|f - h\|_{C^1} \le \varepsilon$.

Therefore $C(f,\varepsilon) \le s(h) = 4\sum_i (2d_i)^n = 4(2k-2)^n \left(\frac{r}{r'}\right)^n$. Substituting the above expression for r', we get

$$C(f,\varepsilon) \le 4(2k-2)^n \left[\frac{kR_k(f)}{r\varepsilon}\right]^{\frac{n}{k-1}} \tag{3.1}$$

By Theorem 2.7 the same expression bounds $M(\varepsilon, \Delta(f))$.

This bound is a little bit weaker than the one of Theorem 1.7. The reason is technical: in the definition of $C(f,\varepsilon)$ we required (in order to simplify a presentation) the same rate of approximation ε for the function f and for its first derivatives. To get a sharp bound of Theorem 1.7, we must allow these rates to be slightly different.

A similar computation gives for a function of one variable $f : [0,1] \to \mathbf{R}^m$:

$$C_2^1(f,\varepsilon) \le 4(k-1)^3 \left(kR_k(f)/\varepsilon\right)^{1/(k-1)} \tag{3.2}$$

Via Theorem 2.15 we get a bound on volume growth, which is a little bit weaker than the one of Theorem 1.1.3. The reason is the same as above.

From (3.1) and (3.2) it may appear that for C^k-functions approximational complexity gives nothing essentially new. But even here it allows us to catch some rather obscure distinctions: for example, let $f(x_1, x_2, x_3, x_4, x_5) = g(x_1, x_2) + h(x_3, x_4, x_5)$, where $g \in C^7 - C^8$, $h \in C^{16}$. Thus f is only C^7 smooth, and by Theorem 1.7 we can conclude only that $M(\varepsilon, \Delta(f)) \sim (1/\varepsilon)^{5/7}$. But the variant of the above computation gives easily that

$$C(f,\varepsilon) \sim \left(\frac{1}{\varepsilon}\right)^{\frac{n_1}{k_1-1}+\frac{n_2}{k_2-1}} = \left(\frac{1}{\varepsilon}\right)^{\frac{1}{3}+\frac{1}{5}} = \left(\frac{1}{\varepsilon}\right)^{\frac{8}{15}},$$

and hence, in fact, $M(\varepsilon, \Delta(f)) \sim \left(\frac{1}{\varepsilon}\right)^{\frac{8}{15}}$.

We pass now to analytic functions. Let $f : Q_r^n \to \mathbf{R}$ be a real analytic function, extendable as a holomorphic one to an h-neighborhood of $Q_r^n \subseteq \mathbf{R}^n \subseteq \mathbf{C}^n$ in $\mathbf{C}^n$. We assume that $\sup |f| \le K$ in this neighborhood $(Q_r^n)_h$. (Compare, e.g., [8]). To approximate f by piecewise-polynomial functions, we proceed as follows: let $\varepsilon > 0$ be given. We subdivide Q_r^n to the equal subcubes Q_i' of the size r', and on each Q_i' we approximate f by an appropriate finite part (say, up to degree q) of its Taylor series. By assumptions, for any $x_0 \in Q_r^n$, f is holomorphic in the ball of radius h in $\mathbf{C}^n$, centered at x_o, and is bounded there by K. The standard estimates of the remainder show then that for $r' \le h/\sqrt{n}$,

$$\|f - P_q^i\|^1 \le c(n) K (\sqrt{n} r'/h)^q (1 - \sqrt{n} r'/h)^{-1} \tag{3.3}$$

where P_q^i is the Taylor polynomial of order q of f at $x_0^i \in Q_i'$.

To simplify computations, we fix $r' = h/e\sqrt{n}$, although this choice is not an optimal one. Thus $\|f - P_q^i\|^1 \le c(n) K (1/e)^q \le \varepsilon$ implies $q = \ln(cK/\varepsilon)$. For g equal to P_q^i on each Q_i', we get $C(f,\varepsilon) \le s(g) = 4(2q)^n \cdot (r/r')^n = (e\sqrt{n}\, r/h)^n \cdot 4 \cdot 2^n (\ln cK/\varepsilon)^n$. Thus

$$C(f,\varepsilon) \le C_1(n,k) \left[\frac{r}{h} \ln\left(\frac{K}{\varepsilon}\right)\right]^n . \tag{3.4}$$

It is interesting to analyze the consequences of the complexity estimates (3.4) for analytic functions. For example, the set of critical values $\Delta(f)$ of any analytic $f : Q_r^n \to \mathbf{R}$ is finite.

However, the bound $M(\varepsilon, \Delta(f)) \le C_1(n,k)\left[\frac{r}{h}\ln\left(\frac{K}{\varepsilon}\right)\right]^n$ is in no way trivial. Indeed, it is easy to see that the number of points in $\Delta(f)$ cannot be bounded in terms of h and K. But the above bound on $M(\varepsilon, \Delta)$ gives essential information on the geometry of Δ. E.g., $\Delta(f)$ cannot contain "too long" finite part of a sequence $A = \{1, 1/2, 1/3, \dots\}$, since $M(\varepsilon, A) \sim 1/\sqrt{\varepsilon}$ grows much faster than $\left[\ln(1/\varepsilon)\right]^n$ as $\varepsilon \to 0$. Sequences which may appear in $\Delta(t)$ (by these metric considerations) have the following form: $a_i = \left(\frac{1}{q}\right)^{i^\alpha}$, where $q > 1$, $\alpha > 0$. Indeed, one can show that for $A = \{a_1, a_2, \dots\}$, $M(\varepsilon, A) \sim \left(\frac{\ln 1/\varepsilon}{\ln q}\right)^{1/\alpha}$. This corresponds to $M(\varepsilon, \Delta(f))$ for f analytic on Q^n, $n \sim 1/\alpha$, and extendable to the h-neighborhood of Q^n in C^n, $h \sim \ln q$.

The same is true for a volume growth under linear dilations. One can easily show that for $f : Q_1^n \to \mathbf{R}^n$ an analytic mapping, $\mathrm{Vol}_n\left(Mf(Q_1^n) \cap B_1^m\right)$ is uniformly bounded for all $M > 0$, but this bound *cannot* be expressed in terms of h and K. Via Theorem 2.15, an estimate (3.4) provides therefore a nontrivial bound on the considered volume for finite values of M.

To bound the complexity of C^k and analytic functions we can use the global polynomial approximation instead of the piecewise Taylor approximation, used above. The bounds obtained in this way essentially coincide with those of (3.1) and (3.4). An important point is that the quasianalytic functions (in the sense of S. Bernstein) turn out to have the same complexity as the analytic ones, while they may have only a finite smoothness or not to be differentiable at all. We do not give here the detailed presentation of this approach.

There is another class of functions whose approximational complexity can be bounded or even exactly computed: those representable as the sums of infinite series of piecewise-polynomial functions. The study of such functions belongs in fact to the field of fractal geometry, so we just outline here the main points.

Let us assume that all the piecewise polynomial (and C^1) functions $h_i \in P$ are normalized in such a way that $\|h_i\|^1 = 1$. Consider $f = \sum_{i=1}^{\infty} \alpha_i h_i$, with $\sum_{i=1}^{\infty} |\alpha_1| < \infty$. Hence $f \in C^1$. Choosing the rate of convergence of α_i to zero, we can provide any desirable rate of approximation of f by the finite sums $\bar{h}_p = \sum_{i=1}^{p} \alpha_i h_i$, and hence, according to section 2 above, any desirable upper bound for the complexity function $C(f, \varepsilon)$.

On the other hand, the second derivative of h_i may be arbitrarily big, so the sequence h_i can be easily chosen in such a way that f is nowhere C^2. This illustrates once more the fact that the differentiability and the complexity properties of functions have quite a different nature.

Finally, in this way we can construct examples of functions with a complexity $C(f, \varepsilon)$ equal to a function $\sigma(\varepsilon)$ prescribed in advance (at least asymptotically, as $\varepsilon \to 0$). Namely, we choose the sequences α_i, h_i in such a way that the critical points and values of $f = \sum_{i=1}^{\infty} \alpha_i h_i$ can be explicitly determined, and that an upper bound $\sigma(f,\varepsilon)$ for $C(f,\varepsilon)$, computed as above, is equal to $\sigma(\varepsilon)$. An important fact then is that f can be constructed in such a way that $M(\varepsilon,\Delta(f))$ behaves asymptotically exactly as $\sigma(f,\varepsilon)$. Theorem 2.7 now shows that $C(f,\varepsilon) \sim \sigma(f,\varepsilon) = \sigma(\varepsilon)$.

4. Complexity of Functions on Infinite-Dimensional Spaces

The approximational complexity techniques, described in the introduction and section 2, apply equally well to functions on infinite-dimensional spaces. Definition 2.1 and Proposition 2.4, which form the main points of the approach, remain valid without any change. However, it is much more difficult to find approximating classes P of "simple" functions.

Roughly speaking, the basic difference between infinite and finite dimensional situations is that the polynomials of an infinite number of variables are no more simple.

This basic fact determines also the following interesting feature of infinite dimensional complexity problems: non-adequacy of the notion of smoothness here becomes quite transparent. Let us start with the example of this kind, belonging to I. Kupka [11].

Let $l^2 = \{x = (x_1,\ldots,x_i,\ldots), \Sigma x_i^2 < \infty\}$ be the standard Hilbert space. We define a function

$$f : l^2 \to \mathbf{R} \quad \text{by} \quad f(x) = \sum_{i=1}^{\infty} \frac{1}{2^i}\varphi(ix_i) ,$$

where φ is the polynomial of degree 3, such that $\varphi(0) = \varphi'(0) = 0$, $\varphi(1) = 1$ and $\varphi'(1) = 0$. Thus φ has exactly two critical points 0 and 1 with the critical values 0 and 1 respectively.

One can easily show that f is infinitely differentiable (in any reasonable definition of differentiability on l^2). In fact f can be considered as a polynomial of degree 3 on l^2.

Now $x = (x_1,\ldots,x_i,\ldots)$ is a critical point of f if and only if for each i, ix_i is a critical point of φ. Thus

$$\Sigma(f) = \left\{\left(a_1,\frac{a_2}{2},\ldots,\frac{a_i}{i},\ldots\right) \;,\; a_i = 0,1\right\} \quad \text{and} \quad \Delta(f) = \left\{\sum_{i=1}^{\infty}\frac{1}{2^i}a_i,\; a_i = 0,1\right\} = [0,1] \,.$$

Hence the critical values of a C^∞-function f cover the interval: the Sard theorem is no more valid on l^2. Notice that the function f can be approximated by "simple" ones, namely by the polynomials, depending only on a finite number of variables: $\sum_{i=1}^{N} \varphi(ix_i)$. Generalizing this remark, we show below that f violates the Sard theorem since the rate of its approximation is not high enough. We present then a wide class of functions on l_2, whose approximational complexity is low enough to imply, in particular, the Sard theorem.

Thus, the class P will consist of polynomials, depending only on a finite number of variables, or more precisely, on a finite number of linear functionals. We need the following result:

Proposition 4.1. *Let V be a Banach space and let $l_1, \dots, l_n$ be linear functions on V. Let $p(x_1, \dots, x_n)$ be a polynomial of degree d. Then for a function $\tilde{p} : B \to \mathbf{R}$, $\tilde{p}(v) = p(l_1(v), \dots, l_n(v))$, $C(\tilde{p}, \varepsilon) \le n \cdot (2d)^n$.*

Here $B \subseteq V$ denotes the unit ball in V, and $C(\tilde{p}, \varepsilon)$, as above, is the minimal number of ε-intervals, covering the set of ε-critical values of $\tilde{p}$ on B.

For $V = l^2$ and $l_1, \dots, l_n$ orthonormal, n can be omitted.

Proof: Consider the mapping $L : V \to \mathbf{R}^n$, $L(v) = (l_1(v), \dots, l_n(v))$. It is well known that there is a Euclidean structure on $\mathbf{R}^n$, such that $B^n_{1/\sqrt{n}} \subseteq L(B) \subseteq B^n_{\sqrt{n}}$, where B^n_r is the ball of radius r in this structure. Therefore, if $v \in B$ is an ε-critical point of $\tilde{p} = p \circ L$, then $L(v) \in \mathbf{R}^n$ is a $\sqrt{n}\varepsilon$-critical point of p. Hence $\Delta(\tilde{p}, \varepsilon, 1) \subseteq \Delta(p, \sqrt{n}\varepsilon, \sqrt{n})$, and the required result follows by Theorem 1.5.

Now, for $V = l^2$ and $l_1, \dots, l_n$ orthonormal, the coefficient $\sqrt{n}$ in the above inequality can be omitted.

Now consider the functions $f : l^2 \to \mathbf{R}$ of the form $f(x) = \sum_{i=1}^{\infty} \alpha_i p_i(x_1, \dots, x_i)$, where $p_i(x_1, \dots, x_i)$ is a polynomial of degree d_i, such that $\max|p_i| \le 1$ on the unit ball in $\mathbf{R}^i$, $i = 1, 2, \dots$. By the Markov inequality (see [5], [10]), $\max\|dp_i\| \le d_i^2$ on the unit ball in $\mathbf{R}^i$. Hence if α_i satisfy $\sum_{i=1}^{\infty} \alpha_i d_i^2 < \infty$, f is continuously differentiable on the unit ball $B \subseteq l^2$.

For f as above we define $\eta(f, \varepsilon)$ as follow: $\eta(f, \varepsilon) = 4[2 \max(d_1, \dots, d_{N(\varepsilon)})]^{N(\varepsilon)}$, where $N(\varepsilon)$ is the smallest natural N, for which the sum $\sum_{i=N+1}^{\infty} \alpha_i d_i^2 \le \varepsilon$.

Theorem 4.2. *For f as above, $C(f, \varepsilon) \le \eta(f, \varepsilon)$. In particular, for any $\varepsilon > 0$, $M(\varepsilon, \Delta(f)) \le \eta(f, \varepsilon)$.*

Proof: We have for $N = N(\varepsilon)$, $\|f - P_N\|^1 \le \varepsilon$, where $P_N = \sum_{i=1}^{N} \alpha_i p_i$. But P_N is a polynomial of the variables $x_1, \dots, x_N$ in l^2 and of the degree $\max(d_1, \dots, d_N)$. Hence, by Proposition 4.1, $C(P_N, \varepsilon) \le (1/4)\eta(f, \varepsilon)$, and the result follows by Proposition 2.4.

In particular, if all the degrees d_i are equal to d, and α_i form the geometric sequence $\alpha_i = \left(\frac{1}{q}\right)^i = \alpha^i$, $\alpha < 1$, we have the following: $\sum_{i=N+1}^{\infty} \alpha^i d = \frac{d\alpha^{N+1}}{1-\alpha}$. For this to be $\le \varepsilon$ we get

$N = [\log_\alpha((1-\alpha)\varepsilon d)]$, and $\eta(f,\varepsilon) \le 4(2d)^{\log_\alpha((1-\alpha)\varepsilon d)} = C(\alpha,d)\left(\frac{1}{\varepsilon}\right)^{\log_q(2d)}$. Thus we have the following corollary:

Corollary 4.3. *For* $f = \sum_{i=1}^{\infty} \left(\frac{1}{q}\right)^i p_i(x_1,\dots,x_i)$, $\deg p_i = d$, $|p_i| \le 1$ *on the unit ball and* $q > 1$, $C(f,\varepsilon) \le C(q,d)\left(\frac{1}{\varepsilon}\right)^{\log_q(2d)}$. *In particular, for* $q > 2d$, f *satisfies the Sard theorem (i.e.,* $m(\Delta(f)) = 0$*).*

Returning to the Kupka example above, we see that it occurs exactly on the boundary: by Corollary 4.3., for any $q > 6$, functions $f = \sum_{i=1}^{\infty} \frac{1}{q_i} p_i(x,\dots,x_i)$ with $\deg p_i = 3$, satisfies the Sard theorem. In the concrete example above it is enough to take $q > 2$.

By formal analogy we can say that the complexity of the function $f : l^2 \to \mathbf{R}$, $f = \Sigma \frac{1}{q^i} p_i(x_1,\dots,x_i)$, $\deg p_i = d$, is the same as the complexity of C^k-functions $g : B^n \to \mathbf{R}$, if $\frac{n}{k-1} = \log_q(2d) = \beta$. In particular, the sequences of the form $1, 1/2^s, 1/3^s, \dots, 1/k^s, \dots$, may appear among the critical values of both f and g only if $1/(s-1) \le \beta$.

Pursuing this formal analogy a little bit further, we find that the functions $f : l^2 \to \mathbf{R}$, $f(x) = \sum_{i=1}^{\infty} \alpha_i p_i(x_1,\dots,x_i)$, have the complexity of the usual analytic functions, if the coefficients α_i tend to zero very fast: $\alpha_i \sim \left(\frac{1}{q}\right)^{b^i}$, where $q > 1$, $b > 1$. Indeed, proceeding exactly as above, we find, that $\|f - P^N\|^1 \le \varepsilon$, if $\left(\frac{1}{q}\right)^{b^N} \le \varepsilon$, or for $N \ge \log\left(\frac{\ln(1/\varepsilon)}{\ln q}\right)$, and for $\deg p_i \equiv d$, we get the following:

Corollary 4.4. *For* $f = \sum_{i=1}^{\infty} \alpha_i p_i(x_1,\dots,x_i)$, *with* $|p_i| \le 1$ *on the unit ball,* $\deg p_i = d$ *and* $\alpha_i = \left(\frac{1}{q}\right)^{b^i}$, $q > 1$, $b > 1$, $C(f,\varepsilon) \le c(b,q,d)\left(\frac{\ln(1/\varepsilon)}{\ln q}\right)^{\log_b(2d)}$.

Comparing this expression with (3.4) we can say formally, that for f as above, $\log_b(2d)$ plays the role of the dimension, and $\ln q$ – the role of the size h of a complex neighborhood, to which the analytic function extends holomorphically. In both the cases of Corollaries 4.3 and 4.4 the examples show that these complexity estimates are essentially sharp.

We conclude with some functions on the spaces $C^k([0,1])$, to show how the geometry of an underlying Banach space enters the complexity estimates.

Let $q > 1$. Consider the sequence of points $x_i = (1/q)^i \in [0,1]$. Let $\varphi_i(x)$ be polynomials of one variable, of degree d, such that $|\varphi_i(x)| \le 1$ for $|x| \le 1$. Define $\Phi : C^k([0,1]) \to \mathbf{R}$ by $\Phi(u) = \sum_{i=1}^{\infty} \alpha_i \varphi_i(u(x_i))$, $u \in C^\kappa([0,1])$. As above, we see that if $\sum_{i=1}^{\infty} \alpha_i d_i^2 < \infty$, then Φ is C^1.

Let $\varepsilon > 0$ be given. To ε-approximate Φ by polynomials of a finite number of variables, we proceed as follows: subdivide $[0,1]$ into two parts: $[0,\delta]$ and $[\delta,1]$. Clearly, $x_i = \left(\frac{1}{q}\right)^i \in$

$[0,\delta]$ for $i \geq [\log_q(\frac{1}{\delta})] + 1 = N(\delta)$. Define $P_1^\delta(u)$ by $P_1^\delta(u) = \sum_{i=1}^{N(\delta)-1} \alpha_i \varphi_i(u(x_i))$. Let $\tilde{u}$ be the Taylor polynomial of $u \in C^k([0,1])$, of degree $k-1$ at the point $0 \in [0,1]$. We define $P_2^\delta(u)$ as $P_2^\delta(u) = \sum_{i=N(\delta)}^{\infty} \alpha_i \varphi_i(\tilde{u}(x_i))$. Finally, we put $P^\delta(u) = P_1^\delta(u) + P_2^\delta(u)$. We notice that $P_2^\delta(u)$ is in fact a polynomial of degree d in k linear functionals on $C^k([0,1])$: $u(0), u'(0), \ldots, u^{(k-1)}(0)$. Thus P^δ is a polynomial of degree d in $N(d)+k-1$ linear functionals $u(x_1), u(x_2), \ldots, u(x_{N(d)-1}), u(0), u'(0), \ldots, u^{(k-1)}(0)$ on $C^k([0,1])$.

We have $\|\Phi - P^\delta\|^1 \leq \sup_{\|u\| \leq 1} (|\Phi(u) - P^\delta(u)| + \|d\Phi(u) - dP^\delta(u)\|)$. Since $\Phi(u) - P^\delta(u) = \sum_{i=N(\delta)}^{\infty} \alpha_i(\varphi_i(u(x_i)) - \varphi_i(\tilde{u}(x_i)))$ and since, by the Taylor formula, $|u(x) - \tilde{u}(x)| \leq \frac{1}{k!}\delta^k$ for $x \in [0,\delta]$, we get $|\Phi(u) - P^\delta(u)| \leq \frac{d^2}{k!}\delta^k \sum_{i=N(\delta)}^{\infty} \alpha_i$. (By the Markov inequality, $\varphi_i' \leq d^2$ on $[-1,1]$.)

For the derivative at u in the direction of the function $v \in C^k([0,1])$, $\|v\| \leq 1$, we get $|d\Phi(u)(v) - dP^\delta(u)(v)| \leq \sum_{i=N(\delta)}^{\infty} \alpha_i |\varphi_i'(u(x_i)) \cdot v(x_i) - \varphi_i'(\tilde{u}(x_i))\tilde{v}(x_i)|$ where as above $\tilde{u}$ and $\tilde{v}$ are the Taylor polynomials of u and v, respectively. Therefore, $|\varphi_i'(u(x_i))v(x_i) - \varphi_i'(\tilde{u}(x_i))\tilde{v}(x_i)| \leq |\varphi_i'(u(x_i)) - \varphi_i'(\tilde{u}(x_i))| \cdot |v(x_i)| + |\varphi_i'(\tilde{u}(x_i))| \cdot |\tilde{v}(X_i) - v(x_i)| \leq d^2(d-1)^2 \frac{1}{k!}\delta^k + d^2 \frac{1}{k!}\delta^k \leq \frac{d^4}{k!}\delta^k$, and $\|d\Phi - dP^\delta\| \leq \frac{d^4}{k!}\delta^k \sum_{i=N(\delta)}^{\infty} \alpha_i$. Finally, we get $\|\Phi - P^\delta\|^1 \leq \frac{2d^4}{k!}\delta^k \sum_{i=N(\delta)}^{\infty} \alpha_i \stackrel{\text{def}}{=} \eta(\delta)$.

Now, let $\delta(\varepsilon)$ be the maximal δ such that $\eta(\delta) \leq \varepsilon$ and let $N(\varepsilon) = N(\delta(\varepsilon))$. Exactly as above and using Proposition 4.1, we get $C(\Phi,\varepsilon) < 4(2d)^{N(\varepsilon)} \cdot N(\varepsilon) \stackrel{\text{def}}{=} \eta(\Phi,\varepsilon)$. Therefore we've proved the following:

Theorem 4.5. *For Φ as above, $C(\Phi,\varepsilon) \leq \eta(\Phi,\varepsilon)$.*

To give explicit bound, assume, e.g., that $\alpha_i = (1/s)^i$. Then $\sum_{i=N(\delta)}^{\infty} \alpha_i \sim (\frac{1}{s})^{N(\delta)} \sim (\frac{1}{s})^{\log_q(1/\delta)} = \delta^{\log_q s}$. Hence $\eta(\delta) \sim \delta^{k+\log_q s}, \delta(\varepsilon) \sim \varepsilon^{1/(k+\log_q s)}$, $N(\varepsilon) \sim \frac{\log_q(1/\varepsilon)}{k+\log_q s}$, and finally, $\eta(\Phi,\varepsilon) = 4(2d)^{N(\varepsilon)} \cdot N \sim (\frac{1}{\varepsilon})^{\frac{\log_q(2d)}{k+\log_q s}} \cdot N$. We have therefore:

Corollary 4.6. *For $\Phi : C^k([0,1]) \to \mathbf{R}$, $\Phi(u) = \sum_{i=1}^{\infty} (\frac{1}{s})^i \varphi_i(u(x_i))$, where $x_i = \left(\frac{1}{q}\right)^i$, $s,q > 1$, where φ_i are the polynomials of one variable of degree d with $|\varphi_i(x)| \leq 1$, $x \in [0,1]$, $C(\Phi,\varepsilon) \leq c(s,q,k,d) (\frac{1}{\varepsilon})^{\frac{\log_q(2d)}{k+\log_q s}} \cdot \log_q(\frac{1}{\varepsilon})$.*

Examples (which can be constructed rather easily, but require some exercises in fractal geometry on $\mathbf{R}$) show that this bound is essentially sharp.

Thus the same function Φ considered on the spaces $C^k([0,1])$ for various k, has various complexities: the higher is k, the simpler is Φ.

5. Concluding Remarks

1. To simplify the presentation we restricted the approximating classes above to polynomials or piecewise-polynomials. In fact the most natural approximating class (in the finite dimensional case) seems to be the class of semialgrebraic functions. In particular, such an extension allows the estimation of complexity of an additional important class of function: functions, representable as $f(x) = \max_{t \in T} h(x,t)$, with $h(x,t)$ smooth, analytic etc., T-compact.

2. In section 2-4 we considered mostly the complexity function $C(f,\varepsilon)$; related to the structure of critical values, and, to some extent, $C_2^1(f,\varepsilon)$, responsible for the volume growth under big dilations. In a similar way other complexity properties can be treated: average complexity of fibers, representation by means of compositions, etc. Computational complexity can be treated similarly. While all these properties concern individual functions, it turns out that the complexity (ε-entropy) of functional spaces, in the spirit of [8], is also closely related to the approximational complexity of functions in the space. *A priori* definitions of all these types of complexity have not much in common, but computations lead to essentially the same formula in each case. One can conjecture the existence of a unifying complexity notion.

3. Although approximational complexity and differentiability are related directly only in one way (degree of smoothness, in finite dimension, bounds complexity), a kind of inverse "Bernstein-type" theorem is true: functions of low approximational complexity have high order differentials at almost every point. More precisely (and more in the spirit of our approach), the set of points, where the high order differential does not exist, is small and "simple", and this set, in turn, can be "stratified" according to the behavior of f on it.

References

1. V.I. Arnold and O.A. Oleinik, Topology of real algebraic varieties. Mosk. Univ. Math. N6 (1980), 7-16.
2. M. Coste, Ensembles semi-algebraique, Lecture Notes Math. 959, 1982, 109-138.
3. M. Gromov, Entropy, homology and semialgebraic geometry (after Y. Yomdin). Seminaire N. Bourbaki, 1985-86, no. 663.
4. L.D. Ivanov, Variazii mnozestv i funkzii, Moscow, Nauka, 1975.
5. O.D. Kellog, On bounded polynomials of several variables, Math. Z.27, 1928, 55-64.
6. A.G. Khovansky, Theorem de Bezout pour les fonctions de Liouville, preprint IHES, 1981.

7. I. Kupka, Counterexample to the Morse-Sard theorem in the case of infinite-dimensional manifolds, Proc. AMS, 16 (1965), 954-957.

8. A.N. Kolmogorov, V.M. Tihomirov, ε-entropy and ε-capacity of sets in functional spaces, Usp. Mat. Nauk 2(86), 14, 1959, 3-86, Am. Math. Soc. Transl. (2), 17, 1961, 277-364.

9. A. Manning, Topological entropy and the first homology group. Springer LNM 468 (1975), 422-429.

10. V. Markov, Uber Polynome, die in einem gegebenem Intervalle Moeglichst wenig von Null abweichen. Math. Ann. 77 (1916), 213-258.

11. M. Misiurewicz and F. Przytyzcki, Topological entropy and degree of smooth mappings. Bull. Acad. Pol. Sci. 25 (1977), 573-574.

12. J.J. Risler, Complexite et Geometrie, (d'apres A. Khovansky), Seminaire Bourbaki, 1984-1985, no. 637.

13. J.J. Risler, Additive complexity and zeroes of real polynomials, preprint.

14. L.A. Santalo, Integral geometry and geometric probability, Addison-Wesley, 1976.

15. A. Sard, The measure of critical values of differentiable maps. Bull. AMS 48 (1942), 883-890.

16. M. Shub, Dynamical systems, filtrations and entropy. Bull. AMS 80 (1974), 27-41.

17. M. Shub and D. Sullivan, Homology theory and dynamical systems. Topology 14 (1975), 109-132.

18. Y. Yomdin, The geometry of critical and near-critical values of differentiable mappings. Math. Ann. 264 (1983), 495-515.

19. Y. Yomdin, The set of zeroes of an "almost polynomial" function, Proceedings of the AMS, 90, N4, (1984), 538-542.

20. Y. Yomdin, Global bounds for the Betti numbers of regular fibers of differentiable mappings, Topology 24, 2,(1985), 145-152.

21. Y. Yomdin, Quantitative version of Kupka-Smale Theorem. Ergodic Theory and Dynamical Systems 5, 3 (1985), 449-472.

22. Y. Yomdin, Volume growth and entropy. Israel J. Math. 57, 3 (1987), 285-300.

23. Y. Yomdin, C^k-resolution of semialgebraic mappings. Israel J. Math. 57, 3 (1987), 301-317.

24. Y. Yomdin, Critical values and representations of functions by means of compositions, SIAM J. on Mathematical Analysis 17, 1 (1986), 236-239.

24. Y. Yomdin, On functions representable as a supremum of smooth functions. SIAM J. on Mathematical Analysis 14, 2 (1983), 239-246.

26. Y. Yomdin, On functions, representable as a supremum of a family of smooth functions, II, SIAM J. on Mathematical Analysis 17, 1986.

27. Y. Yomdin, Metric properties of semialgebraic sets and mappings and their applications in smooth analysis, Proceedings of the Second International Conference on Algebraic Geometry, La Rabiba, Spain, 1984, J.M. Aroca, T. Sanchez-Geralda, J.L. Vicente, eds. Travaux en Course, Hermann, Paris 1987.

28. Y. Yomdin, Metric semialgebraic geometry with applications in smooth analysis, to appear.

MINKOWSKI SUMS AND SYMMETRIZATIONS

J. Bourgain IHES, France

J. Lindenstrauss Hebrew University

V.D. Milman Tel Aviv University

0. Introduction

It was found out recently that the empirical distribution method from probability theory has interesting (and perhaps even surprising) consequences in finite-dimensional convexity theory. The relevance of this method to the problem of approximating zonoids by zonotopes with few summands was first pointed out by Schechtman [S]. In [B.L.M] we refined the method and results of Schechtman. In section 6 of [B.L.M] we pointed out that the empirical distribution method gives interesting results if we consider instead of Minkowski sums of segments (i.e., zonotopes) Minkowski sums of more general convex sets. Here we intend to explore this direction in more detail.

In the first section we present results whose proofs are relatively easy modulo known results in the literature. We present those results because of their intrinsic interest from the geometric point of view. There are obviously many variants of the results which can be proved along similar lines. One of the interesting aspects of some of the results of this section, as well as the result of the second section, is that they combine the empirical distribution method with the concentration of measure technique in the spirit of [F.L.M] or more generally [M.S].

In the second section we consider Minkowski symmetrizations (sometimes also called Blaschke symmetrizations) of convex sets. We obtain there a rather sharp estimate on the number of steps needed in order to obtain approximately a ball if we start from an arbitrary convex body and choose the directions of symmetrization randomly and independently. Technically, section 2 is more involved than section 1.

1. Minkowski Sums

We start by recalling the basic tool from the empirical distribution method namely the so called Bernstein's inequality. We shall state here three versions of this basic inequality as all three will be used in the sequel. The proofs of all versions are simple and very similar (and of course also classic). We shall present here the proof of the third version (the proofs of (i) and (ii) were recalled in [B.L.M]; a martingale version of (ii) will be proved in the next section). Before stating the result, we recall the definition of Orlicz norms which enter in its statement.

Let (Ω, μ) be a probability space and let $\Phi(t)$ be a strictly increasing convex function on $[0, \infty)$ so that $\Phi(0) = 0$ and $\lim_{t\to\infty} \Phi(t) = \infty$. We denote by $L_\Phi(\mu)$ the space of all real valued measurable functions on Ω so that $\int_\Omega \Phi(|f|/\lambda)d\mu < \infty$ for some $\lambda > 0$ and put

$$\|f\|_{L_\Phi(\mu)} = \inf\{\lambda > 0 \,;\, \int_\Omega \Phi(|f|/\lambda)d\mu \le 1 .\} \tag{1}$$

We shall be concerned here only with the two functions

$$\psi_1(t) = e^t - 1 \quad , \quad \psi_2(t) = e^{t^2} - 1 \tag{2}$$

(besides of course the functions t^p, $1 \le p < \infty$, which give rise to the usual L_p spaces).

Proposition 1. *Let $\{g_j\}_{j=1}^N$ are independent random variables with mean 0 on some probability space (Ω, μ).*

(i) Assume that the $\{g_j\}$ are all bounded and that $\|g_j\|_1 \le 2$ and $\|g_j\|_\infty \le A$ for all j and some constant A; then, for $0 < \varepsilon < 1$,

$$\text{Prob}\,\{|\sum_{j=1}^N g_j| > \varepsilon N\} \le 2\exp(-\varepsilon^2 N/8A) . \tag{3}$$

(ii) Assume that $\{g_j\}$ belong to $L_{\psi_1}(\mu)$ and that $\|g_j\|_{L_{\psi_1}} \le A$ for all j and some constant A. Then, for $0 < \varepsilon < 4A$,

$$\text{Prob}\,\{|\sum_{j=1}^N g_j| > \varepsilon N\} \le 2\exp(-\varepsilon^2 N/16A^2) . \tag{4}$$

(iii) Assume that the $\{g_j\}$ belong to $L_{\psi_2}(\mu)$ and that $\|g_j\|_{L_{\psi_2}} \le A$ for all j and some constant A. Then, for all $\varepsilon > 0$,

$$\text{Prob}\,\{|\sum_{j=1}^N g_j| > \varepsilon N\} \le 2\exp(-\varepsilon^2 N/8A^2) . \tag{5}$$

Note that in contrast to (i) and (ii) we do not have in (iii) any restriction of the size of ε. We will in fact use (iii) below also for large ε.

Proof of (iii). Without loss of generality we may assume that $A = 1$ and thus, for $1 \le j \le N$,

$$\int_\Omega \exp(g_j^2) d\mu) \le 2 .$$

In particular

$$\int_\Omega |g_j|^{2k} d\mu \le 2 \cdot k! , \qquad 1 \le k < \infty,$$

and hence also $\int_\Omega |g_j|^{2k-1} d\mu \le 2 \cdot k!$ for $k \ge 1$. If $\lambda \ge 1$ we deduce that

$$\int_\Omega \exp(\lambda g_j) d\mu \le 1 + \sum_{k=2}^{\infty} \int_\Omega |\lambda g_j|^k d\mu / k!$$

$$\le 1 + \lambda^2 + 2 \sum_{k=2}^{\infty} \lambda^{2k} k! ((2k!)^{-1} + (2k-1)!^{-1}) \le e^{2\lambda^2} .$$

This estimate holds also for $0 < \lambda \le 1$. Consequently,

$$\text{Prob}\{ \sum_{j=1}^{N} g_j > \varepsilon N \} e^{\varepsilon \lambda N} \le \int_\Omega \exp(\lambda \sum_{j=1}^{N} g_j) d\mu$$

$$= \prod_{j=1}^{N} \int_\Omega \exp(\lambda g_j) d\mu \le e^{2\lambda^2 N} .$$

By taking $\lambda = \varepsilon/4$ we deduce (5). □

In order to apply in a significant way Proposition 1 in convexity theory it is useful to have at our disposal distribution inequalities which ensure that for some functions g which are of geometric interest the norm $\|g\|_1$ is comparable to $\|g\|_{L_{\psi_1}}$ or even $\|g\|_{L_{\psi_2}}$. The basic result in this direction is Khintchine's inequality which states the following: There is an absolute constant c so that for every choice of reals $\{u_i\}_{i=1}^n$ we have

(6)
$$(\text{Average} |\sum_{i=1}^{n} \pm u_i|^p)^{1/p} \le c p^{1/2} (\sum_{i=1}^{n} u_i^2)^{1/2} , \qquad 1 \le p < \infty$$

(7)
$$(\sum_{i=1}^{n} u_i^2)^{1/2} \le \sqrt{2}\, \text{Average} |\sum_{i=1}^{n} \pm u_i| ,$$

where the averages are taken over all 2^n possible choice of signs. The best constants in (6) and (7) were found in [Sz] and [H] but their exact values are not important to us. The only thing which matters is that the constant in the right hand side of (6) behaves like $p^{1/2}$ as $p \to \infty$ and this is not hard to verify. Another form of stating (6) and (7) is that

$$(8) \qquad \Big(\int_0^1 \Big|\sum_{i=1}^n r_i(t)u_i\Big|^p dt\Big)^{1/p} \le cp^{1/2}\Big(\sum_{i=1}^n |u_i|^2\Big)^{1/2} \le c_1 p^{1/2} \int_0^1 \Big|\sum r_i(t)u_i\Big| dt$$

where $\{r_i\}$ denote the Rademacher functions. By expanding $\exp(x^2)$ into power series we deduce from (8) that for some constant c (we use here and below the same notation for different absolute constants in different formulas)

$$(9) \qquad \Big\|\sum_{i=1}^n r_i(t)u_i\Big\|_{L_{\psi_2}(0,1)} \le c\Big\|\sum_{i=1}^n r_i(t)u_i\Big\|_{L_1(0,1)} .$$

Khintchine's inequality has been generalized in many directions and several of these generalizations will be useful to us below.

Kahane [Ka] and Kwapien [Kw] (cf. also [L.T, p.74] for a proof and [Ta] for a new approach) showed that (8) and (9) remain valid if we replace the scalars $\{u_i\}_{i=1}^n$ by vectors in a Banach space X (the constants being independent of X as well as of the particular choice of n and $\{u_i\}_{i=1}^n$). Thus we have

$$(10) \qquad \Big\|\sum_{i=1}^n r_i(t)u_i\Big\|_{L_{\psi_2}((0,1),X)} \le c\Big\|\sum_{i=1}^n r_i(t)u_i\Big\|_{L_1((0,1),X)} .$$

A second generalization of (9) is Borell's inequality [Bo] (cf. [G.M] for the current point of view) which follows from the Brunn Minkowski inequality. Let K be a convex body symmetric with respect to the origin in $\mathbb{R}^n$ with volume 1. Let $y \in \mathbb{R}^n$ be a vector of Euclidean norm 1 and let $\varphi(t)$ be the $(n-1)$-dimensional volume of $K \cap \{x; \langle x, y\rangle = t\}$. Then for some absolute constants c and C

$$(11) \qquad \varphi(t) \le C\varphi(0)\exp\big(-c\varphi(0)|t|\big)$$

and consequently if μ is the Lebesgue measure restricted to K then

$$(12) \qquad \|\langle\cdot, y\rangle\|_{L_{\psi_1}(\mu)} \le c_1\|\langle\cdot, y\rangle\|_{L_1(\mu)}$$

In this case we cannot replace in general L_{ψ_1} by L_{ψ_2} (e.g., for the unit ball in ℓ_1^n, suitably normalized). For K the cube (12) becomes a weak form of Khintchine inequality. A discussion of the relation between Borell's inequality and Kahane's inequality is presented in Appendix II of [M.S].

Another direction in which (9) or (10) can be generalized is by replacing the Rademachers by more general random variables. In particular these inequalities hold if the Rademachers are replaced by independent normalized Gaussian variables [L.S]. Marcus and Pisier [Ma.P] discussed this question in detail and deduced from the Gaussian version of Khintchine's inequality the following. Let $\| \ \|$ be any norm on $\mathbb{R}^n$, let $x \in \mathbb{R}^n$ and let μ_n the normalized Haar measure on the orthogonal group O^n. Then for some absolute c

(13) $$\|Ux\|_{L_{\psi_2}(\mu_n)} < c\|Ux\|_{L_1(\mu_n)} .$$

After presenting these preliminaries we can turn to the main subject of this section, i.e., results concerning Minkowski sums of convex bodies. We start by recalling a result from [B.L.M] (theorem 6.3. there).

Theorem 2. *Let K be a compact convex set in $\mathbb{R}^n$ and let $0 < \varepsilon < 1/2$. There is a constant $r(K) = r$ and there are orthogonal transformations $\{U_j\}_{j=1}^N$ on $\mathbb{R}^n$ with*

(14) $$N \le cn\varepsilon^{-2}\log\varepsilon^{-1}$$

so that

(15) $$(1-\varepsilon)rB^n \subset N^{-1}\sum_{j=1}^{N} U_j K \subset (1+\varepsilon)rB^n$$

Here B^n denote the Euclidean ball in $\mathbb{R}^n$ and c is, as usual, an absolute constant. Since the method of proof of Theorem 2 serves as a model to several of the theorems below we recall briefly its proof.

Proof: Without loss of generality we may assume that K is symmetric with respect to the origin (otherwise replace K by $K + (-K)$) and that it has non-empty interior. By duality (15) is equivalent to

(16) $$(1-\varepsilon)r|||x||| \le N^{-1}\sum_{j=1}^{N}\|U_j^* x\|_* \le (1+\varepsilon)r|||x|||$$

where $|||\cdot|||$ denote the Euclidean norm and $\| \ \|_*$ is the norm whose unit ball is the polar to K.

We choose

(17) $$r = \int_{S^{n-1}} \|x\|_* d\sigma_n(x) = \int_{O^n} \|Ux\|_* d\mu_n(U)$$

where σ_n is the normalized rotation invariant measure on S^{n-1} and μ_n is the Haar measure on O^n. Let $\{x_i\}_{i=1}^m$ be an ε-net with respect to $|||\cdot|||$ on S^{n-1} with $m \le (4/\varepsilon)^n$. For each i we consider the random variables $\{g_{i,j}\}_{j=1}^N$ on (O^n, μ_n) defined by

$$g_{i,j}(U) = \|Ux_i\|_* - r \qquad 1 \le j \le N$$

where N will be determined shortly. Clearly, for every i and j, $g_{i,j}$ has mean 0 and $\|g_{i,j}\|_{L_1(\mu_n)} \le 2r$. It follows from (13) that $\|g_{i,j}\|_{L_{\psi_2}(\mu_n)} \le c_1$ for some absolute constant c_1. From (5) we deduce that for every $\varepsilon > 0$ and every $1 \le i \le m$

$$\text{Prob}\left\{(U_1,\dots,U_N)\ ;\ \left|N^{-1}\sum_{j=1}^{N}\|U_jx_i\|_* - r\right| < \varepsilon r\right\} \le 2\exp(-\varepsilon^2 N/8c_1^2)\ .$$

Consequently if $2\exp(-\varepsilon^2 N/8c_1^2) \le 1/m$, i.e., if N satisfies (14) there exist $\{U_j\}_{j=1}^N$ so that (16) holds if we take as x any element of the ε-net $\{x_i\}_{i=1}^m$. This implies that (16) holds for every x with ε replaced by $\Delta = 3\varepsilon/(1-\varepsilon)$ in view of the following simple and well known lemma.

Lemma 3. *Let T be a bounded linear map from a Banach space X into a Banach space Y. Let $0 < \varepsilon < 1$ and assume that for an ε-net $\mathcal{F}$ of the unit sphere of X, $\big|\|Tx\| - \|x\|\big| < \varepsilon$ for every $x \in \mathcal{F}$. Then $\big|\|Tx\| - \|x\|\big| < 3\varepsilon\|x\|/(1-\varepsilon)$ for every $x \in X$.*

Proof: Put $\Delta = \sup\left\{\big|\|Tx\| - \|x\|\big|\ ,\ \|x\| = 1\right\}$. For every $x \in X$ with $\|x\| = 1$ there is a $u \in \mathcal{F}$ wo that $\|x - u\| \le \varepsilon$. We have

$$\big|\|Tx\| - \|x\|\big| \le \big|\|Tu\| - \|u\|\big| + \big|\|T(x-u)\| - \|x-u\|\big| + 2\|x-y\|\ .$$

Hence $\Delta \le \varepsilon + \varepsilon\Delta + 2\varepsilon$ or $\Delta \le 3\varepsilon/(1-\varepsilon)$. This concludes the proof of the Lemma as well as of Theorem 2. □

Remark. The dependence of N in (14) on n and ε is close to being optimal in the case K is a segment, especially for large n. This question was discussed in some detail in section 6 of [B.L.M]. It is known that if K is a segment one can replace (14) by $N \le c\varepsilon^{-2}n$ ([G]). It is likely that the $|\log\varepsilon|$ factor can be dropped also for general K, we did not check this point.

Our next result is of a very similar nature to Theorem 2. It uses instead of the orthogonal group Kahane's theorem. The point in this application is to show that the fact that a relatively small number N is needed in (14) is not a consequence of some deep group theoretical properties of the orthogonal group. A similar phenomenon happens if we consider the simpler and obviously discrete setting of the operation of changing signs.

Let $\| \ \|$ be a norm in $\mathbb{R}^n$ and let $\{e_i\}_{i=1}^n$ denote the unit vector basis. We define a new norm $\|x\|_0$ in $\mathbb{R}^n$ by putting for $x = \sum_{i=1}^n \alpha_i e_i$

$$\|x\|_0 = \operatorname*{Average}_{\pm} \Big\| \sum_{i=1}^n \pm \alpha_i e_i \Big\| = \int_0^1 \Big\| \sum_{i=1}^n r_i(t) \alpha_i e_i \Big\| dt \,. \tag{18}$$

The unit vectors $\{e_i\}_{i=1}^n$ form a 1-unconditional basis with respect to $\|\cdot\|_0$ (if they are already a 1-unconditional basis of $\| \ \|$ then of course $\|x\| = \|x\|_0$ for all x).

Proposition 4. *Let $\|\cdot\|_0$ be defined as in (18). Then for every $\frac{1}{2} > \varepsilon > 0$ there are N choices of n-tuples of signs $\{\theta_1^j, \ldots, \theta_n^j\}_{j=1}^N$ with N given by (14) so that*

$$(1-\varepsilon)\|x\|_0 \le N^{-1} \sum_{j=1}^N \|\Sigma \theta_i^j \alpha_i e_i\| \le (1-\varepsilon)\|x\|_0 \,, \qquad x = \sum_{i=1}^n \alpha_i e_i \,. \tag{19}$$

Proof: Use exactly the same argument as in the proof of Theorem 2, just replace the use of (13) by that of (10). □

We now pass to a situation where L_{ψ_2} estimates are not available and we have to use L_{ψ_1} estimates. Given a convex symmetric body K in $\mathbb{R}^n$ normalized so that its volume is 1. We associate to it a zonoid $M(K)$ (called the centroid or moment zonoid of K) whose support functional (i.e., the norm determined by its polar) is given by

$$h_{M(K)}(u) = \int_K |\langle x, u\rangle| d\mu(x)$$

where μ denotes here the usual Lebesgue meaure. The zonoid $M(K)$ should not be confused with another zonoid associated to K, namely the projection body of K (which will be discussed in some detail in the paper [B.L] in this volume). In contrast to the projection body which can be an arbitrary zonoid the zonoid $M(K)$ defines a norm which is up to an absolute constant $\tilde{c}$ (independent of K or n) Euclidean. This is an immediate consequence of (12) above ((12) also implies of course that $\|\langle \cdot \ y\rangle\|_{L_1(\mu)}$ is equivalent to $\|\langle \cdot \ y\rangle\|_{L_2(\mu)}$). Like any zonoid which is of type 2 it can be (in view of [B.L.M]) approximated up to ε by a zonotope having $c(\varepsilon)n$ summands. The argument used above combined with (12) gives however a little more specific information.

Proposition 5. *Let K be a symmetric convex body with volume 1 in $\mathbb{R}^n$. Then for $\varepsilon < 1/2$ and for N satisfying (14), we have with probability $1 - o(\varepsilon)$ that*

$$(1-\varepsilon)M(K) \subset N^{-1}\sum_{j=1}^{N}[-x_j, x_j] \subset (1+\varepsilon)M(K) \tag{20}$$

if the $\{x_j\}_{j=1}^N$ are chosen randomly (with uniform distribution on K) and independently.

Proof: By duality (20) is equivalent to

$$\Big|1 - N^{-1}\sum_{j=1}^{N}|\langle u, x_j\rangle|\Big| < \varepsilon \tag{20'}$$

whenever $\int_K |\langle u,x\rangle| d\mu = h_{M(K)}(u) = 1$. We consider now an ε-net $\{u_i\}_{i=1}^m$ in the set $\{u \; ; h_{M(K)}(u) = 1\}$ and consider the random variables $|\langle u_i, \cdot\rangle| - 1$ on K. The same argument as that of the proof of Theorem 2 (using now (4) and (12)) shows that (20') holds with a large probability if the $\{x_j\}_{j=1}^N$ are chosen independently and uniformly distributed on K. □

We come back to the theme of Theorem 2 namely the representation of the Euclidean ball as a Minkowski sum. We shall show, by combining the proof of Theorem 2 with some results from [F.L.M], that the estimate on N given in (14) can be improved somewhat if K has an interior point and the Euclidean norm is properly chosen (or equivalently, we first apply a proper affine transformation on K). For some K the improvement on the estimate of N can be very substantial.

Theorem 6. *Let K be a convex symmetric body in $\mathbb{R}^n$ so that the Euclidean ball B^n is the ellipsoid of minimal volume containing K. Then (15) holds with N satisfying*

$$N \le cn(\log n)^{-1}\varepsilon^{-2}|\log\varepsilon| \,. \tag{21}$$

If moreover the space $\mathbb{R}^n$ normed by the polar K° of K has cotype q for some $q < \infty$ with cotype constant $C_q(K^\circ)$ then we can replace (21) by

$$N \le c_1 C_q(K^\circ)^2 n^{1-2/q}\varepsilon^{-2}|\log\varepsilon| \,. \tag{22}$$

Proof: By the well known result of F. John [J] we have that $n^{-1/2}B^n \subset K \subset B^n$, i.e.,

$$n^{-1/2}|||x||| \le \|x\|_* \le |||x||| \qquad x \in \mathbb{R}^n$$

where $|||\cdot|||$ and $\|\cdot\|_*$ denote the norms whose unit balls are B^n and K° respectively. It was shown in [F.L.M] that for some positive absolute constant c_2

$$(23) \qquad r(K) = \int_{S^{n-1}} \|x\|_* d\sigma_n(x) \geq c_2(\log n/n)^{1/2}$$

and that $\|x\|_*$ is strongly concentrated aroung $r(K)$ in the sense that for $m = 0, 1, 2, \ldots,$

$$(24) \qquad \sigma_n\{x \in S^{n-1}, mn^{-1/2} \leq |\|x\|_* - r(K)| \leq (m+1)n^{-1/2}\} \leq 4\exp\left(-(m-c_3)^2/2\right)$$

where c_3 is an absolute constant. It follows from (23) and (24) that for x with $|||x||| = 1$

$$\left|\|Ux\| - r(K)\right|_{L_{\psi_2}(O^n,\mu_n)} \leq c_4 n^{-1/2} \leq c_4 r(K)/c_2(\log n)^{1/2} .$$

By the reasoning in the proof of Theorem 2 we get that for every $\varepsilon > 0$ there exist $\{U_j\}_{j=1}^N$ in O^n with N satisfying (21) so that

$$|N^{-1}\sum_{j=1}^{N} \|U_j x\|_* - r(K)| < \varepsilon r(K)$$

for every x in an ε-net of S^{n-1}. An application of Lemma 3 and duality conclude the proof for general K. If we have information on $C_q(K^\circ)$ for some $q < \infty$ then as shown in [F.L.M] we can improve (23) to

$$r(K) \geq c_5 C_q(K^\circ)^{-1} n^{1/q-1/2}$$

and this yields the estimate (22) on N. □

Remarks. 1. The estmates (21) and (22) are the best possible in general, at least as far as the dependence on n is concerned. We shall explain this in the case of (21). Let K be the octahedron in $\mathbb{R}^n$ and let $\{T_j K\}_{j=1}^N$ be affine images of K (not necessarily rotations!) such that $\sum_{j=1}^{N} T_j K$ is up to 2 say an Euclidean ball. Then $N \geq cn/\log n$. Indeed, by passing to the duals the assumption implies that the Banach Mazur distance from ℓ_2^n to a subspace of $Z = (\ell_\infty^n \oplus \cdots \oplus \ell_\infty^n)_1$ (with N summands) is at most 2. By counting extreme points in the unit ball of Z^* we get that Z embeds isometrically in $\ell_\infty^{(2n)^N}$. Hence, as observed in [F.L.M], $c_1 \log\left((2n)^N\right) \geq n$ which is the desired result. The same reasoning gives the following corollary of (21) in Banach space terminology. For every n dimensional Banach space X, the space ℓ_2^n is of Banach Mazur distance ≤ 2 from a subspace of $(X \oplus X \oplus \cdots \oplus X)_1$ with $cn/\log n$ summands. This fact however, also follows directly from [F.L.M] without applying the empirical distribution method.

2. One should note that in the proof above we had to use Proposition 1 (iii) for large ε. In the proof of Theorem 2 itself, we used only small ε and thus we could have used there only L_{ψ_1} estimates and Proposition 1 (ii).

The previous result becomes of special interest if we consider bodies K for which $C_2(K^\circ)$ is bounded (e.g., K the n-dimensional cube). In this case we get in (22) an estimate on N which depends only on ε but not on n. A variant of this observation is the following known fact.

Proposition 7. *Let K be a convex symmetric body in $\mathbb{R}^n$ so that the Euclidean ball B^n is the ellipsoid of minimal volume containing K. Then there is an orthogonal transformation U so that*

$$\lambda^{-1} B^n \subset \tfrac{1}{2}(K + UK) \subset B^n \tag{25}$$

where

$$\lambda \le c C_2(K^\circ) \ln\left(C_2(K^\circ) + 1\right) \tag{26}$$

Proof: The right hand inclusion in (25) is clear for every choice of U so we have only to check the other inclusion relation. We denote the norm induced by B^n, respectively K°, by $|||\cdot|||$ and $\|\ \|_*$. It was proved in [K] in the case of the cube and in [M], [M.P] for a general K that the set of subspaces Y of $\mathbb{R}^n$ of dimension $[(n+1)/2]$ for which

$$|||y||| \le c_1 \|y\|_* C_2(K^\circ) \ln\left(C_2(K^\circ) + 1\right) \qquad y \in Y \tag{27}$$

has measure $> \frac{1}{2}$ (with respect to the usual normalized measure on the Grassmanian). Hence there is a subspace $Y \subset \mathbb{R}^n$ so that (27) holds for Y as well as its orthogonal complement (with respect to $|||\ \ |||$) $Y^\perp$. Choose now U to be the orthogonal map defined by

$$Uy = y\ ,\ y \in Y \quad , \quad Uy = -y\ ,\ y \in Y^\perp\ .$$

Then for $x = y + z$ with $y \in Y$ and $z \in Y^\perp$ we have

$$\|Ux\|_* + \|x\|_* \ge \|y\|_* + \|z\|_* \ge \left(c C_2(K^\circ) \ln C_2(K^\circ)\right)^{-1} |||x||| \tag{28}$$

and this proves (25). □

Remark. It is of some interest to look at (28) also in the following form. By definition of cotype 2 we have for every choice of $\{x_i\}_{i=1}^m$ in $\mathbb{R}^n$

$$C_2(K^\circ)\int_0^1 \|\sum_{i=1}^m r_i(t)x_i\|_* dt \geq \left(\sum_{i=1}^m \|x_i\|^2\right)^{1/2} .$$

Inequality (28) shows that also an inequality in the other direction holds (of course not the obvious reverse inequality since this would imply type 2 and thus that $\| \;\|_*$ is uniformly equivalent to an inner product norm). The inequality we get is that

$$\int_0^1 \|\sum_{i=1}^m r_i(t)x_i)\|_* dt \leq \tilde{c}C_2(K^\circ)\ln 2C_2(K^\circ)\left(\left(\sum_{i=1}^m \|x_i\|_*^2\right)^{1/2} + \left(\sum_{i=1}^m \|Ux_i\|_*^2\right)^{1/2}\right) . \tag{29}$$

Indeed, we have

$$\int_0^1 \|\sum_{i=1}^m r_i(t)x_i\|_* dt \leq \int_0^1 |||\sum r_i(t)x_i|||dt \leq c_1\left(\sum_{i=1}^m |||x_i|||^2\right)^{1/2}$$

and by (28) we deduce (29).

We now give a result on a very general Minkowski sum. The proof is straightforward generalization of the argument of Schechtman in [S].

Theorem 8. *Let $\{K_i\}_{i=1}^m$ be symmetric convex bodies in $\mathbb{R}^n$ and let $K = \sum_{i=1}^m K_i$. Then for every $0 < \varepsilon < 1/2$ there is subset $\{i_j\}_{j=1}^N$ of $\{1,\ldots,m\}$ with*

$$N \leq cn^2\varepsilon^{-2}|\log\varepsilon| \tag{30}$$

and scalars $\{\lambda_j\}_{j=1}^N$ so that

$$(1-\varepsilon)K \subset \sum_{j=1}^N \lambda_j K_{i_j} \subset (1+\varepsilon)K . \tag{31}$$

Proof: We pass again to the duals. Let $\|\cdot\|_i$ be the norm in $\mathbb{R}^n$ whose unit ball is K_i°. Then $\|x\| = \sum_{i=1}^m \|x\|_i$ is the norm whose unit ball is K°. Let $\{x_k\}_{k=1}^n$ be an Auerbach basis of $\mathbb{R}^n$ with respect to $\|\cdot\|$. In other words $\|x_k\| = 1$ for $1 \leq k \leq n$ and for any $x \in \mathbb{R}^n$ with $\|x\| \leq 1$ we have $x = \sum_{k=1}^n a_k x_k$ with $|a_k| \leq 1$ for all k Let $\alpha_i = \sum_{k=1}^n \|x_k\|_i$, $1 \leq i \leq m$. Then $\sum_{i=1}^m \alpha_i = \sum_{k=1}^n \|x_k\| = n$ and for every x with $\|x\| = 1$ we have $\|x\|_i \leq \alpha_i$. Let ν be the probability measure on $\Omega = \{1,2,\ldots,m\}$ defined by $\nu(i) = \alpha_i/n$. Let $\{u_t\}_{t=1}^s$ be an ε-net with respect to

$\|\cdot\|$ of the unit sphere in $\mathbb{R}^n$ with the same norm and with $s \le (4/\varepsilon)^n$. The functions $\{g_t\}_{t=1}^s$ on Ω defined by

$$g_t(i) = (n\|u_t\|_i/\alpha_i) - 1 \qquad 1 \le i \le m .$$

These are functions of mean 0 with $\|g_t\|_{L_\infty(\nu)} \le n$ and $\|g_t\|_{L_1(\nu)} \le 1$. By applying (3) we can find $\{i_j\}_{j=1}^N$ so that for every $1 \le t \le s$

$$\left| N^{-1} \sum_{j=1}^N n\|u_t\|_{i_j}/\alpha_{i_j} - 1 \right| < \varepsilon \tag{32}$$

provided that $2\log s = 2n|\log 4/\varepsilon| \le \varepsilon^2 N/8n$, i.e., that (30) holds. Now by Lemma 3 it follows from (32) that

$$(1-\Delta)K \subset \frac{1}{N}\sum_{j=1}^N n\alpha_{i_j}^{-1}K_{i_j} \subset (1+\Delta)K$$

where $\Delta = 3\varepsilon/(1-\varepsilon)$. □

It is likely that at least some of the refinements of Schechtman's result which were worked out in [B.L.M] carry over to the setting of Theorem 8. We did not pursue this point.

2. Minkowski symmetrizations

Let K be a compact convex set in $\mathbb{R}^n$ and let u be any vector in $S^{n-1} = \{u \,;\, |||u||| = 1\}$, where as before $|||\cdot|||$ denotes the Euclidean norm. We denote by π_u the reflection with respect to the hyperplane through 0 orthogonal to u, i.e.,

$$\pi_u x = x - 2\langle x, u\rangle u . \tag{33}$$

Obviously π_u is an orthogonal transformation in $\mathbb{R}^n$. The Minkowski symmetrization K with respect to u (or with respect to hyperplane orthogonal to u) is defined to be the convex set $\frac{1}{2}(\pi_u K + K)$. This bype of symmetrization was apparently used first by Blaschke in his book "Kreis and Kugel" and is therefore also called Blaschke symmetrization. The main result in this section is a proof of the fact that for every $\varepsilon > 0$ we get from K after performing $cn\log n + c(\varepsilon)n$ random Minkowski symmetrizations a body which is (with a large probability) up to ε an Euclidean ball. By a random Minkowski symmetrization we mean an operation in which u is chosen randomly on S^{n-1} with the usual rotation invariant distribution. If we perform several operations the u's in different steps are chosen independently. After N symmetrizations by vectors $u_1, \ldots, u_N$, the resulting body may be written in the form

$$2^{-N} \sum_{D \subset \{1,\ldots,N\}} \prod_{i \in D} \pi_{u_i}(K)$$

where the sum is taken over all subsets $D \subset \{1, \ldots, N\}$. As in section 1, we want this Minkowski sum to approximate the euclidean ball for a random choice of vectors $u_1, \ldots, u_N$ in the sphere S^{n-1}, where N should be taken as small as possible.

This result on Minkowski symmetrizations is in some sense related to the fact discovered by Diaconis and Shashahani (cf. for example [DS]) that the product of $\frac{1}{2} n \log n + cn$ random reflections π_u produce a random orthogonal transformation on $\mathbb{R}^n$. It does not seem however that this fact can be used for simplifying the proof of our main result. The methods we use here are different from the tools employed in [DS]. In particular we do not use here results from the theory of group representations.

In the proof given here, the methods of the previous section have to be combined with additional techniques, given in Lemmas 10,11,12. We will make use of the concentration phenomena on the orthogonal group (see Lemma 13) besides the concentration on the sphere.

The version of Bernstein's inequality we use in this section involves martingales rather than independent random variables. More specifically we use the following martingale version of Proposition 1(ii).

Proposition 9. *Let $(\Omega, \mathcal{B}, \mu)$ be a probability measure space and let $\mathcal{B}_1 \subset \mathcal{B}_2 \subset \cdots \subset \mathcal{B}_N$ be sub-σ-algebras of $\mathcal{B}$. Let f be a $\mathcal{B}_N$ measurable real valued function and let put $E(f \mid \mathcal{B}_i) = f_i$, $1 \le i \le N$. Assume further that for some constant A*

$$E\big(\exp(|f_i - f_{i-1}|/A \mid \mathcal{B}_{i-1}\big) \le 2\,, \qquad 1 \le i \le N \tag{34}$$

($\mathcal{B}_0$ is the trivial σ-field and $f_0 = E(f \mid \mathcal{B}_0) = \int_\Omega f(\omega)d\mu$). Then for $0 < \varepsilon < 4A$

$$\text{Prob}\Big\{\big|f - \int_\Omega f\,d\mu\big| \ge \varepsilon N\Big\} \le 2\exp(-\varepsilon^2 N/16A^2)\,. \tag{35}$$

Note that the assumption (34) is stronger than $\|f_i - f_{i-1}\|_{L_{\psi_1}} \le A$; it is a pointwise condition on the conditional expectation.

Proof: There is no loss of generality to assume that $A = 1$ and that $\int_\Omega f\,d\mu = 0$. Put $d_i = f_i - f_{i-1}$ and note that $E(d_i \mid \mathcal{B}_{i-1}) = 0$. Also by (34) $E\big(|d_i|^k \,|\, \mathcal{B}_{i-1}\big) \le 2 \cdot k!$ for

$1 \le i \le N$ and all k. Hence for $0 < \lambda < 1/2$

$$E\big(\exp(\lambda\, d_i) \mid \mathcal{B}_{i-1}\big) \le 1 + 2\sum_{k=2}^{\infty} \lambda^k \le 1 + 4\lambda^2\,, \qquad 1 \le i \le N\,,\ \omega \in \Omega\,.$$

Consequently for $i \le N$

$$\int_\Omega \exp(\lambda f_i)d\mu = \int_\Omega E\big(\exp(\lambda f_i) \mid \mathcal{B}_{i-1}\big)d\mu$$
$$= \int_\Omega \exp(\lambda f_{i-1}) \cdot E\big(\exp(\lambda d_i) \mid \mathcal{B}_{i-1}\big)d\mu \le (1+4\lambda^2)\int_\Omega \exp(\lambda f_{i-1})d\mu\,.$$

It follows that

$$\mathrm{Prob}\{f > \varepsilon N\}\exp(\varepsilon\lambda N) \le \int_\Omega \exp(\lambda f)d\mu \le (1+4\lambda^2)^N \le \exp(4\lambda^2 N)\,.$$

By taking $\lambda = \varepsilon/8$ $(\le 1/2)$ we get

$$\mathrm{Prob}\{f_N > \varepsilon N\} \le \exp(-\varepsilon^2 N/16)$$

and this implies (35)

We shall apply Proposition 9 in the proof of

Lemma 10. *For $x \in S^{n-1}$ and $u = (u_1, u_2, \ldots, u_N) \in (S^{n-1})^N$ put*

$$\varphi_u(x) = \sup_{y \in S^{n-1}} 2^{-N} \sum_D \Big|\Big\langle \prod_{i\in D} \pi_{u_i} x, y\Big\rangle\Big| \tag{36}$$

where the sum is taken over all subsets D of $\{1,\ldots,N\}$. Then for fixed x, $0 < \delta < \delta_0$, $n > n(\delta)$ and $N = c_1 n|\log\delta|$ we have

$$\mathrm{Prob}\big\{u\ ;\ \varphi_u(x) > \delta\big\} \le \exp(-c_2\delta^2 n/|\log\delta|) \tag{37}$$

where δ_0, c_1 and c_2 are absolute constants.

Proof: Clearly $\varphi_u(x) = 2^{-N} = 2^{-N}\||\sum_D \theta_D \prod_{i\in D} \pi_{u_i} x|\||$ for suitable signs θ_D (i.d., $\theta_D = \pm 1$). It follows that

$$\varphi_u(x)^2 \le 4^{-N} \sum_D \sum_{D'} \Big|\Big\langle \prod_{i\in D} \pi_{u_i} x, \prod_{i\in D'} \pi_{u_i} x\Big\rangle\Big| \tag{38}$$

Every summand in the right hand side of (38) is at most 1. The number of summands for which the symmetric difference $D\Delta D'$ has cardinality $\overline{\overline{D\Delta D'}} \le N/4$ is $o(4^N)$ and thus for $n > n(\delta)$

is at most $\delta 4^N/4$. We consider the other summands and take a choice of D and D' so that $\overline{\overline{D\Delta D'}} \geq N/4$. We consider

$$f_{D,D'}(u) = f(u) = \langle \prod_{i\in D} \pi_{u_i} x \, , \, \prod_{i\in D'} \pi_{u_i} x \rangle$$

as a martingale taking as the sub-σ-algebras $\mathcal{B}_i$ those subsets of $(S^{n-1})^N$ which depend only on the u_j with $j > N - i$. The indices i which contribute a non-zero summand to the presentation of f as a sum of martingale differences are exactly those which belong to $D\Delta D'$. We explain this by considering $i = 1$. If $1 \notin D \cup D'$, then obviously f is already $\mathcal{B}_{N-1}$ measurable. The same is true if $1 \in D \cap D'$ since π_{u_1} being orthogonal implies that

$$f(u) = \langle \prod_{i\in D\sim\{1\}} \pi_{u_i} x \, , \prod_{i\in D'\sim\{1\}} \pi_{u_i} x \rangle \, .$$

Assume that $1 \in D$ but $1 \notin D'$. Put $v = \prod_{i\in D\sim\{1\}} \pi_{u_i} x$, $w = \prod_{i\in D'} \pi_{u_i} x$. Then

$$f(u) = \langle \pi_{u_1} v, w \rangle = \langle v, w \rangle - 2\langle v, u_1 \rangle \langle w, u_1 \rangle \, .$$

Put

$$f_{N-1}(u) = E\{f \mid \mathcal{B}_{N-1}\} = \langle v, w \rangle - 2 \int_{S^{n-1}} \langle v, u \rangle \langle w, u \rangle d\sigma_n(u)$$

$$= \left(1 - \frac{2}{n}\right) \langle v, w \rangle \, .$$

Note that

$$f - f_{N-1} = \frac{2}{n} \langle v, w \rangle - 2\langle v, u_1 \rangle \langle w, u_1 \rangle$$

and that $\|\langle \cdot \, v \rangle^2 \|_{L_{\psi_1}(S^{n-1})} \approx n^{-1}$. Hence, for some $c_3 > 0$ and all v and w

$$E\big(\exp(c_3 n |f - f_{N-1}|) \mid \mathcal{B}_{N-1} \big) \leq 2 \, .$$

The function f_{N-1} is of the same form as f only with the additional factor $\left(1 - \frac{2}{n}\right)$. We can now repeat the same reasoning with $i = 2$ and so on. We get in particular that

$$E(f) = \int_{(S^{n-1})^N} f(d\sigma_n)^N = \left(1 - \frac{2}{n}\right)^{N'} \quad \text{where} \quad N' = \overline{\overline{D\Delta D'}} \, .$$

Since we assume that $N' \geq N/4$ we can, by a suitable choice of c_1 (in the definition of N), ensure that $|E(f)| < \delta/4$. Since $\delta/2N \leq 4/c_3 n$ for $\delta < \delta_0$ we may apply Proposition 9 and get that for some absolute $c_4 > 0$

$$\begin{aligned} \operatorname{Prob}\{|f| \geq \delta/2\} &\leq \operatorname{Prob}\big\{|f - E(f)| \geq \delta/4\big\} \\ &\leq 2\exp\big(- \delta^2 (c_3 n)^2 / 16 \cdot 4 \cdot N' \big) \leq \exp(-c_4 \delta^2 n / |\log \delta|) \, . \end{aligned}$$

For D and D' with $\overline{\overline{D\Delta D'}} \geq N/4$ put

$$\chi_{D,D'}(u) = \begin{cases} 1 & \text{if } |f_{D,D'}(u)| \geq \delta/2 \\ 0 & \text{otherwise .} \end{cases}$$

Then

$$\text{Prob}\left\{u \; ; \; \chi_{D,D'}(u) = 1 \text{ for at least } \delta \cdot 4^N/4 \text{ pairs } D, D'\right\} \leq 4(4^N\delta)^{-1} E\left(\sum_D \sum_{D'} \chi_{D,D'}\right) \leq$$
$$\leq 4\delta^{-1} \exp(-c_4\delta^2 n/|\log \delta|) \; .$$

If u is such that for at most $\delta \cdot 4^N/4$ pairs D, D' with $\overline{\overline{D\Delta D'}} \geq N/4$ we have $|f_{D,D'}(u)| > \delta/2$ then clearly

$$\varphi_u(x) < \delta/4 + \delta/4 + \delta/2 = \delta \; .$$ □

Our next goal is to strengthen Lemma 10 by showing that $\varphi_u(x)$ is small for most u and all x and not only a single x.

Lemma 11. *Let $\varphi_u(x)$ be as in (36). Then for $N = c_5\delta^{-2}|\log \delta|^3 n$, $0 < \delta < \delta_0$ and $n > n(\delta)$*

(39) $$\text{Prob}\left\{u \; ; \; \varphi_u(x) > \delta \text{ for some } x \in S^{n-1}\right\} \leq \exp(-c_6 n|\log \delta|) \; .$$

Proof: Let k and N_1 be integers, let $u \in (S^{n-1})^{N_1}$ and $v \in (S^{n-1})^{kN_1}$. We consider (u, v) as an element of $(S^{n-1})^{(k+1)N_1}$ and there is an obvious meaning to $\varphi_{u,v}(x)$ for $x \in S^{n-1}$. Note that $\varphi_{u,v}(x) \leq \varphi_v(x)$ for all u, v and x. We assume that ε, δ and N_1 are such that for all $x \in S^{n-1}$

(40) $$\text{Prob}\left\{u \in (S^{n-1})^{N_1} \; ; \; \varphi_u(x) \geq \delta/2\right\} \leq \varepsilon \; .$$

Since, for each fixed v_0, $\varphi_{u,v_0}(x)$ is bounded from above by the arithmetic average of 2^{kN_1} of expressions of the form $\varphi_u(z)$ for suitable vectors z in S^{n-1} we get by the same reasoning as that used at the end of the proof of Lemma 10 that

$$\text{Prob}\left\{u \in (S^{n-1})^{N_1} \; ; \; \varphi_{u,v_0}(x) \geq \delta\right\} \leq 2\varepsilon/\delta \; .$$

Hence, for all $x \in S^{n-1}$,

(41) $$\text{Prob}\left\{(u, v) \; ; \; \varphi_{u,v}(x) \geq \delta\right\} \leq \text{Prob}\left\{v \; ; \; \varphi_v(x) \geq \delta\right\} \cdot 2\varepsilon/\delta \; .$$

By induction on k we deduce that

(42) $$\text{Prob}\left\{v \in (S^{n-1})^{kN_1} \; ; \; \varphi_v(x) \geq \delta\right\} \leq (2\varepsilon/\delta)^k \; .$$

By Lemma 10, (40) holds with $N_1 = c_1 n|\log\delta|$ and $\varepsilon = \exp(-8c_2\delta^2 n/|\log\delta|)$. Hence by taking $k \sim |\log\delta|^2\delta^{-2}$ in (42) we get that for $N \approx n\delta^{-2}|\log\delta|^3$, $x \in S^{n-1}$, $0 < \delta < \delta_0$ and $n > n(\delta)$

$$\mathrm{Prob}\left\{v \in (S^{n-1})^N \; ; \; \varphi_v(x) \geq \delta\right\} < (\delta/4)^{2n} .$$

Let $\{x_\alpha\}_{\alpha\in A}$ be a δ net in S^{n-1} with $\overline{\overline{A}} \leq (4/\delta)^n$. Then

$$\mathrm{Prob}\left\{v \; ; \varphi_v(x_\alpha) \geq \delta \;\; \text{for some } \alpha \in A\right\} < (\delta/4)^n . \tag{43}$$

A trivial argument, similar to the proof of Lemma 3, shows that if $\varphi_v(x_\alpha) \leq \delta$ for all x_α then $\varphi_v(x) \leq \delta/(1-\delta) \leq 2\delta$ for all $x \in S^{n-1}$ and this concludes the proof of our assertion. □

Remark. It is evident from the proof that if we take a larger value of N we get (39) with a smaller value for the probability. This holds also for the results below, in particular for Theorem 14.

Our next lemma is a simple exhaustion argument.

Lemma 12. *Assume that $\{w_\alpha\}_{\alpha\in A}$ is a family of vectors in S^{n-1} so that for some $\delta > 0$*

$$\sup_{y\in S^{n-1}} \overline{\overline{A}}^{-1} \sum_{\alpha\in A} |\langle w_\alpha, y\rangle| \leq \delta .$$

Let $0 < \lambda < 1$ and let k be an integer. Then there are disjoint families $\mathcal{F}_\beta = \{\beta_i\}_{i=1}^k$, $\beta \in B$, of elements of A so that $\overline{\overline{\bigcup_{\beta\in B} \mathcal{F}_\beta}} \geq (1-\lambda)\overline{\overline{A}} - k$ and so that for every $\beta \in B$ there is an orthonormal set of vectors $\{v_{\beta_i}\}_{i=1}^k$ satisfying

$$|||v_{\beta_i} - w_{\beta_i}||| \leq \delta 4^k/\lambda , \qquad 1 \leq i \leq k . \tag{44}$$

Proof: Assume that we already have families $\mathcal{F}_\beta$ as above and that $\overline{\overline{\bigcup_\beta \mathcal{F}_\beta}} \leq (1-\lambda)\overline{\overline{A}} - k$. Then we can construct by induction on i distinct elements $\{a_i\}_{i=1}^k$ in A which do not belong to the $\mathcal{F}_\beta$ and for which

$$\sum_{j=1}^{i} |\langle w_{a_j}, w_{a_{i+1}}\rangle| \leq i\delta/\lambda , \qquad 1 \leq i \leq k-1 . \tag{45}$$

Indeed, put $\widetilde{A} = A \sim \left(\{a_j\}_{j=1}^i \cup \bigcup_\beta \mathcal{F}_\beta\right)$ then $\overline{\overline{\widetilde{A}}} \geq \lambda\overline{\overline{A}}$ and by our assumption

$$\sum_{j=1}^{i} \sum_{\alpha\in\widetilde{A}} |\langle w_{a_j}, w_\alpha\rangle| \leq \sum_{j=1}^{i} \sum_{\alpha\in A} |\langle w_{a_j}, w_\alpha\rangle| \leq i\delta\overline{\overline{A}} .$$

Hence there is an $\alpha \in \widetilde{A}$ which can serve as a_{i+1} in (45). Let $\{v_{a_i}\}_{i=1}^k$ be the system obtained from $\{w_{a_i}\}_{i=1}^k$ by the Gram Schmidt orthogonalization procedure. Then

$$\begin{aligned}|||v_{a_{i+1}} - w_{a_{i+1}}||| &\leq 2\sum_{j=1}^{i} |\langle w_{a_{i+1}}, v_{a_j}\rangle| \leq \\ &\leq 2\sum_{j=1}^{i} |\langle w_{a_{i+1}}, w_{a_j}\rangle| + 2\sum_{j=1}^{i} |||w_{a_j} - v_{a_j}||| \leq \\ &\leq 2\delta i/\lambda + 2\sum_{j=1}^{i} |||w_{a_j} - v_{a_j}||| .\end{aligned}$$

From this inequality we deduce by induction on i that

$$|||v_{a_i} - w_{a_i}||| \leq \delta \cdot 4^i/\lambda , \qquad 1 \leq i \leq k .$$

□

The next lemma is a concentration of measure result on the orthogonal group.

Lemma 13. *Let $\|\cdot\|$ be a norm on $\mathbb{R}^n$ so that $\|x\| \leq |||x|||$ for all x. Let $\{x_i\}_{i=1}^k$ be orthonormal vectors (with respect to $|||\ \ |||$), and let μ_n be Haar measure on the orthogonal group O^n. Then*

$$\mu_n\{U \in O^n ;\ |k^{-1}\sum_{i=1}^{k} \|Ux_i\| - \int_{O^n} \|Ux\| d\mu_n(U)| \geq \varepsilon\} \leq \exp(-c\varepsilon^2 nk) . \tag{46}$$

Proof: We use the following known fact (cf. [MS], p. 29). Let F be a function from O^n to $\mathbb{R}$ which has Lipschitz constant λ if O^n is taken in the Hilbert-Schmidt metric, i.e.,

$$|F(U) - F(V)| \leq \lambda |||U - V|||_{\mathrm{H.S.}}$$

Then

$$\mu_n\{U \in O^n ;\ |F(U) - \int_{O^n} F(U) d\mu_n(U)| \geq \varepsilon\} \leq \exp(-c\varepsilon^2 n/\lambda^2) .$$

We use here the function $F(U) = k^{-1} \sum_{i=1}^{k} \|Ux_i\|$. Note that

$$\begin{aligned}|F(U) - F(V)| &\leq k^{-1}\sum_{i=1}^{k} \|(U-V)x_i\| \leq \\ &\leq k^{-1}\sum_{i=1}^{k} |||(U-V)x_i||| \leq k^{-1/2}\Big(\sum_{i=1}^{k} |||(U-V)x_i|||^2\Big)^{1/2} \leq \\ &\leq k^{-1/2} |||U - V|||_{\mathrm{H.S.}}\end{aligned}$$

Thus we may take $\lambda = k^{-1/2}$ in the estimate above and this proves (46). □

We are now ready to prove the main result.

Theorem 14. *Let K be a symmetric convex body in $\mathbb{R}^n$ and let $\varepsilon > 0$. If $n > n_0(\varepsilon)$ and if we perform $N = cn \log n + c(\varepsilon)n$ random Minkowski symmetrizations on K we obtain with probability $1 - \exp\big(-\tilde{c}(\varepsilon)n\big)$ a body $\tilde{K}$ which satisfies*

$$(1-\varepsilon)rB^n \subset \tilde{K} \subset (1+\varepsilon)rB^n$$

for a suitable $r = r(K)$.

The meaning of random Minkowski symmetrizations was explained in the beginning of this section. As usual c denotes an absolute constant while $c(\varepsilon)$ and $\tilde{c}(\varepsilon)$ denote positive constants depending only on ε.

As was the case in section 1 it is more convenient to carry out the proof in the dual space because of the simple behaviour of the support function of a Minkowski sum. We first formulate the theorem in the dual form

Theorem 14′. *Let $\|\cdot\|$ be a norm in $\mathbb{R}^n$ and let $N = cn\log n + c(\varepsilon)n$ (where $\varepsilon > 0$ and $n > n_0(\varepsilon)$). Then for all $\{u_i\}_{i=1}^N$ in $(S^{n-1})^N$ with the exception of a set of $(\sigma_n)^N$ measure at most $1 - \exp\big(-\tilde{c}(\varepsilon)n\big)$ we have*

$$(47)\qquad (1-\varepsilon)r|||x||| \le 2^{-N}\sum_D \Big\|\prod_{i\in D}\pi_{u_i}x\Big\| \le (1+\varepsilon)r|||x||| , \qquad x \in X$$

where the sum is taken over all subsets D of $\{1,\dots,N\}$ and

$$(48)\qquad r = \int_{S^{n-1}} \|x\|d\sigma_n(x) = \int_{O^n} \|Ux\|d\mu_n(U) .$$

Proof: We assume as we may that $\|x\| \le |||x|||$ for all x and that for some vector e, $\|e\| = |||e||| = 1$. Then clearly $\|x\| \ge |\langle e,x\rangle|$ and thus

$$(49)\qquad 1 \ge r \ge \int_{S^{n-1}} |\langle e,x\rangle|d\sigma_n(x) \approx 1/\sqrt{n} .$$

We shall perform the operations in steps. We shall show below that the following holds.

Statement a. *If $r < 1/8$, then with $N_1 < c_1 n$ we have*

$$(50)\qquad \|x\|_2 - 2^{-N_1}\sum_D \Big\|\prod_{i\in D}\pi_{u_i}x\Big\| \le \tfrac{1}{2}|||x||| , \qquad x \in \mathbb{R}^n$$

with probability of at least $1 - \exp(-x_2 n)$.

In proving Statement a, we shall use Lemma 11 for a certain absolute value of δ. Note that because of the rotation invariance of σ_n and the fact that the π_u are unitary operators we have that $r = \int_{S^{n-1}} \|x\|_2 d\sigma_n(x)$, i.e., r does not change in this operation. If $r < 1/16$ then by applying Statement a to the norm $2\|x\|_2$ we obtain a norm $2\|x\|_3$ which satisfies $\|x\|_3 \leq \frac{1}{4}|||x|||$. (Statement a uses only the normalizing condition $\|x\| \leq |||x|||$. The requirement on the existence of e with $\|e\| = |||e||| = 1$ is used only to obtain a lower estimate on r in (49). As mentioned, r is fixed throughout the whole procedure.) Note that because of the special form of our operation if follows that $\|x\|_3$ can be obtained directly from $\|x\|$ via the formula.

$$\|x\|_3 = 2^{-2N_1} \sum_D \| \prod_{i \in D} \pi_{u_i} x\|$$

where here of course, the sum is over all subsets D of $\{1, 2, \ldots, 2N_1\}$. In view of (49) we will get, after $\frac{1}{2} \log n$ steps, to a norm for which the ratio between its maximum and average on S^{n-1} is at most 8. So far we have performed $\frac{1}{2} c_1 n \log n$ symmetrizations and the probability that we fail to get a good norm is at most $\frac{1}{2} \log n \exp(-c_2 n) \leq \exp(-c_3 n)$ (for $n \geq n_0$). Thus to conclude the proof of the theorem it suffices to verify besides Statement a also

Statement b. *If $f \geq 1/8$, then for all $\varepsilon > 0$ and $n \geq n(\varepsilon)$ and with $N_2 = c(\varepsilon) n$*

$$(51) \qquad (1-\varepsilon) r |||x||| \leq 2^{-N_2} \sum_D \| \prod_{i \in D} \pi_{u_i} x\| \leq (1+\varepsilon) r |||x||| , \qquad x \in X$$

with probability of at least $1 - \exp\big(-c_1(\varepsilon) n\big)$.

Proof of Statement a. We shall use as δ, λ and k absolute constants which will be determined shortly. By Lemmas 11 and 12 we get with probability $1 - (\delta/4)^n$ (see (43)) that if $N_1 = c_5 \delta^{-2} |\log \delta|^3 n$ we can decompose, for every $x \in S^{n-1}$, the set consisting of the 2^{N_1} vectors $\prod_{i \in D} \pi_{u_i} x$ into disjoint families $\{\mathcal{F}_\beta\}_{\beta \in B}$ consisting of k elements each and a remainder of size $\lambda 2^{N_1}$. For each such family $\mathcal{F} = \{w_j\}_{j=1}^k$ there are orthonormal vectors $\{v_j\}_{j=1}^k$ so that $|||w_j - v_j||| \leq \delta \cdot 4^k / \lambda$ for all j. By Lemma 13

$$(52) \qquad \mu_n \{U \in O^n \; ; \; k^{-1} \sum_{j=1}^k \|U v_j\| - r \geq 1/8\} \leq \exp(-c_4 k n) \; .$$

Hence

$$\mu_n \{U \in O^n \; ; \; k^{-1} \sum_{j=1}^k \|U w_j\| - r \geq 1/8 + \delta \cdot 4^k / \lambda\} \leq \exp(-c_4 k n) \; .$$

Consequently the set of $U \in O^n$ for which

$$k^{-1} \sum_{j=1}^k \|U w_{\beta_j}\| \leq r + 1/8 + \delta \cdot 4^k / \lambda$$

for at least $(1-\lambda)\overline{\overline{B}}$ of the families $\mathcal{F}_\beta$ have a measure not smaller than $1-\lambda^{-1}\exp(-c_4kn)$ (use the same argument as that at the end of the proof of Lemma 10). We deduce from this that there is a set of $\{u_i\}_{i=1}^{N_1}$ of measure at least $1-(\delta/4)^n$ so that for all $x \in S^{n-1}$ the set of $U \in O^n$ for which

$$2^{-N_1}\sum_D \|U\prod_{i\in D}\pi_{u_i}x\| \le r+\frac{1}{8}+\delta\cdot 4^k/\lambda+\lambda+\lambda \tag{53}$$

has measure $\ge 1-\lambda^{-1}\exp(-c_4kn)$. By Fubini's theorem we deduce that for some $U_0 \in O^n$, (53) holds for all $\{u_i\}_{i=1}^{N_1}$ in a set of measure $\ge 1-(\delta/4)^n-\lambda^{-1}\exp(-c_4kn)$ and all x in a set of measure $\ge 1-\lambda^{-1}\exp(-c_4kn)$. Since

$$U\prod_{i\in D}\pi_{u_i}x - \prod_{i\in D}\pi_{Uu_i}Ux$$

and the map $\{u_i\}_{i=1}^{N_1} \to \{Uu_i\}_{i=1}^{N_1}$ is measure preserving there is no loss of generality to assume that $U_0=$ identity. We choose λ,δ and k so that $2\lambda=1/16$, $\delta 4^k/\lambda=1/16$ and then $\lambda^{-1}\exp(-c_4kn)$ takes the form $\exp(-c_5|\log\delta|n)$ and we are still free to choose δ as small as we please. For δ sufficiently small a set on S^{n-1} with measure $\ge 1-\exp(-c_6|\log\delta|n)$ contains a $1/4$ net. Thus for a $1/4$ net on S^{n-1} and a set of $\{u_i\}_{i=1}^{N_1}$ of measure $\ge 1-2\exp(-c_6|\log\delta|n)$ we have

$$2^{-N_1}\sum_D \|\prod_{i\in D}\pi_{u_i}x\| \le 3/8\ . \tag{54}$$

This implies that (54) holds for the same $\{u_i\}_{i=1}^{N_1}$ and all $x \in S^{n-1}$ with $3/8$ replaced by $3/8\cdot\left(1-\frac{1}{4}\right)^{-1}=\frac{1}{2}$ and this proves (50).

Proof of Statement b. The proof is almost identical to the proof of Statement a only now k,δ and λ will depend on the given ε.

Instead of (52) we will get now

$$\mu_n\{U\in O^n\ ;\ |k^{-1}\sum_{j=1}^{k}\|Uv_j\|-r|\ge \varepsilon r/4\}\le \exp(-c_7\varepsilon^2kn)\ .$$

We shall choose now k and λ and δ so that $2\lambda=\varepsilon/32$ ($\le \varepsilon r/4$) and $\delta\cdot 4^k/\lambda=\varepsilon/32$. This gives for the measure of the exceptional set (of $\{u_i\}_{i=1}^{N_1}$ or of x) an estimate of $\exp\left(-c_8\varepsilon^2|\log(\varepsilon/\delta)|n\right)$. For non exceptional $\{u_i\}_{i=1}^{N_1}$ and $x\in S^{n-1}$ we get that

$$|2^{-N_1}\sum_D \|\prod_{i\in D}\pi_{u_i}x\|-r|\le 3\varepsilon r/4\ . \tag{55}$$

We choose now $\delta(\varepsilon)$ small enough so that by an argument using nets on S^{n-1} we can ensure that (55) holds for all $x\in S^{n-1}$ with $3\varepsilon r/4$ replaced by εr. □

Remarks. 1. The theorem and its proof are valid also for non symmetric convex bodies K which contains O in their interior. We just have to work throughout with norms such that $\|-x\| \neq \|x\|$.

2. The dependence of the $c(\varepsilon)$ (appearing in the statement of Theorem 14) on ε which we get is $c(\varepsilon) \approx \exp(\alpha\varepsilon^{-2}|\log\varepsilon|)$.

3. The dependence of N on n in Theorem 14 is optimal. Let e be some vector in $\mathbb{R}^n$ with $|||e||| = 1$ and consider the norm $\|x\| = |\langle e, x\rangle| + |||x|||/n^{1/2}$. For this norm $r = \int_{S^{n-1}} \|x\| d\sigma_n(x) \approx n^{-1/2}$. After k random symmetrizations we get from the given norm a norm whose expected value at x is $\geq \left(1 - \frac{2}{n}\right)^k |\langle e, x\rangle| + |||x|||/n^{1/2}$. Thus unless $k \geq \frac{1}{2} n \log n$ we get a norm whose expected value for e is significantly larger than its average (i.e., r). If we consider also deterministic symmetrizations then it is no longer clear to us that cn symmetrizations will not suffice always. For the deterministic case it is evident that in general we need at least $n - c\log n$ steps. Indeed, if we symmetrize with respect to $\{u_i\}_{i=1}^k$ we do not change the norm at all of vectors x which are orthogonal to all the $\{u_i\}_{i=1}^k$. For some norms (e.g., the norm in ℓ_∞^n) it is known that there are no subspaces of dimension $\geq c\log n$ which are Euclidean up to 2,say.

References

[B.L] J. Bourgain and J. Lindenstrauss, Projection bodies, This volume.

[B.L.M] J. Bourgain, J. Lindenstrauss and V. Milman, Approximation of zonoids by zonotopes, Preprint I.H.E.S., September 1987, 62pp, to appear in Acta Math.

[Bo] C. Borell, The Brunn-Minkowski inequality in Gauss spaces, Inventiones Math. 30 (1975), 207-216.

[D.S] P. Diaconis and M. Shahshahani, Products of random matrices as they arise in the study of random walks on groups, Contemporary Math. 50 (1986), 183-195.

[F.L.M] T. Figiel, J. Lindenstrauss and V.D. Milman, The dimensions of almost spherical sections of convex bodies, Acta Math. 139 (1977), 53-94.

[G] Y. Gordon, Some inequalities for Gaussian processes and applications, Israel J. Math. 50 (1985), 265-289.

[G.M] M. Gromov and V.D. Milman, Brunn theorem and a concentration of volume of convex bodies, GAFA Seminar Notes, Israel 1983-1984.

[H] U. Haagerup, The best constants in the Khinchine inequality, Studia Math, 70 (1982), 231-283.

[J] F. John, Extremum problems with inequalities as subsidiary conditions, Courant Anniversary Volume, Interscience, New York, 1948, 187-204.

[K] B.S. Kashin, Sections of some finite dimensional sets and classes of smooth functions, Izv. ANSSSR, ser. mat. 41 (1977), 334-351 (Russian).

[Ka] J.P. Kahane, Series of Random Functions, Heath Math. Monographs, Lexington, Mass., Heath & Co., 1968.

[Kw] S. Kwapień, Isomorphic characterizations of inner product spaces by orthogonal series with vector valued coefficients, Studia Math. 44 (1972), 583-595.

[L.T] J. Lindenstrauss and L. Tzafriri, Classical Banach spaces, vol II, Function spaces, Ergebnisse der Math., v. 97, Springer Verlag 1979.

[L.S] H. Landau and L. Shepp, On the supremum of a Gaussian process, Sankhya A32 (1970), 369-378.

[Ma.P] M.B. Marcus and G. Pisier, Random Fourier series with applications to harmonic analysis, Ann. Math. Studies, 101, Princeton 1981.

[M] V.D. Milman, Random subspaces of proportional dimension of finite dimensional normed spaces: Approach through the isoperimetric inequality, Banach spaces, Proc. Missouri Conference 1984, Springer Lecture Notes #1166, 106-115.

[M.P] V.D. Milman and G. Pisier, Banach spaces with a weak cotype 2 property, Israel J. Math., 54 (1986), 139-158.

[M.S] V.D. Milman and G. Schechtman, Asymptotic theory of finite dimensional normed spaces, Springer Lecture Notes #1200 (1986).

[S] G. Schechtman, More on embedding subspaces of L_p in ℓ_r^n, Compositio Math. 61 (1987), 159-170.

[Sz] S.J. Szarek, On the best constant in the Khinchine inequality, Studia Math. 58 (1976), 197-208.

[Ta] M. Talagrand, An isoperimetric theorem on the cube and the Khinchine-Kahane inequalities, Preprint.

ON TWO THEOREMS OF LOZANOVSKII CONCERNING INTERMEDIATE BANACH LATTICES*

Shlomo Reisner
Department of Mathematics and School of Education of the Kibbutz Movement
University of Haifa
Haifa, Israel

1. Introduction, Notations and Statements of the Main Theorems

We deal in this work with two theorems of G.Ya. Lozanovskii ([7] and [9]). In the above-mentioned papers Lozanovskii investigated methods of constructing Banach lattices which are intermediate "between" two given Banach lattices. The constructions follow and generalize those of Calderon [2].

The two above mentioned theorems characterize lattices dual in certain senses to these intermediate lattices. Lozanovskii formulated and proved the theorems in a very general context (cf. also [8]) and it is not completely trival to follow the proofs and interpret the results in the case of Banach lattices of measurable functions. Also the terminology used in [7] and [9] is the Russian terminology which differs from the one usually used in the West. Therefore, it seems useful to present a detailed proof of the theorems in the setting of Banach lattices of measurable functions. Our approach is somewhat different from that of the original papers. In particular the theorems are shown to be basically expressions of minimax phenomena.

We obtain here exact formulas for the norm in the duals of some function spaces (e.g., Orlicz spaces). These formulas seem to be new. In Lozanovskii's papers (and in other places in the literature) the norms are computed only up to equivalence.

A Köthe function space on a complete σ-finite measure space (Ω, Σ, μ) (cf. [6] Def. 1.b.17) is a Banach space L, consisting of equivalnce classes, modulo equality a.e., of locally integrable, real (or complex) valued functions on Ω, verifying

(1) $|f| \leq |g|$, f measurable and $g \in L$ implies $f \in L$ and $\|f\| \leq \|g\|$.

(2) For all $\sigma \in \Sigma$ with $\mu(\sigma) < \infty$ the characteristic function χ_σ of σ is in L.

---

* This talk was delivered at a special session of the Seminar which was dedicated to the memory of G.Ya. Lozanovskii (1937-1976)

If L is a Köthe function space we denote by L' the space of those elements θ in the dual L^* of L of the form $\theta(f) = \int_\Omega fg d\mu$ for some measurable g and we identify θ with g. The space L' in the norm induced from L^* is also a Köthe function space.

Let X, Y be Köthe function spaces on the same measure space and $0 < s < 1$. A new Köthe function space $X^{1-s}Y^s$ is defined by Calderon's construction:

(3) $z \in X^{1-s}Y^s$ iff $|z| \leq x^{1-s}y^s$ for some $x \in X_+$, $y \in Y_+$.

The norm of z in $X^{1-s}Y^s$ is defined by

(4) $\|z\| = \inf\{\|x\|_X^{1-s}\|y\|_Y^s \mid x, y \text{ as in (3)}\}$.

Let U_2^0 be the set of all real-valued, concave functions φ on $\mathbb{R}_+^2$ which are positive homogeneous and satisfy

(5) $\forall \xi, \eta \geq 0,\ \varphi(\xi, 0) = \varphi(0, \eta) = 0$.

(6) $\forall \xi, \eta > 0,\ \lim_{\alpha \to \infty} \varphi(\xi, \alpha) = \lim_{\beta \to \infty} \varphi(\beta, \eta) = \infty$.

For $\varphi \in U_2^0$ the function $\hat{\varphi}$ conjugate to φ is defined on $\mathbb{R}_+^2$ by

$$\hat{\varphi}(\xi, \eta) = \inf_{\alpha, \beta > 0} \frac{\alpha\xi + \beta\eta}{\varphi(\alpha, \beta)},$$

we have $\hat{\varphi} \in U_2^0$ and $\hat{\hat{\varphi}} = \varphi$.

Let X, Y be as above. For $1 \leq p \leq \infty$ we denote by $X \oplus_p Y$ the direct sum of X and Y with the norm

$$\|(x, y)\|_p = \|(x, y)\|_{X \oplus_p Y} = (\|x\|_X^p + \|y\|_Y^p)^{1/p}$$

$$\left(\max(\|x\|_X, \|y\|_Y) \quad \text{if} \quad p = \infty\right).$$

For $\varphi \in U_2^0$ a Köthe function space $\varphi(X, Y)$ is defined by:

(7) $z \in \varphi(X, Y)$ iff $|z| \leq \varphi(x, y)$ for some $x \in X_+$, $y \in Y_+$.

For $1 \leq p \leq \infty$ norms ${}_p\|\cdot\|_{\varphi(X,Y)}$ are defined in $\varphi(X, Y)$ by

(8) ${}_p\|z\|_{\varphi(X,Y)} = \inf\{\|(x, y)\|_p \mid x, y \text{ as in (7)}\}$.

All the norms ${}_p\|\cdot\|_{\varphi(X,Y)}$ are, of course, mutually equivalent. In particular

$${}_\infty\|z\|_{\varphi(X,Y)} \leq {}_1\|z\|_{\varphi(X,Y)} \leq 2\,{}_\infty\|z\|_{\varphi(X,Y)}$$

(The norm ${}_\infty\|z\|_{\varphi(X,Y)}$ is the one defined in $\varphi(X, Y)$ by Lozanovskii.)

In the special case $\varphi(\xi, \eta) = \xi^{1-s}\eta^s$ we have

(9) $_{\infty}\|z\|_{\varphi(X,Y)} = \|z\|_{X^{1-s}Y^s}$.

In this case $\widehat{\varphi}(\xi,\eta) = \dfrac{\xi^{1-s}\eta^s}{(1-s)^{1-s}s^s}$.

Using the formula

$$\xi^{1-s}\eta^s = \inf_{t>0}\left[t^{1/(1-s)}(1-s)\xi + t^{-1/s}s\eta\right]$$

(for $\xi, \eta > 0$) we get easily

(10) $$_1\|z\|_{\widehat{\varphi}(X,Y)} = \|z\|_{X^{1-s}Y^s} .$$

Denote by $_p\varphi(X,Y)$ the space $\varphi(X,Y)$ equipped with the norm $_p\|\cdot\|_{\varphi(X,Y)}$. In all the statements which follow, equality of normed spaces means also equality of the norms.

Theorem 1. *([7]). Let X, Y be two Köthe function spaces on the same measure space, then for all $0 < s < 1$*

$$(X^{1-s}Y^s)' = (X')^{1-s}(Y')^s .$$

Theorem 2. (mainly [9]). *Let X, Y be as in Theorem 1 and $\varphi \in U_2^0$. Let $\psi = \widehat{\varphi}$, then for $1 \le p \le \infty$ and p' such that $\frac{1}{p} + \frac{1}{p'} = 1$*

$$_p\varphi(X,Y)' = {}_{p'}\psi(X',Y') .$$

In view of (9) and (10), Theorem 1 is a special case of Theorem 2. Nevertheless we include in section 2 a direct proof of Theorem 1 in the finite dimensional case. We do this because this proof demonstrates that Theorem 1 is basically a consequence of Ky-Fan's minimax theorem for convex-concave-like functions.

In section 3 we prove Theorem 2 in the general case. In the finite dimensional case the proof is simple, along the same lines as that of Theorem 1. The general case, however, requires more work. In section 4 we bring some corollaries and special cases.

Let L be a Köthe function space on (Ω, Σ, μ). For $f \in L^*$ and $x \in L$ we denote by $f\langle x\rangle$ the result of the action of f on x. In particular, if $f \in L'$

$$f\langle x\rangle = \int_\Omega f(w)x(w)d\mu .$$

We shall extend the use of the notation $f\langle x\rangle$ to denote $\int_\Omega f(w)x(w)d\mu$ whenever f and x are measurable functions such that $f(\cdot)x(\cdot)$ is integrable. An element $0 \le f \in L^*$ is said to be *singular* if whenever $g \in L'$ satisfies $0 \leq g \leq f$ it follows that $g = 0$. An arbitrary element $f \in L^*$ is singular if f_+ and f_- are both singular.

Every $f \in L^*$ has a unique decomposition $f = f_I + f_S$ with $f_I \in L'$ and $f_S \in L^*$ is singular. If $0 \leq f \in L^*$ then $f_I, f_S \geq 0$. We have for $0 \leq f \in L^*$

(11) $\forall x \in L_+,\ f_I(x) = \inf\{\lim f(x_n) \mid 0 \leq x_n \uparrow x\}$

(cf. [10] ch. 15 exc. 70.1).

f_I is called the *integral part of* f.

Finally we recall Ky-Fan's minimax theorem. Let Γ, Λ be non-empty sets. A real function f on $\Gamma \times \Lambda$ is said to be *convex-concave-like* if for all $0 < \alpha < 1$ we have the following

(12) For all $\gamma_1, \gamma_2 \in \Gamma$ there exists $\gamma_3 \in \Gamma$ so that for all $\lambda \in \Lambda$, $f(\gamma_3, \lambda) \leq \alpha f(\gamma_1, \lambda) + (1 - \alpha) f(\gamma_2, \lambda)$.

(13) For all $\lambda_1, \lambda_2 \in \Lambda$ there exists $\lambda_3 \in \Lambda$ so that for all $\gamma \in \Gamma$, $f(\gamma, \lambda_3) \geq \alpha f(\gamma, \lambda_1) + (1 - \alpha) f(\gamma, \lambda_2)$.

Theorem A. (Fan's minimax th. [3]). *Let f be convex-concave-like on $\Gamma \times \Lambda$. Suppose also that Λ is compact in some Hausdorff topology and $f(\gamma, \cdot)$ is upper semi-continuous on Λ for every $\gamma \in \Gamma$. Then*

$$\inf_{\Gamma} \max_{\Lambda} f(\gamma, \lambda) = \max_{\Lambda} \inf_{\Gamma} f(\gamma, \lambda) .$$

(cf. [1] for a short and simple proof).

Throughout the paper we adopt the convention $\frac{0}{0} = 0$.

2. A proof of Theorem 1 in the finite dimensional case

We assume that $\dim X = \dim Y = n$ and that the unit vectors' basis $\{e_i\}_{i=1}^n$ is a 1-unconditional basis in X and in Y. We have to show that for $0 < s < 1$ and $b = (\beta_i)_{i=1}^n \in \mathbb{R}_+^n$

(1) $$\|b\|_{(X^{1-s}Y^s)'} = \|b\|_{(X')^{1-s}(Y')^s} .$$

By definition we have

(2) $$\|b\|_{(X^{1-s}Y^s)'} = \sup_{\substack{x,y \in \mathbb{R}_+^n \\ \|x\|_X \leq 1,\ \|y\|_Y \leq 1}} \sum_{i=1}^n \beta_i x_i^{1-s} y_i^s$$

(3) $$\|b\|_{(X')^{1-s}(Y')^s} = \inf_{\substack{\gamma,\delta \in \mathbb{R}_+^n \\ \gamma_i^{1-s}\delta_i^s = \beta_i}} \sup_{\substack{x,y \in \mathbb{R}_+^n \\ \|x\|_Y \leq 1,\ \|y\| \leq 1}} (\Sigma \gamma_i x_i)^{1-s} (\Sigma \delta_i y_i)^s$$

Let $\Gamma = \{(\gamma,\delta) \in \mathbb{R}^n_+ \times \mathbb{R}^n_+ \mid \forall i,\ \gamma_i^{1-s}\delta_i^s = \beta_i\}$
$\Lambda = \{(x,y) \in \mathbb{R}^n_+ \times \mathbb{R}^n_+ ; \|x\|_X \le 1 ,\ \|y\|_Y \le 1\}$

and $f : \Gamma \times \Lambda \to \mathbb{R}$ be defined by

$$f((\gamma,\delta),(x,y)) = (\Sigma\gamma_i x_i)^{1-s}(\Sigma\delta_i y_i)^s .$$

Clearly, for all $(\gamma,\delta) \in \Gamma$, $f((\gamma,\delta),(\cdot,\cdot))$ is concave and continuous on the compact convex set Λ. Also, f is convex-like on Γ. In fact, given $(\gamma_1,\delta_1),(\gamma_2,\delta_2)$ in Γ and $0 < \alpha < 1$, let $\gamma_3 = \gamma_1^{1-\alpha}\gamma_2^\alpha$, $\delta_3 = \delta_1^{1-\alpha}\delta_2^\alpha$ then clearly for all $(x,y) \in \Lambda$ we have

$$f((\gamma_3,\delta_3),(x,y)) \le (1-\alpha)f((\gamma_1,\delta_1),(x,y)) + \alpha f((\gamma_2,\delta_2),(x,y))$$

(use Hölder's and the arithmetic-geometric-mean inequalities). It follows from Theorem A and from (3) that

$$\|b\|_{(X')^{1-s}(Y')^s} = \sup_{(x,y)\in\Lambda}\ \inf_{(\gamma,\delta)\in\Gamma} (\Sigma\gamma_i x_i)^{1-s}(\Sigma\delta_i y_i)^s \tag{4}$$

It is now easily checked (using the case of equality in Hölder's inequality and approximating in case $x_i = 0$ or $y_i = 0$ for some i) that for all $(x,y) \in \Lambda$ we have

$$\inf_{(\gamma,\delta)\in\Gamma} (\Sigma\gamma_i x_i)^{1-s}(\Sigma\delta_i y_i)^s = \Sigma\beta_i x_i^{1-s} y_i^s . \tag{5}$$

Of course, (2), (4), and (5) imply (1). □

A proof of the general result - Theorem 2, in the finite dimensional case could be given along similar lines.

3. The Proof of Theorem 2 – General Case

The proof is composed of two parts.

Step a. Consists of proving that

$$\psi(X',Y') \subset \varphi(X,Y)'$$

and for all $f \in \psi(X',Y')$

$$\|f\|_{\psi(X',Y')} = \|f\|_{\varphi(X,Y)'}$$

Step b. A proof that actually $\psi(X',Y') = \varphi(X,Y)'$.

Lemma 1. *Let $0 \le z \in \varphi(X,Y)_+$ be such that $z \le \varphi(x,y)$, $(x \in X_+ \,,\; y \in Y_+)$. Then there are $0 \le x_1 \le x$, $0 \le y_1 \le y$ such that $z = \varphi(x_1,y_1)$.*

Proof: Put $\lambda(\omega) = z(\omega)/\varphi(x(\omega),y(\omega))$, $\omega \in \Omega$, and $x_1 = \lambda x, y_1 = \lambda y$. By homogeneity, $z = \varphi(x_1,y_1)$. □

Lemma 2. *There exists $z \in X \cap Y$ such that $z > 0$ a.e. on Ω.*

Proof: Write $\Omega = \cup_{n=1}^{\infty}\Omega_n$ with $\mu(\Omega_n) < \infty$ for all n and $\{\Omega_n\}$ disjoint. Let $z = \sum_{n=1}^{\infty} \gamma_n \chi_{\Omega_n}$ where $\{\gamma_n\}$ are positive and $\Sigma\gamma_n\|\chi_{\Omega_n}\|_X < \infty$, $\Sigma\gamma_n\|\chi_{\Omega_n}\|_Y < \infty$. □

Lemma 3. a) *If $g \in X'_+$, $h \in Y'_+$ and $0 \le f \le \psi(g,h)$ then for all $x \in X_+$, $y \in Y_+$*

$$f\langle\varphi(x,y)\rangle \le g\langle x\rangle + h\langle y\rangle \,.$$

b) Let f be a non-negative, locally integrable function on Ω, if $g \in X_+^*$, $h \in Y_+^*$ and for all $x \in X_+$, $y \in Y_+$ we have $f\langle\varphi(x,y)\rangle \le g\langle x\rangle + h\langle y\rangle$ then $f \le \psi(g_I,h_I)$ (where g_I, h_I are the integral parts of g,h respectively).

Proof: a) By the definition of ψ, we have for ω such that $\varphi(x(\omega),y(\omega)) \ne 0$

$$\psi(g(\omega),h(\omega)) \le \frac{g(\omega)x(\omega) + h(\omega)y(\omega)}{\varphi(x(\omega),y(\omega))}$$

from which follows

$$f\langle\varphi(x,y)\rangle \le \psi(g,h)\langle\varphi(x,y)\rangle =$$
$$= \int_\Omega \psi(g(\omega),h(\omega))\varphi(x(\omega),y(\omega))d\mu \le g\langle x\rangle + h\langle y\rangle \,.$$

b) Assume that $f\langle\varphi(x,y)\rangle \le g\langle x\rangle + h\langle y\rangle$ for all $x \in X_+$, $y \in Y_+$ $(g \in X_+^*, h \in Y_+^*)$, but $f \not\le \psi(g_I,h_I)$. Let $E \in \Sigma$ be such that $0 < \mu(E) < \infty$ and $f > \psi(g_I,h_I)$ on E.

Case i. $\mu(E \cap \operatorname{supp} g_I \cap \operatorname{supp} h_I) = 0$. Then we may assume, e.g., that $\mu(E_1) > 0$ where $E_1 = E \setminus \operatorname{supp} h_I$. Take $x = \varepsilon\chi_{E_1}$, $y = M\chi_{E_1}$. We get

(1) $$f\langle\varphi(x,y)\rangle = \varphi(\varepsilon,M)\int_{E_1} f d\mu$$

(2) $$g_I\langle x\rangle + h_I\langle y\rangle = \int_{E_1} g_I x d\mu = \varepsilon\int_{E_1} g_I d\mu$$

If we choose $\varepsilon > 0$ so small that $\varepsilon\int_{E_1} g_I d\mu < \int_{E_1} f d\mu$ and then M such that $\varphi(\varepsilon,M) > 1$ we get

(3) $$f\langle\varphi(x,y)\rangle > g_I\langle x\rangle + h_I\langle y\rangle \,.$$

Case ii. $\mu(E \cap \operatorname{supp} g_I \cap \operatorname{supp} h_I) > 0$.

We may assume that $0 < \delta \le g_I, h_I \le K$ on E and that $f > \psi(g_I, h_I) + \theta$ on E, for some $\theta > 0$.

Let $0 < \varepsilon \ll 1$. By Egoroff's theorem there exists $E_1 \in \Sigma$, $E_1 \subset E$ with $\mu(E_1) > 0$ and real numbers $\delta/2 \le a, b \le 2K$ so that on E_1 we have $|g_I - a| < \varepsilon$, $|h_I - b| < \varepsilon$.

Define $x = \alpha \chi_{E_1}$, $y = (1-\alpha)\chi_{E_1}$ $(0 < \alpha < 1)$. We have

(4) $$f\langle\varphi(x, y)\rangle = \varphi(\alpha, 1-\alpha) \int_{E_1} f\,d\mu .$$

Also

(5) $$\begin{aligned} g_I\langle x\rangle + h_I\langle y\rangle &= \int_{E_1} \left[\alpha g_I + (1-\alpha)h_I\right] d\mu \le \\ &\le \varepsilon\mu(E_1) + \left[\alpha a + (1-\alpha)b\right]\mu(E_1) . \end{aligned}$$

By the definition (and homogeneity) of ψ, we may choose $0 < \alpha < 1$ such that

$$\alpha a + (1-\alpha)b = \psi(a, b)\varphi(\alpha, 1-\alpha) .$$

Notice also that

(6) $$\varphi(\alpha, 1-\alpha) = \frac{\alpha a + (1-\alpha)b}{\psi(a, b)} \ge \frac{\delta/2}{\max\limits_{\delta/2 \le c,d \le 2K} \psi(c, d)} = \varphi_0 > 0 .$$

Since a, b and $g_I(\omega), h_I(\omega)$ $(\omega \in E_1)$ are in $J = [\delta/2, 2K]$, we have, due to uniform continuity of ψ on $J \times J$:

(7) $$\psi(a, b) \le \psi(g_I(\omega), h_I(\omega)) + \delta(\varepsilon) , \qquad (\omega \in E_1) ,$$

where $\delta(\varepsilon) \to 0$ as $\varepsilon \to 0$.

Hence (5) yields

$$\begin{aligned} g_I\langle x\rangle + h_I\langle y\rangle &\le \left[\varepsilon + \delta(\varepsilon)\varphi(\alpha, 1-\alpha)\right]\mu(E_1) + \varphi(\alpha, 1-\alpha) \int_{E_1} \psi(g_I(\omega), h_I(\omega))\,d\mu \\ &< \left[\frac{1}{\varphi_0}\varepsilon + \delta(\varepsilon)\right] \varphi(\alpha, 1-\alpha)\mu(E_1) + \varphi(\alpha, 1-\alpha) \int_{E_1} f\,d\mu - \theta\varphi(\alpha, 1-\alpha)\mu(E_1) . \end{aligned}$$

If $\varepsilon > 0$ is so small that $(1/\varphi_0)\varepsilon + \delta(\varepsilon) < \theta$ we get by (4)

(8) $$g_I\langle x\rangle + h_I\langle y\rangle < f\langle\varphi(x, y)\rangle$$

So, in both cases – i) and ii), we found $x \in X_+$, $y \in Y_+$ so that (8) holds.

By (11) in section 1, we may find sequences $(x_n) \subset X_+$, $(y_n) \subset Y_+$, $x_n \uparrow x$, $y_n \uparrow y$, so that for given $\varepsilon > 0$,

(9)
$$\begin{aligned} g_I(x) &\leq \lim g(x_n) \leq g_I(x) + \varepsilon \\ h_I(y) &\leq \lim h(y_n) \leq h_I(y) + \varepsilon \,. \end{aligned}$$

By (8) and the monotone convergence theorem we conclude that for n big enough

(10)
$$f(\varphi(x_n, y_n)) > g(x_n) + h(y_n)$$

which clearly contradicts the assumption of b). □

We are now ready to prove Step a.

Assume that ${}_{p'}\|f\|_{\psi(X',Y')} = 1$ and define

$$\Gamma = \{(g,h) \mid g \in X_+^* \,,\; h \in Y_+^* \,;\; f \leq \psi(g_I, h_I)\} \,.$$
$$\Lambda = \{(x,y) \mid x \in X_+ \,,\; y \in Y_+ \,;\, \|(x,y)\|_{X \oplus_p Y} \leq 1\} \,.$$

We have

(11)
$$\|f\|_{{}_p\varphi(X,Y)'} = \sup_{(x,y)\in\Lambda} f(\varphi(x,y))$$

(12)
$$\begin{aligned} 1 = {}_{p'}\|f\|_{\psi(X',Y')} &= \inf_{(g,h)\in\Gamma} \|(g_I,h_I)\|_{X' \oplus_{p'} Y'} \\ &= \inf_{(g,h)\in\Gamma} \sup_{(x,y)\in\Lambda} \left[g(x) + h(y)\right] \end{aligned}$$

(a-priori we should have written g_I and h_I instead of g and h inside the square brackets, but a simple check shows that anyway the inf. in (12) is actually on the integral elements).

Γ is a convex subset of $X^* \oplus_{p'} Y^*$ and, in view of Lemma 3, is ω^*-closed. Notice that if $(g,h) \in \Gamma$ then $\|(g,h)\|_{p'} \geq 1$, so that for all $0 < \varepsilon < 1$, Γ is disjoint from the ω^*-compact set $(1-\varepsilon)B(X^* \oplus_{p'} Y^*)$ ($B(E)$ denotes the unit ball of the Banach space E). It follows that for every $0 < \varepsilon < 1$ there exists a ω^*-continuous functional $(x_\varepsilon, y_\varepsilon)$ with $\|(x_\varepsilon, y_\varepsilon)\|_{X \oplus_p Y} = 1$, separating these two sets.

In other words

$$1 - \varepsilon \leq g(x_\varepsilon) + h(y_\varepsilon) \quad \text{for all } (g,h) \in \Gamma \,.$$

Hence $1 \leq \sup_\Lambda \inf_\Gamma [g(x) + h(y)]$ and since we always have $\sup\inf \leq \inf\sup$ we get from (12):

(13)
$$1 = {}_{p'}\|f\|_{\psi(X',Y')} = \sup_{(x,y)\in\Lambda} \inf_{(g,h)\in\Gamma} [g(x) + h(y)] \,.$$

Let $(x,y) \in \Lambda$, by Lemma 3 a), we have for all $(g,h) \in \Gamma$

$$g\langle x\rangle + h\langle y\rangle \geq g_I\langle x\rangle + h_I\langle y\rangle \geq f\langle\varphi(x,y)\rangle .$$

Hence, the right-hand side of (13) is $\geq$ the right-hand side of (11).

Let $k \in X_+ \cap Y_+$, $k > 0$ a.e. be as in Lemma 2. Replacing (x,y) in Λ by

$$\left(\frac{x+\varepsilon k}{C}, \frac{y+\varepsilon k}{C}\right), C = \|(x+\varepsilon k, y+\varepsilon k)\|_p$$

with $\varepsilon > 0$ small, it is easily checked that the "sup" in (13) and in (11) may be taken over $(x,y) \in \Lambda$ such that $x > 0$ and $y > 0$ a.e. We shall show that for every such $(x,y) \in \Lambda$ there is, for every $\delta > 0$, a $(\tilde{g},\tilde{h}) \in \Gamma$ verifying

$$\tilde{g}\langle x\rangle + \tilde{h}\langle y\rangle \leq f\langle\varphi(x,y)\rangle + 6\delta . \tag{14}$$

This will show that the right-hand side of (13) and (11) are equal and Step a) will be completed.

Let $g \in X'_+, h \in Y'_+$ with $\|g\|_{X'} \leq 2$, $\|h\|_{Y'} \leq 2$ be such that $f = \psi(g,h)$ (cf. Lemma 1). We intend to find non-negative measurable functions G, H so that

$$\psi(Gg, Hh) = \psi(g,h) = f , \tag{15}$$

$$\int_\Omega Ggx d\mu + \int_\Omega Hhy d\mu = \int_\Omega \psi(Gg,Hh)\varphi(x,y)d\mu = f\langle\varphi(x,y)\rangle . \tag{16}$$

We have for $\omega \in \Omega$ with $g(\omega) \neq 0$, $h(\omega) \neq 0$

$$\varphi(x(\omega),y(\omega)) = \min_{0<\kappa<1} \frac{\kappa g(\omega)x(\omega) + (1-\kappa)h(\omega)y(\omega)}{\psi(\kappa g(\omega),(1-\kappa)h(\omega))} .$$

For such ω define

$$k(\omega) = \min\left\{\kappa \mid 0<\kappa<1 ; \frac{\kappa g(\omega)x(\omega) + (1-\kappa)h(\omega)y(\omega)}{\psi(\kappa g(\omega),(1-\kappa)h(\omega))} = \right.$$
$$\left. = \varphi(x(\omega),y(\omega))\right\} .$$

It is not difficult to check that $k(\omega)$ is a measurable function on $\Omega_+ = \{\omega \in \Omega \mid g(\omega) \neq 0$ and $h(\omega) \neq 0\}$. Define $G(\omega) = H(\omega) = 0$ for $\omega \notin \Omega_+$. For $\omega \in \Omega_+$ let

$$\lambda(\omega) = \frac{\psi(g(\omega),h(\omega))}{\psi(k(\omega)g(\omega),(1-k(\omega))h(\omega))}$$

and define

$$G(\omega) = \lambda(\omega)k(\omega) \quad , \quad H(\omega) = \lambda(\omega)[1-k(\omega)] .$$

Then (15) and (16) are clearly satisfied.

Since everything "lives" on Ω_+ we assume now that $h > 0$ and $g > 0$ on Ω. We then have $gx, hy > 0$ a.e. and they are integrable functions. Due to absolute continuity with respect to μ of the measures defined by gx, hy, Ggx, Hhy we can, for a given $\delta > 0$, find $\varepsilon > 0$ such that the set

$$\Omega_\varepsilon = \left\{\omega \mid \varepsilon \leq g(\omega) \leq \frac{1}{\varepsilon} \quad \text{and} \quad \varepsilon \leq h(\omega) \leq \frac{1}{\varepsilon}\right\}$$

satisfies

$$\int_{\Omega\setminus\Omega_\varepsilon} gx d\mu < \delta\ , \quad \int_{\Omega\setminus\Omega_\varepsilon} hy d\mu < \delta\ , \quad \int_{\Omega\setminus\Omega_\varepsilon} Ggx d\mu < \delta\ , \quad \int_{\Omega\setminus\Omega_\varepsilon} Hhy d\mu < \delta\ .$$

Given two real numbers σ, τ, $\varepsilon \leq \tau, \sigma \leq 1/\varepsilon$, let γ vary and η vary accordingly so that the equality

$$\psi(\sigma, \tau) - \psi(\gamma\sigma, \eta\tau)$$

will remain true. We have then

(17) $$\psi(\gamma\varepsilon, \eta\varepsilon) \leq \psi\left(\frac{1}{\varepsilon}, \frac{1}{\varepsilon}\right) \quad \text{for all such} \quad \gamma, \eta\ .$$

(17) implies that there exists $M > 0$ (which depends on ε, δ) such that

(18) $$\begin{cases} \text{a)} & \gamma \geq M \longrightarrow \eta \leq \delta \\ \text{b)} & \gamma \geq \delta \Longrightarrow \eta \leq M \end{cases}$$

For M as above define functions $G_\varepsilon, H_\varepsilon$ by $G_\varepsilon = H_\varepsilon = 1$ on $\Omega\setminus\Omega_\varepsilon$, and for $\omega \in \Omega_\varepsilon$

$$G_\varepsilon(\omega) = \begin{cases} M & \text{if} \quad G(\omega \geq M \\ G(\omega) & \text{if} \quad \delta \leq G(\omega) \leq M \\ \delta & \text{if} \quad G(\omega) \leq \delta \end{cases}$$

and $H_\varepsilon(\omega)$ is determined by the equation

(19) $$\psi(G_\varepsilon g, H_\varepsilon h) = \psi(g, h) = f$$

(H_ε is clearly measurable).

From (18) we conclude

(20) $$H_\varepsilon(\omega) \begin{cases} \leq \delta & \text{if} \quad G(\omega) > M \\ = H(\omega) & \text{if} \quad \delta \leq G(\omega) \leq M \\ \leq H(\omega) & \text{if} \quad G(\omega) \leq \delta \end{cases}$$

Also G_ε and H_ε are L_∞ functions (G_ε by definition and H_ε by (18)b)).

We get

$$(21)\quad (G_\epsilon g)\langle x\rangle + (H_\epsilon h)\langle y\rangle = \int_{\Omega\setminus\Omega_\epsilon} gx\,d\mu + \int_{\Omega\setminus\Omega_\epsilon} hy\,d\mu + \int_{\Omega_\epsilon} G_\epsilon gx\,d\mu + \int_{\Omega_\epsilon} H_\epsilon hy\,d\mu$$

$$\leq 2\delta + \int_{\{\delta\leq G\leq M\}\cap\Omega_\epsilon} [Ggx + Hhy]\,d\mu + \delta \int_{\{G<\delta\}\cap\Omega_\epsilon} gx\,d\mu +$$

$$+ \delta \int_{\{G>M\}\cap\Omega_\epsilon} hy\,d\mu + \int_{\{G>M\}\cap\Omega_\epsilon} Ggx\,d\mu + \int_{\{G<\delta\}\cap\Omega_\epsilon} Hhy\,d\mu$$

$$\leq \int_\Omega Ggx\,d\mu + \int_\Omega Hhy\,d\mu + 6\delta$$

$$= f\langle\varphi(x,y)\rangle + 6\delta\ .$$

Let $\tilde{g} = G_\epsilon g$, $\tilde{h} = H_\epsilon h$. By (19) we have $\psi(\tilde{g},\tilde{h}) = f$ and by (20) we have (14). Thus the proof of Step a) is completed.

For Step b) – the proof that $\varphi(X,Y)' = \psi(X',Y')$, we need the following introduction.

Let L be a Köthe function space. Define $\tau = \tau(L,L')$ to be the locally convex Hausdorff topology on L', generated by the Riesz semi-norms $\{p_x \mid x \in L_+\}$, where for $x \in L_+$, p_x is defined by

$$p_x(f) = |f|\langle x\rangle \quad \text{for} \quad f \in L'\ .$$

Clearly τ is weaker than the norm topology in L' and stronger than the week-* topology restricted to L'. Denote by $(L'_\tau)^*$ the dual space of the l.c.s. space (L',τ). Since τ is weaker than the norm we have

$$(L'_\tau)^* \subset (L')^*\ .$$

Notice also that since $\max(p_{x_1}, p_{x_2}) \leq p_{x_1\vee x_2}$, $\{p_x\}$ is a fundamental system of semi-norms.

Lemma 4. $(L'_\tau)^* = L$.

Proof: Since $\{p_x\}$ is fundamental, a linear functional $\theta \in (L')^*$ is τ-continuous iff there is at least one seminorm p_x s.t. for all $f \in L'$

$$(22).\qquad |\theta\langle f\rangle| \leq p_x(f)\ .$$

If $\theta = z \in L$ then clearly $x = |z|$ satisfies (22) hence $\theta \in (L'_\tau)^*$. On the other hand, suppose $\theta \in (L'_\tau)^*$. Then $\theta \in (L')^*$ and there is $x \in L_+$ which satisfies (22). We have for all $0 \leq f \in L'$

$$|\theta|\langle f\rangle = \sup\{\theta\langle g\rangle - \theta\langle f-g\rangle \mid 0 \leq g \leq f\}\ .$$

For all $0 \leq g \leq f$ we have by (22)

$$|\theta\langle g\rangle| \leq g\langle x\rangle \quad \text{and} \quad |\theta\langle f-g\rangle| \leq (f-g)\langle x\rangle \ .$$

So

$$|\theta|\langle f\rangle \leq f\langle x\rangle \quad \text{for all} \quad f \in L'_+ \ .$$

Hence $|\theta| \leq x$ in $(L')^*$, but x is an integral in $(L')^*$ hence ([10,p. 464) $|\theta|$ is also an integral and since $|\theta| \leq x \in L$ we get $\theta \in L$. □

A straightforward consequence of Lemma 4 and the Hahn-Banach theorem is

Corollary 5. *Let $Y \subset L'$ be a subspace which is total on L then Y is τ-dense in L'.*

From the inequalities

$$\left.\begin{array}{l} |f_1 \vee g_1 - f_2 \vee g_2| \\ |f_1 \wedge g_1 - f_2 \wedge g_2| \end{array}\right\} \leq |f_1 - f_2| + |g_1 - g_2|$$

we get

Lemma 6. *The operations $(f,g) \to f \vee g$, $(f,g) \to f \wedge g$ are τ-continuous from $L' \times L'$ into L'. In particular $f \to |f|$ is τ-continuous in L'.*

We are now ready to prove that $\varphi(X,Y)' = \psi(X',Y')$. By step a) $\psi(X',Y')$ is a subspace of $\varphi(X,Y)'$, it is total on $\varphi(X,Y)$ because it contains the characteristic functions of finite-measure sets. By Corollary 5, $\psi(X',Y')$ is τ-dense in $\varphi(X,Y)'$.

Let $f \in \varphi(X,Y)'$ and let (f_α) be a net in $\psi(X',Y')$ s.t. $f_\alpha \xrightarrow{\tau} f$. By Lemma 6 $|f_\alpha| \xrightarrow{\tau} |f|$. Moreover, if $g_\alpha = |f_\alpha| \wedge |f|$ we have $g_\alpha \xrightarrow{\tau} |f|$. $g_\alpha \leq |f_\alpha|$ so $g_\alpha \in \psi(X',Y')$. Also $g_\alpha \leq |f|$ hence by Step a)

$$_{p'}\|g_\alpha\|_{\psi(X',Y')} = \|g_\alpha\|_{{}_p\varphi(X,Y)'} \leq \|f\|_{{}_p\varphi(X,Y)'} \ .$$

So for some $M > 0$ we can represent for all α, $g_\alpha = \psi(h_\alpha, k_\alpha)$ with $h_\alpha, k_\alpha \geq 0$, $\|h_\alpha\|_{X'}$, $\|k_\alpha\|_{Y'} \leq M$.

There exists a subnet (β) of (α) such that

$$h_\beta \xrightarrow{w^*} h \in X^*_+ \quad , \quad k_\beta \xrightarrow{w^*} k \in Y^*_+ \ .$$

Let $x \in X_+$, $y \in Y_+$ then

$$\begin{aligned} |f|\langle\varphi(x,y)\rangle &= \lim g_\beta\langle\varphi(x,y)\rangle \\ &= \lim \psi(h_\beta,k_\beta)\langle\varphi(x,y)\rangle \\ &\leq \lim\left[h_\beta\langle x\rangle + k_\beta\langle y\rangle\right] = h\langle x\rangle + k\langle y\rangle \ . \end{aligned}$$

By Lemma 3 b) we conclude now that $|f| \leq \psi(h_I, k_I)$ where h_I, k_I are the integral parts of h, k resp. Hence $f \in \psi(X', Y')$. □

4. Some Special Cases and Corollaries

The results of this section are, like the preceeding results, basically taken from [7] and [9]. The possibility of describing exactly the norm dual to $\varphi(X,Y)$ provides a somewhat more elegant presentation of some of these results.

Some special properties of Köthe function spaces are (cf. [6])

a) X has the *Fatou property* if $0 \leq f_n \uparrow f$ a.e., $(f_n)_{n=1}^{\infty} \subset X$ and $\sup_n \|f_n\| < \infty$ implies $f \in X$ and $\|f\| = \lim_n \|f_n\|$. X has the Fatou property if and only if $X = X''$.

b) X is *σ-order continuous* if $f_n \downarrow 0$ a.e., $(f_n)_{n=1}^{\infty} \subset X$ implies $\|f_n\| \to 0$. X is σ-order continuous if and only if $X^* = X'$.

Proposition 1. *If X and Y have the Fatou property then ${}_p\varphi(X,Y)$ has the Fatou property. Also, in this case every $z \in \varphi(X,Y)$ has a representation*

$$|z| = \varphi(x,y) \quad , \quad x \in X_+ \, , \, y \in Y_+$$

with ${}_p\|z\|_{\varphi(X,Y)} = \|(x,y)\|_{X \oplus_p Y}$.

Proof: The first assertion is a straightforward consequence of Theorem 2:

$${}_p\varphi(X,Y)'' = {}_{p'}\psi(X',Y')' = {}_p\varphi(X'',Y'') = {}_p\varphi(X,Y) \, .$$

In order to prove the second assertion, let $0 \leq z \in \varphi(X,Y)$ with ${}_p\|z\|_{\varphi(X,Y)} = 1$. For every $n \in I\!N$ we find $x_n \in X_+$, $y_n \in Y_+$ with $\|(x_n, y_n)\|_{X \oplus_p Y} \leq 1 + (1/n)$ such that $z = \varphi(x_n, y_n)$. There is a subnet (n_α) of $I\!N$ so that $(x_{n_\alpha}, y_{n_\alpha}) \xrightarrow{\omega^*} (\xi, \eta) \in (X^{**} \oplus_p Y^{**})_+$. We have of course $\|(\xi,\eta)\|_{X^{**} \oplus_p Y^{**}} \leq 1$. For all $f \in X'_+$, $g \in Y'_+$ we have by Lemma 3 of section 3

$$\psi(f,g)\langle z \rangle \leq \xi\langle f \rangle + \eta\langle g \rangle \, .$$

We conclude, using the same lemma, that $z \leq \varphi(\xi_I, \eta_I)$ where ξ_I, η_I are the integral parts of ξ, η resp. Also $\|\xi_I\|_X = \|\xi_I\|_{X''} \leq \|\xi\|_{X^{**}}$, $\|\eta_I\|_Y = \|\eta_I\|_{Y''} \leq \|\eta\|_{Y^{**}}$, so $\|(\xi_I, \eta_I)\|_{X \oplus_p Y} \leq 1$. □

Definition 2. Let $\varphi \in U_2^0$. We say that φ satisfies the *Left* (Right)Δ_2-condition (in short $L-\Delta_2$ or $R-\Delta_2$) if there exists a constant $C \geq 1$ such that for all ξ, η

$$(1)_L \quad \varphi(2\xi, 2\eta) \leq \varphi(C\xi, \eta)$$

$$(\text{resp.} \quad (1)_R \quad \varphi(2\xi, 2\eta) \leq \varphi(\xi, C\eta)) \ .$$

Remark 3. Every $\varphi \in U_2^0$ is associated with two Orlicz functions M_L and M_R via the identity

$$\varphi(\xi,\eta) = \begin{cases} 0 & \text{if } \eta = 0 \\ \eta M_L^{-1}(\xi/\eta) & \text{if } \eta > 0 \end{cases} = \begin{cases} 0 & \text{if } \xi = 0 \\ \xi M_R^{-1}(\eta/\xi) & \text{if } \xi > 0 \end{cases}$$

The $L\ (R) - \Delta_2$ condition for φ is equivalent to the Δ_2-condition for $M_L\ (M_R)$.

Proposition 4. *If X or Y is σ-order continuous and φ satisfies the appropriate one sided Δ_2-condition, then ${}_p\varphi(X,Y)$ is σ-order continuous and*

$${}_p\varphi(X,Y)^* = {}_{p'}\psi(X',Y') \ .$$

Proof: Assume that $X' = X^*$ and φ satisfies the $L - \Delta_2$ condition. Let $z_n \in \varphi(X,Y)$ be such that $z_n \downarrow 0$ a.e.

We write $z_1 = \varphi(x_1, y_1)$, $x_1 \in X_+$, $y_1 \in Y_+$.

Let x_n be defined by the equation $z_n = \varphi(x_n, y_1)$. Notice that since $z_n \downarrow 0$ we have $x_n \leq x_1$ so $x_n \in X$ and also $x_n \downarrow 0$ a.e. It follows that $\|x_n\|_X \to 0$ (since X is σ-order continuous). By the $L - \Delta_2$ condition we have for all $m, n \in \mathbb{N}$:

$$z_n \leq \varphi\left(\left(\frac{C}{2}\right)^m x_n \ , \ \frac{1}{2^m} y_1\right) \ . \tag{2}$$

Choose now a sequence $(m(n))_{n=1}^\infty$ of positive integers satsifying $m(n) \to \infty$ as $n \to \infty$ and $\left(\frac{C}{2}\right)^{m(n)} \|x_n\|^{1/2} \leq K$ for some fixed constant K. Then

$$\left\| \left(\frac{C}{2}\right)^{m(n)} x_n \right\|_X \longrightarrow 0 \ , \quad \left\| \frac{1}{2^{m(n)}} y_1 \right\|_Y \longrightarrow 0 \ .$$

Hence (2) yields ${}_p\|z_n\|_{\varphi(X,Y)} \to 0$. We have thus shown that ${}_p\varphi(X,Y)$ is σ-order continuous. We conclude that

$${}_p\varphi(X,Y)^* = {}_p\varphi(X,Y)' = {}_{p'}\psi(X',Y') \ . \qquad \square$$

Corollary 5. *Assume that either X or Y has the Fatou property and is σ-order continuous. Assume that φ satisfies the appropriate one sided Δ_2-condition. Assume also, independently* *of the previous assumption, that either X' or Y' is σ-order continuous and that ψ satisfies the appropriate one-sided Δ_2-condition. Then ${}_p\varphi(X,Y)$ is reflexive.*

Proof: By Proposition 4 and the first assumption we have: ${}_p\varphi(X,Y)^* = {}_p\psi(X',Y')$. The second assumption implies now ${}_p\varphi(X,Y)^{**} = {}_p\varphi(X'',Y'')$. Assume, e.g., that the first assumption was true for X, so $X = X''$, hence ${}_p\varphi(X,Y)^{**} = {}_p\varphi(X,Y'')$. Repeating the first step we get

$${}_p\varphi(X,Y)^{***} = {}_p\varphi(X,Y'')^* = {}_p\psi(X',Y''') = {}_p\psi(X',Y') = {}_p\varphi(X,Y)^* .$$

Hence ${}_p\varphi(X,Y)^*$ is reflexive and so is ${}_p\varphi(X,Y)$. □

Proposition 6. ([7], cf. also [5] and [4]). *Let X be a Köthe function space on (Ω,Σ,μ). For every $0 \le f \in L_1(\mu)$ and $\varepsilon > 0$ there exists $0 \le g \in X$, $0 \le h \in X'$ such that $f = gh$ and*

$$\|g\|_X \|h\|_{X'} \le (1+\varepsilon)\|f\|_{L_1} . \tag{3}$$

If X has the Fatou property we may take $\varepsilon = 0$ in (3).

Proof: Let $Y = X^{1/2}(X')^{1/2}$. By Theorem 1 $Y' = (X')^{1/2}(X'')^{1/2}$.

$$Y'' = (X'')^{1/2}(X''')^{1/2} = (X'')^{1/2}(X')^{1/2} = Y' .$$

From the equality $Y'' = Y'$ it follows that $Y' = L_2(\mu)$, so also $Y = L_2(\mu)$, from which the result follows.

The case that X has the Fatou property follows from Proposition 1. □

Example 7. Orlicz spaces. Let M be an Orlicz function on $\mathbb{R}_+$, i.e., M is convex, strictly increasing and $M(0) = 0$. The complementary Orlicz function M^* is defined by

$$M^*(\xi) = \max_{\eta\ge 0}\left[\xi\eta - M(\eta)\right] .$$

With an Orlicz function M, a function $\varphi \in U_2^0$ is associated by

$$\varphi(\xi,\eta) = \begin{cases} 0 & \text{if } \eta = 0 \\ \eta M^{-1}(\xi/\eta) & \text{if } \eta > 0 \end{cases}$$

the complementary function M^* is then associated with $\psi = \widehat{\varphi}$ by

$$\psi(\eta,\xi) = \begin{cases} 0 & \text{if } \xi = 0 \\ \xi M^{*-1}(\eta/\xi) & \text{if } \xi > 0 . \end{cases}$$

Let (Ω,Σ,μ) be a σ-finite measure space, denote by $L_M = L_M(\mu)$ the space $\varphi(L_1,L_\infty)$. When equipped with the norm ${}_p\|\cdot\|_{\varphi(L_1,L_\infty)}$, we shall denote L_M by ${}_pL_M$ and the norm in it will be denoted by ${}_p\|\ \|_M$ $(1 \le p \le \infty)$.

Claim 8.

(4) $$_p\|f\|_M = \inf_{0<\alpha<1} \inf\left\{\lambda > 0 \;\middle|\; \left\|M\left(\frac{|f|}{\lambda\alpha}\right)\right\|_{L_1} \leq \left(\frac{1}{\alpha^p} - 1\right)^{1/p}\right\} \; ; \; 1 \leq p < \infty ,$$

(5) $$_\infty\|f\|_M = \inf\left\{\lambda > 0 \;\middle|\; \left\|M\left(\frac{|f|}{\lambda}\right)\right\|_{L_1} \leq 1\right\} .$$

Proof: From the definition it follows that

(6) $$_p\|f\|_M = \inf_{\substack{0 \leq h \in L_\infty, \|h\|_\infty \leq 1 \\ \operatorname{supp} f \subset \operatorname{supp} h}} \inf\left\{\lambda > 0 \;\middle|\; \left\|hM\left(\frac{|f|}{\lambda h}\right)\right\|_{L_1}^p + \|h\|_\infty^p \leq 1\right\}$$

Let $0 \leq h \in L_\infty$ be such that $h(\omega) = 0 \Longrightarrow f(\omega) = 0$ and for some $\lambda > 0$ $\left\|hM\left(\frac{|f|}{\lambda h}\right)\right\|_{L_1}^p +$ $\|h\|_\infty^p \leq 1$. Assume $\|h\|_\infty = \alpha > 0$ then convexity of M yields for almost all $\omega \in \Omega$ with $f(\omega) \neq 0$:

$$h(\omega)M\left(\frac{f(\omega)}{\lambda h(\omega)}\right) > \alpha M\left(\frac{|f(\omega)|}{\lambda\alpha}\right)$$

So $\left\|\alpha M\left(\frac{|f|}{\lambda\alpha}\right)\right\|_{L_1}^p + \alpha^p < 1$. Since the functions $h \equiv \alpha$ are included in the game, the right hand sides of (6) and (4) are the same. □

Remarks. Formula (5) shows of course that $_\infty\|\cdot\|_M$ is the usual Luxemburg norm in L_M. By Theorem (2) we have

$$_p\varphi(L_1, L_\infty)' = {}_{p'}\psi(L_\infty, L_1)$$

so $({}_pL_M)' = {}_{p'}L_{M^*}$ with equality of norms. In particular, we get

$$(L_M)' = ({}_\infty L_M)' = {}_1L_{M^*} .$$

So (4) provides an exact formula for the norm in $(L_M)'$:

(7) $$\|f\|_{(L_M)'} = \inf_{0<\alpha<1} \left\{\lambda > 0 \;\middle|\; \left\|M^*\left(\frac{|f|}{\lambda\alpha}\right)\right\|_{L_1} \leq \frac{1}{\alpha} - 1\right\} .$$

It is of course stated often in the literature that the norms in $(L_M)'$ and L_{M^*} are equivalent, but the exact formula seems to be new. Also, since L_1 is σ-order continuous, we get that if M satisfies the Δ_2-condition then $(L_M)^* = {}_1L_{M^*}$ and the norm dual to L_M is given by (7).

The well known criteria for reflexivity of Orlicz spaces are corollaries of Corollary 5. In general, if M and M^* satisfy the Δ_2-condition then L_M is reflexive. If, however, (Ω, Σ, μ) is a finite, or respectively a discrete measure space with a positive lower bound to the measure of its atoms, then a simple check of the proof of Corollary 5 shows that it is enough to assume the Δ_2 condition at ∞ or respectively at 0.

5. A Remark on Intermediate Construction for More than Two Lattices

Let $\varphi \in U_n^0$, i.e., φ is a positive-homogeneous function on $\mathbb{R}_+^n$ which is concave and satisfies $\varphi(\xi_1, \ldots, \xi_n) = 0$ whenever $\xi_i = 0$ for some i and for all i, $\varphi(\xi_1, \ldots, \xi_n) \to \infty$ as $\xi_i \to \infty$ if $\xi_j > 0$ for all $j \neq i$. If $X_1, \ldots, X_n$ are Köthe function spaces on the same σ-finite measure space, an intermediate Köthe function space ${}_p\varphi(X_1, \ldots, X_n)$ can be constructed in a completely analogous way to the one discussed here. The proof of Theorem 2 can be modified with only minor changes to yield

$$ {}_p\varphi(X_1, \ldots, X_n)' = {}_{p'}\widehat{\varphi}(X_1', \ldots, X_n') $$

where

$$ \widehat{\varphi}(\xi_1, \ldots, \xi_n) = \inf_{\eta_i > 0} \frac{\sum_{i=1}^{n} \xi_i \eta_i}{\varphi(\eta_1, \ldots, \eta_n)} . $$

(Here we get an exact value of the dual norm rather than equivalence with constant n which is obtained by using ${}_\infty\widehat{\varphi}$.)

References

[1] J.M. Borwein and D. Zhung, On Fan's minimax theorem, Math. Prog. 34 (1986), 232-234.

[2] A.P. Calderon, Intermediate spaces and interpolation, the complex method, Studia Math. 24 (1964), 113-190.

[3] K. Fan, Minimax theorems, Proc. Nat. Acad. Sc. USA 39 (1953), 42-47.

[4] T.A. Gillespie, Factorization in Banach function spaces, Indag. Math. 43 (1981), 287-300.

[5] R.E. Jamison and W.H. Ruckle, Factoring absolutely convergent series, Math. Ann. 224 (1976), 143-148.

[6] J. Lindenstrauss and L. Tzafriri, Classical Banach spaces II, Springer-Verlag, Berlin-Heidelberg-New York, 1979.

[7] G.Ya. Lozanovskii, On some Banach lattices, Siberian Math. J. 10 (1969), 419-431 (English translation).

[8] G.Ya. Lozanovskii, On some Banach lattices III, Siberian Math. J. 13 (1972), 910-916 (English translation).

[9] G.Ya. Lozanovskii, On some Banach lattices IV, Siberian Math. J. 14 (1973), 97-108 (English translation).

[10] A.C. Zaanen, Integration, North-Holland, Amsterdam, 1967.

ON MILMAN'S INEQUALITY AND RANDOM SUBSPACES WHICH ESCAPE THROUGH A MESH IN $\mathbb{R}^n$

Y. Gordon*

Technion - Israel Institute of Technology
Haifa, Israel

Abstract

Let S be a subset in the Euclidean space $\mathbb{R}^n$ and $1 \leq k < n$. We find sufficient conditions which guarantee the existence and even with probability close to 1, of k-codimensional subspaces which miss S. As a consequence we derive a sharp form of Milman's inequality and discuss some applications to Banach spaces.

Introduction

Consider the following theorem which we label as the "escape" phenomenon of random k-codimensional subspaces through a mesh S.

Theorem. *Let S be a subset of the unit sphere S_2^{n-1} of R^n and associate to S the number $s = a_n \int_{S_2^{n-1}} \sup_{x \in S} \langle x, u \rangle m_{n-1}(du)$, where m_{n-1} is the normalized rotation invariant measure on the sphere and*

$$a_n = \sqrt{2}\Gamma\left(\frac{n+1}{2}\right) \Big/ \Gamma\left(\frac{n}{2}\right) = \sqrt{n}\left(1 - \frac{1}{4n} + 0(n^{-2})\right) .$$

If $s < a_k$ $(1 \leq k < n)$ then there is a k-codimensional subspace which does not intersect S.

This theorem and other results developed here, are based on two theorems A and B on Gaussian processes, which were originally proved in [G1] and generalizations proved in [G3]. Using Theorem A, we obtain in section 1, as a simple corollary to the "escape" phenomenon, a sharp new proof of an inequality of V. Milman and its refinements due to Pajor and Tomczak. The inequality, stated here in Theorem 1.5, has some important applications to Banach space theory. One of them is the theorem due to Milman [M1, M2]:

---

* Supported in part by the USA-Israel Binational Science Foundation (BSF) grant #86-00074.
Supported in part by the Fund for the Promotion of Research at the Technion #100-700 and the V.P.R. grant #100-677.

Theorem. *For any $0 < \lambda < 1$, any Banach space X of sufficiently large dimension n has a subspace of a quotient of X of dimension $\geq \lambda n$, which is $f(\lambda)$ isomorphic to a Hilbert space.*

As we shall see in section 1, this and other results can easily be derived from the inequality.

In section 2 we combine the escape phenomenon and Lévy's isoperimetric inequality to prove the following sharper form of a fundamental result stated in section 2 of [MS] and originally proved in [M5].

Theorem. *Given a function $f(x) \in C(S_2^{n-1})$ let M_f and $\omega_f(\varepsilon)$ denote the median and modulus of continuity of f. If $0 < \varepsilon \leq 1$ and $\ell = [n\varepsilon^2/4] - 1 \geq 0$, then there is a subspace E of dimension ℓ such that $|f(x) - M_f| \leq \omega_f(2\varepsilon)$ for every $x \in E \cap S_2^{n-1}$.*

This eliminates the log term in the estimate $\ell = [n\varepsilon^2/(2\log 4/\varepsilon)]$ of Theorem 2.4 of [MS].

In order to obtain measure estimates as well, it is necessary to replace Theorem A by its quantitative analogue Theorem B, and combine it, with Lévy's theorem or Pisier's theorem, or as we shall do here, with Theorem 3.2 due to Maurey on the tail distribution of a real Lipschitz function defined on the Gaussian probability space $(\mathbb{R}^n, P)$. Thus we are able in section 3 to estimate the measure P of the subset of all k-codimensional subspaces of the Grassman manifold $\mathcal{G}_{n,n-k}$ which miss the set $S + K$, S being an arbitrary closed subset of S_2^{n-1} and K a closed convex subset in $\mathbb{R}^n$ which contains the origin in its relative interior. A special case is $K = rB_2^n$, the Euclidean ball of radius r. P is then the probability that a random k-codimensional subspace will be at Euclidean distance greater than r from S; if $r = 0$ this simply means that ξ will miss S. We shall see for example that if $r = 0$, P is very close to 1 if $s < \sqrt{k} - c$ for large k and suitable large constant c.

In section 4 we consider a finite sequence of N arbitrary convex sets $\{S_i\}_{i=1}^N$ situated at various locations in $\mathbb{R}^n$. We are interested in estimating the probability that a randomly chosen k-codimensional subspace will for all $i = 1, \ldots, N$ pass at a distance greater than d_i (≥ 0) from S_i. This probability can be estimated by an expression which involves only the numbers d_i, t_i, T_i and s_i $(i = 1, \ldots, N)$, where t_i and T_i are the smallest and largest distance of the points of S_i to the origin and s_i is the number s we associate above with the set $S = S_i$.

We shall see in section 5 that some of these results admit generalizations. In order to do that we shall also need to extend Theorems A and B. For example, we consider a finite sequence of L arbitrary subsets of the unit sphere S_2^{n-1} and we find conditions which imply that there exists a subspace of codimension k which misses at least one of these sets and estimate the probability of this event.

1. The Escape Phenomenon and Milman's Inequality

Theorem A below is the key to the existence theorems contained in sections 1 and 2 and was originally proved in [G1]. We shall see in section 5 that Theorem A is a particular case of Theorem C, which was proved in [G3] by a different and simpler method.

Theorem A. *Let $\{X_{ij}\}$ and $\{Y_{ij}\}$, $i = 1, 2, \ldots, n$, $j = 1, 2, \ldots, m$, be two centered Gaussian processes which satisfy the following inequalities for all choices of indices:*

(i) $\mathbb{E}|X_{ij} - X_{ik}|^2 \le \mathbb{E}|Y_{ij} - Y_{ik}|^2$,

(ii) $\mathbb{E}|X_{ij} - X_{\ell k}|^2 \ge \mathbb{E}|Y_{ij} - Y_{\ell k}|^2$, *if* $i \ne \ell$.

Then, $\mathbb{E}\min_i \max_j X_{ij} \le \mathbb{E}\min_i \max_j Y_{ij}$.

Taking $n = 1$ in Theorem A we obviously obtain the Fernique, Sudakov inequality [F1,F2].

Corollary 1.1. *Let $\{X_j\}$ and $\{Y_j\}$, $1 \le j \le m$, be two centered Gaussian processes such that $\mathbb{E}|X_j - X_k|^2 \le \mathbb{E}|Y_j - Y_k|^2$ for all j, k. Then, $\mathbb{E}\max_j X_j \le \mathbb{E}\max_j Y_j$.*

Throughout, we shall denote by $\{g_{ij}\}, \{h_i\}, \{g_j\}, \{g\}$ independent sets of orthonormal Gaussian r.v.'s and

$$G = G(\omega) = \sum_{i=1}^{n}\sum_{j=1}^{k} g_{ij}(\omega) e_i \otimes e_j$$

will denote the random Gaussian operator from $\mathbb{R}^n$ to $\mathbb{R}^k$, where $\{e_i\}_{i=1}^n$ is the usual unit vector basis of $\mathbb{R}^n$, and $\|\bullet\|_2$ stands for the Euclidean norm on $\mathbb{R}^n$. The letter c will denote constants.

Corollary 1.2. *Let $S \subset \mathbb{R}^n$ be a closed subset, $1 \le k \le n$. Then*

$$\mathbb{E}\big(\min_{x\in S}\|G(\omega)(x)\|_2\big) \ge \mathbb{E}\Big(\min_{x\in S}\Big\{\|x\|_2\Big(\sum_1^k g_j^2\Big)^{1/2} + \sum_1^n h_i x_i\Big\}\Big) \tag{1.1}$$

$$\ge a_k \min_{x\in S}\|x\|_2 - \mathbb{E}\Big(\max_{x\in S}\sum_1^n h_i x_i\Big)\ ,$$

$$\mathbb{E}\big(\max_{x\in S}\|G(\omega)(x)\|_2\big) \le \mathbb{E}\Big(\max_{x\in S}\Big\{\|x\|_2\Big(\sum_1^k g_j^2\Big)^{1/2} + \sum_1^n h_i x_i\Big\}\Big) \tag{1.2}$$

$$\le a_k \max_{x\in S}\|x\|_2 + \mathbb{E}\Big(\max_{x\in S}\sum_1^n h_i x_i\Big)\ .$$

where $a_k = \mathbb{E}\big(\sum_1^k g_j^2\big)^{1/2} = \sqrt{2}\,\Gamma\left(\frac{k+1}{2}\right)/\Gamma\left(\frac{k}{2}\right)$ satisfies $k/\sqrt{k+1} < a_k < \sqrt{k}$.

Proof: For $x \in S$ and $y \in S_2^{k-1} = \{y \in \mathbb{R}^k \; ; \; \|y\|_2 = 1\}$, we define the two Gaussian processes

$$X_{x,y} = \|x\|_2 \sum_{j=1}^{k} g_j y_j + \sum_{i=1}^{n} h_i x_i$$

and

$$Y_{x,y} = \langle G(\omega)(x), y\rangle = \sum_{i=1}^{n}\sum_{j=1}^{m} g_{ij} x_i y_j \, .$$

It follows then that for all $x, x' \in S$ and $y, y' \in S_2^{k-1}$, we have

$$\begin{aligned}
&\mathbb{E}|X_{x,y} - X_{x',y'}|^2 - \mathbb{E}|Y_{x,y} - Y_{x',y'}|^2 \\
&= \|x\|_2^2 + \|x'\|_2^2 - 2\|x\|_2\|x'\|_2\langle y, y'\rangle - 2\langle x, x'\rangle(1 - \langle y, y'\rangle) \\
&\geq \|x\|_2^2 + \|x'\|_2^2 - 2\|x\|_2\|x'\|_2\langle y, y'\rangle - 2\|x\|_2\|x'\|_2(1 - \langle y, y'\rangle) \\
&\geq 0 \, .
\end{aligned}$$

with equality if $x = x'$. By Theorem A this implies that

$$\mathbb{E}(\min_{x\in S} \max_{y\in S_2^{k-1}} X_{x,y}) \leq \mathbb{E}(\min_{x\in S} \max_{y\in S_2^{k-1}} Y_{x,y})$$

and by Corollary 1.1

$$\mathbb{E}(\max_{x,y} X_{x,y}) \geq \mathbb{E}(\max_{x,y} Y_{x,y}) \, .$$

The last two inequalities imply inequalities (1.1) and (1.2). The estimate for a_k follows from $a_k^2 \leq \mathbb{E}(\sum_1^k g_j^2) = k$ and $a_k\sqrt{k+1} \geq a_k a_{k+1} = k$. □

Remark (1.1). Using the identity $a_k a_{k+1} = k$, it follows by induction that

$$a_k = \sqrt{k}\left(1 - \frac{1}{4k} + \frac{1}{32k^2} + 0(k^{-3})\right) \, .$$ □

Corollary 1.3. *Let S be a closed subset of the unit sphere*

$$S_2^{n-1} = \{x \in \mathbb{R}^n \; ; \; \|x\|_2 = 1\} \quad \text{and} \quad s = \mathbb{E}(\max_{x\in S}\sum_1^n h_i x_i) \, .$$

If $1 \leq k < n$ satisfies $a_k > s$, then there is an operator $T : \ell_2^n \to \ell_2^k$ such that

$$\frac{\|Tx\|_2}{\|Ty\|_2} \leq \frac{a_k + s}{a_k - s} \quad \text{for all} \quad x, y \in S \, . \tag{1.3}$$

Proof: By Corollary 1.2, $\mathbb{E}(\min_{x\in S}\|G(\omega)(x)\|_2) \geq a_k - s > 0$ and $\mathbb{E}(\max_{x\in S}\|G(\omega)(x)\|_2) \leq a_k + s$. Hence, there exists ω_0 such that $T = G(\omega_0)$ satisfies (1.3). □

There is a natural upper bound for s which is obtained by taking $S = S_2^{n-1}$, thus $s \le a_n$. Integrating over the unit sphere we get that s can be expressed by $s = a_n \int_{S_2^{n-1}} \max_{x\in S}\langle x, u\rangle m_{n-1}(du)$ where m_{n-1} is the normalized rotation invariant measure defined on S_2^{n-1}.

Another estimate for s can be obtained, for example, by using Dudley's theorem [D]: *Let $N(\varepsilon)$ be the number of elements in a set $S \subset \mathbb{R}^n$ which forms an ε net in the $\|\bullet\|_2$ norm for S. Then, $s \le c\int_0^\infty \sqrt{\log N(\varepsilon)}d\varepsilon$.* Thus, s can be replaced by $c\int_0^\infty \sqrt{\log N(\varepsilon)}d\varepsilon$ in Corollary 1.3.

The escape phenomenon mentioned in the introduction is summarized in the next theorem.

Theorem 1.4. *Let $S \subset \mathbb{R}^n$ be a compact subset, $1 \le k < n$ and $G(\omega) = \sum_{i=1}^{n}\sum_{j=1}^{k} g_{ij}(\omega)e_i \otimes e_j$ be the Gaussian map from $\mathbb{R}^n$ to $\mathbb{R}^k$.*

(1) $\mathbb{E}\min_{x\in S}\|G(\omega)(x)\|_2 > 0$ iff there exists a subspace of codimension k which misses S.

(2) If $S \subset S_2^{n-1}$ and $a_k > s = \mathbb{E}\big(\max_{x\in S}\sum_1^n x_i h_i\big)$, then $\mathbb{E}\min_{x\in S}\|G(\omega)(x)\|_2 > 0$.

Proof: The subspace $E_\omega = G(\omega)^{-1}(0)$ has codimension k a.e., hence if $\mathbb{E}\min_{x\in S}\|G(\omega)(x)\|_2 > 0$ then there is an E_{ω_0} of codimension k which misses S. Conversely, every k-codimensional subspace has the form E_ω and if E_{ω_0} misses S, then compactness of S implies that E_ω misses S, i.e., $\min_{x\in S}\|G(\omega)(x)\|_2 > 0$, on a set of positive measure, this proves (1). Statement (2) follows from inequality (1.1) □

Given a closed convex body K in $\mathbb{R}^n$ which contains the origin in its interior, we associate with K the dual body $K^* = \{x \in \mathbb{R}^n \ ; \ \langle x, y\rangle \le 1 \text{ for all } y \in K\}$ and the "norm" $\|x\|_K = \inf\{t > 0\ ;\ x \in tK\}$. $\mathbb{E}(K)$ will denote $\mathbb{E}\|\sum_1^n g_i e_i\|_K$ and $\varepsilon_2(K) = \max\{\|x\|_2\ ;\ x \in K^*\}$. If K is centrally symmetric, then $\|\bullet\|_K$ is a norm and the normed space $X = (\mathbb{R}^n, \|\bullet\|_K)$ has K for its unit ball; we shall then refer to $\mathbb{E}(K)$ by $\mathbb{E}(X)$ and $\varepsilon_2(X) = \varepsilon_2(K)$.

Theorem 1.5. *Let K be a closed convex body in $\mathbb{R}^n$, $0 \in \operatorname{int} K$, $1 \le k \le n$ and let $K_r = r^{-1}K \cap B_2^n$ $(r > 0)$. Then there is a subspace E of codimension k such that*

$$\|x\|_2 \le r_0\|x\|_K \quad \text{for all} \quad x \in E\ ,$$

where $r_0 = \inf\{r > 0\ ;\ a_k > \mathbb{E}(K_r^*)\} \le a_k^{-1}\mathbb{E}(K^*)$.

Proof: Let $S = S_2^{n-1} \cap K_r$. By Theorem 1.4, if $a_k > s = \mathbb{E}\big(\sup_{x\in S}\sum x_i h_i\big)$ then there is a k-codimensional subspace E which misses S, hence if $x \in E \cap S_2^{n-1}$ then $x \notin r^{-1}K$, that is $r\|x\|_K \ge \|x\|_2$ for all $x \in E$. Notice that $s \le \mathbb{E}\big(\sup_{x\in K_r}\sum x_i h_i\big) = \mathbb{E}(K_r^*)$. To prove the inequality for r_0, since $K_r \subseteq r^{-1}K$, $K_r^* \supseteq rK^*$, therefore $r\|x\|_{K_r^*} \le \|x\|_{K^*}$ and $\mathbb{E}(K_r^*) \le r^{-1}\mathbb{E}(K^*)$, which implies $r_0 \le a_k^{-1}\mathbb{E}(K^*)$. □

Remark (1.2). Originally V. Milman was the first to prove that there is a function ψ : $(0,1) \to \mathbb{R}_+$ with the property that in every Banach space X of dimension n there is a subspace E of codimension $< \varepsilon n$ such that $\|x\|_2 \le \psi(\varepsilon) n^{-1/2} \mathbb{E}(X^*) \|x\|$ for all $x \in E$ and the estimate $\psi(\varepsilon) \le c\varepsilon^{-1}$ was proved in [M1]. Pajor and Tomczak [PT2] improved the estimate to $\psi(\varepsilon) \le C\varepsilon^{-1/2}$ where C is a universal constant. Their original proof is based on Lévy's isoperimetric inequality [MS] and Sudakov's minoration theorem [F2], an earlier version of which appears in [PT1]. It is worthwhile to recall here some consequences of Theorem 1.5. We first give the operator formulation of the theorem.

Let $u : X \to Y$ be an operator between the Banach spaces X and Y. The kth Gelfand number of u is defined by $c_k(u) = \inf\{\|u \mid E\| \,;\, E \subset X \,,\, \operatorname{codim} E < k\}$, $k = 1, 2, \ldots$. For $u : \ell_2^n \to Y$, $\ell(u) = \mathbb{E}\|\sum_1^n g_i u(e_i)\|$ (usually the ℓ norm of u, $\ell(u)$ is defined by $\left(\mathbb{E}\|\sum_1^n g_i u(e_i)\|^2\right)^{1/2}$, but by Kahane's inequality [MS] the two expressions are equivalent). For $u : \ell_2 \to Y$, $\ell(u)$ is defined as $\sup\left\{\ell(uv) \,;\, \|v : \ell_2^n \to \ell_2\| \le 1\right\}$. Theorem 1.5. can be reformulated as follows:

Theorem 1.5′. *Let $u : X \to \ell_2^n$ and $X_r = (X, \|\bullet\|_r)$ be the space normed by $\|x\|_r = \max\{r\|x\| \,,\, \|ux\|_2\}$ $(r > 0)$. Let $1 \le k \le n$, then $c_{k+1}(u) \le \tilde{r}_0 = \inf\left\{r > 0 \,;\, \ell(u^* : \ell_2^n \to X_r^*) \le a_k\right\}$. In particular, $a_k c_{k+1}(u) \le \ell(u^* : \ell_2^n \to X^*)$.*

Proof: Without loss of generality we can assume that u is invertible and define on $\mathbb{R}^n$ the norm $|x| = \|u^{-1}x\|$, then $u : X \to (\mathbb{R}^n, |\bullet|)$ is an isometry. Now let $|x|_r = \max\{r|x| \,,\, \|x\|_2\}$ and apply Theorem 1.5 to the unit ball K_r of the normed space $(\mathbb{R}^n, |\bullet|_r)$, with K equal to the unit ball of $(\mathbb{R}^n, |\bullet|)$. We thus obtain a k-codimensional subspace E for which $\|ux\|_2 \le \tilde{r}_0\|x\|$ for all $x \in E$, where

$$\tilde{r}_0 = \inf\left\{r > 0 \,;\, a_k \ge \mathbb{E}(K_r^*)\right\} = \inf\left\{r > 0 \,;\, \ell(u^* : \ell_2^n \to X_r^* \le a_k\right\} . \qquad \square$$

Theorem 1.6. *Let $X = (\mathbb{R}^n, \|\bullet\|)$ be an n-dimensional Banach space and $T_2(X^*)$ and $C_2(X)$ be the Gaussian type 2 and cotype 2 constants of X^* and X respectively. Then for every $\lambda \in (0,1)$, there exists a subspace E of X of dimension $[\lambda n]$ such that*

(i) $d(E, \ell_2^{[\lambda n]}) \le (1 - \lambda - n^{-1})^{-1/2} T_2(X^*)$, *if* $n^{-1} \le \lambda \le 1 - n^{-1}$;

(ii) $d(E, \ell_2^{[\lambda n]}) \le c(1-\lambda)^{-1/2} C_2(X) \log\left(1 + c(1-\lambda)^{-1/2} C_2(X)\right)$, *if* $n \ge N_\lambda$.

Proof: (i) without loss of generality assume B_2^n, the unit ball of ℓ_2^n, is the ellipsoid of maximal volume contained in B_X, the unit ball of X, and let $k + 1 = n - [\lambda n]$. By Theorem 1.5 there exists a subspace E of codimension $k+1$ such that

$$\|x\| \le \|x\|_2 \le a_{k+1}^{-1} \mathbb{E}(X^*) \|x\| \quad \text{for all} \quad x \in E\,.$$

Now, $a_{k+1} \geq \sqrt{k} \geq \sqrt{n}(1-\lambda-n^{-1})^{1/2}$ and by Lemma 4.11 of [BG], $\mathbb{E}(X^*) \leq \sqrt{n}T_2(X^*)$. This proves (i).

(ii) Since $T_2(X^*) \leq C_2(X)K(X) \leq cC_2(X)\log\left(d(X,\ell_2^n)+1\right)$ where $K(X)$ is the K convexity constant of X (cf. [P1] or [MS] and Remark 2.2 [P2]), the required result follows by applying the iteration procedure of Milman [M3] a finite number of times (with an appropriate sequence of λ_i's in $(0,1)$) to a subspace of a subspace ..., etc., of X. □

Remark (1.3). The fact that (ii) has linear dependence on the cotype 2 constant $C_2(X)$, except for a log term, was first proved in [M1, M3]. In the form stated above, (ii) was established in [PT2] by means of the sharp estimate $\psi(\varepsilon) < c\varepsilon^{-1/2}$ obtained there. The following theorem is due to Milman and appears in the form stated here in [M4].

Theorem 1.7. *Let $f(\lambda) = c_1(1-\lambda)^{-1}\log\left(c_2/(1-\lambda)\right)$, $\lambda \in \left[\frac{1}{2},1\right)$. Then, given any $\lambda \in \left(\frac{1}{2},1\right)$, every Banach space X of dimension $n(> N_\lambda)$ contains a subspace of a quotient of X, E, of dimension $[\lambda n]$, such that $d(E,\ell_2^{[\lambda n]}) \leq f(\lambda)$.*

Proof: Given any $\frac{1}{2} \leq \lambda_1 < 1$, we apply Theorem 1.5 to find a subspace E_1 of X^* of dimension $> [\lambda_1 n]$ such that

$$\|x\|_2 \leq \frac{\mathbb{E}(X)}{\sqrt{n}\sqrt{1-\lambda_1}}\|x\| \quad \text{for all} \quad x \in E_1 .$$

Dualizing this inequality and applying Theorem 1.5 again, we find a subspace $E_2 \subset E_1^*$ of dimension $> [\lambda_1^2 n]$ such that

$$\|x\|_2 \leq \frac{\mathbb{E}(E_1)}{\sqrt{\lambda_1 n - \lambda_1^2 n}}\|x\| \quad \text{for all} \quad x \in E_2 ,$$

both inequalities imply

$$d(E_2,\ell_2^{\dim E_2}) \leq \frac{\mathbb{E}(E_1)\mathbb{E}(X)}{n\sqrt{\lambda_1}(1-\lambda_1)} \leq \frac{2\mathbb{E}(X)\mathbb{E}(X^*)}{n(1-\lambda_1)} .$$

By a result due to Lewis [L], there exists $u : \ell_2^n \to X$ for which $\ell(u)\ell^*(u^{-1}) = n$, and by [FT] $\ell(u^{*-1}) \leq c\ell^*(u^{-1})K(X)$ and now applying the inequality $K(X) \leq c\log\left(d(X,\ell_2^n)+1\right)$, we obtain by choosing the proper ellipsoid in $\mathbb{R}^n$, namely the Lewis ellipsoid

$$d(E_2,\ell_2^{\dim E_2}) \leq \frac{c\log\left(d(X,\ell_2^n)+1\right)}{1-\lambda_1} ,$$

where E_2 is a subspace of a quotient of X of dimension $\geq \lambda_1^2 n$. Replacing X above by E_2 and continuing in this manner, we obtain the result after a finite number of iterations and the proper choice of the sequence λ_i. □

In order to obtain a subspace of a quotient of X which is $\frac{1+\varepsilon}{1-\varepsilon}$ isomorphic to a Hilbert space for a given $\varepsilon \in (0,1)$, we shall use the following theorem proved in [G1], see also [G2].

Theorem 1.8. *Let $Y = (\mathbb{R}^n, \|\bullet\|)$ be a Banach space and $\varepsilon_2(Y) = \max_{y\neq 0}(\|y\|/\|y\|_2)$. If $1 < m < n$ satisfies $\mathbb{E}(Y) > \sqrt{m}\varepsilon_2(Y)$, then Y contains an m-dimensional subspace F with*

$$d(F, \ell_2^m) \leq \frac{\mathbb{E}(Y) + \sqrt{m}\varepsilon_2(Y)}{\mathbb{E}(Y) - \sqrt{m}\varepsilon_2(Y)} .$$

Theorem 1.8 combined with Theorem 1.7 provides immediately and obviously the $\frac{1+\varepsilon}{1-\varepsilon}$ version of Theorem 1.7.

1.9. *Let $0 < \varepsilon < 1$. There exists N such that for any $n(\geq N)$ dimensional Banach space X, there exists a subspace of a quotient of X of dimension $> c\varepsilon^2 n$ which is $\frac{1+\varepsilon}{1-\varepsilon}$ isomorphic to a Hilbert space.*

Proof: Take $\lambda = 1/2$ in Theorem 1.7 and apply Theorem 1.8 to the space E of dimension $[n/2]$, to find a subspace $F \subset E$ of dimension m which will be $\frac{1+\varepsilon}{1-\varepsilon}$ isomorphic to ℓ_2^m provided $\sqrt{m}\varepsilon_2(E) < \varepsilon\mathbb{E}(E)$. But since $d(E, \ell_2^{[n/2]}) \leq f(\frac{1}{2})$, it follows that $\varepsilon_2(E)/\mathbb{E}(E) \leq f(\frac{1}{2})/\mathbb{E}(\ell_2^{[n/2]}) \sim cn^{-1/2}$ and this yields the estimate for m. □

Remark (1.4). In a similar manner we can obtain $\frac{1+\varepsilon}{1+\varepsilon}$ versions of Theorem 1.6.

2. Applications to Continuous Functions on the Sphere

Given a set $A \subset S_2^{n-1}$ and $\varepsilon > 0$ we denote by A_ε the set $\{x \in S_2^{n-1} \ ; \ \rho(x, A) \leq \varepsilon\}$, where ρ is the geodesic metric on S_2^{n-1}. The median M_f of a function $f \in C(S_2^{n-1})$ is a number such that $m_{n-1}(f \leq M_f) \geq \frac{1}{2}$ and $m_{n-1}(f \geq M_f) \geq \frac{1}{2}$, where m_{n-1} is the normalized rotation invariant measure of S_2^{n-1}. Lévy's classical isoperimetric inequality states (cf. [MS]):

(2.1.) *Let $f \in C(S_2^{n-1})$ and $A = \{x \in S_2^{n-1} \ ; \ f(x) = M_f\}$. Then $m_{n-1}(A_\varepsilon) \geq 1 - \sqrt{\pi/2}\exp\left(-\frac{1}{2}(n-1)\varepsilon^2\right)$.*

Let $\omega_f(\varepsilon) = \sup\{|f(x) - f(y)| \ ; \ \rho(x,y) \leq \varepsilon\}$ be the modulus of continuity of f. It is clear from the definition of A in (2.1) that $|f(x) - M_f| \leq \omega_f(\varepsilon)$ for every $x \in A_\varepsilon$. Thus by Lévy's isoperimetric inequality, $f(x)$ is close to M_f on the set A_ε which has large measure if $n\varepsilon^2$ is big.

Theorem 2.1. *Let $A \subset S_2^{n-1}$, $0 < \varepsilon \leq 1$ and $\ell = \left[\frac{1}{2}n\varepsilon^2 m_{n-1}(A)\right] - 1 \geq 0$. Then there exists a subspace E of dimension ℓ such that $E \cap S_2^{n-1} \subset A_\varepsilon$.*

Proof: We apply Theorem 1.4 to $S = (A_\varepsilon)^C$, the complement of A_ε. By standard integration

$$s = E\big(\sup_{x\in S}\sum_1^n x_i h_i\big) = a_n \int_{S_2^{n-1}} \sup_{x\in S}\langle x,u\rangle m_{n-1}(du) .$$

Notice that if $u \in A$ and $x \in S$, then $\langle x,u\rangle \le \cos\varepsilon$ and for all other $u \in S_2^{n-1}$ $\langle x,u\rangle \le 1$, hence $s \le a_n\big(m_{n-1}((A)^C) + m_{n-1}(A)\cos\varepsilon\big) = a_n\big(1-(1-\cos\varepsilon)m_{n-1}(A)\big)$.

Let $k = n - \ell$, then

$$\begin{aligned}(2.2)\qquad a_k/a_n &> \sqrt{(k-1)/n} \ge \big(1 - \tfrac{1}{2}\varepsilon^2 m_{n-1}(A)\big)^{1/2}\\ &\ge \left(1 - \frac{2}{3}\varepsilon^2 m_{n-1}(A) + \frac{\varepsilon^4}{9} m_{n-1}(A)\right)^{1/2}\\ &\ge 1 - \frac{1}{3}\varepsilon^2 m_{n-1}(A) > 1 - (1-\cos\varepsilon)m_{n-1}(A) \ge s/a_n ,\end{aligned}$$

therefore $a_k > s$. By Theorem 1.4 there exists a subspace E of dimension ℓ which misses S, hence $E \cap S_2^{n-1} \subset A_\varepsilon$. □

Theorem 2.1, combined with the isoperimetric inequality (2.1), sharpens the estimate obtained in Theorem 2.4 of [MS].

Corollary 2.2. *Let $0 < \varepsilon \le 1$ and $n\varepsilon^2 \ge 4$. Then for any function $f \in C(S_2^{n-1})$, there exists a subspace E of dimension $\ge [n\varepsilon^2/4] - 1$ such that*

$$(2.3)\qquad |f(x) - M_f| < \omega_f(2\varepsilon) \quad \text{for every} \quad x \in E \cap S_2^{n-1} .$$

Proof: If $B = f^{-1}(M_f)$, then by the isoperimetric inequality, $m_{n-1}(B_\varepsilon) \ge 1 - \sqrt{2/\pi}\exp\big(-(n-1)\varepsilon^2/2\big) > \frac{1}{2}$ since by our assumption $n\varepsilon^2 \ge 4$. Hence taking $A = B_\varepsilon$ the dimension ℓ of the subspace E in Theorem 2.1 is $\ge [n\varepsilon^2/4] - 1$ and $E \cap S_2^{n-1} \subset B_{2\varepsilon}$. □

3. An Isoperimetric Inequality on the Sphere of $\mathbb{R}^n$

In this section we shall make use of Theorem B proved in [G1]. Theorem B turns out to be essential for obtaining the measure estimates in this section. A simpler proof of this theorem appeared recently also in [K], and other related results, as well as generalizations in various directions, may be found in [G3].

Theorem B. *Let $\{X_{ij}\}$ and $\{Y_{ij}\}$, $1 \le i \le n$, $1 \le j \le m$, be two centered Gaussian processes which satisfy the following inequalities for all choices of indices*

(i) $\mathbb{E}(X_{ij}^2) = \mathbb{E}(Y_{ij}^2)$

(ii) $\mathbb{E}(X_{ij}X_{ik}) \ge \mathbb{E}(Y_{ij}Y_{ik})$

(iii) $\mathbb{E}(X_{ij}X_{\ell k}) \le \mathbb{E}(Y_{ij}Y_{\ell k})$ *if* $i \ne \ell$.

Then, $P(\bigcap_i \bigcup_j [X_{ij} \ge \lambda_{ij}]) \le P(\bigcap_i \bigcup_j [Y_{ij} \ge \lambda_{ij}])$ for all choices of $\lambda_{ij} \in \mathbb{R}^1$.

Lemma 3.1. *Let $G(\omega) = \sum_{i=1}^{n}\sum_{j=1}^{k} g_{ij}(\omega) e_i \otimes e_j : \mathbb{R}^n \to \mathbb{R}^k$ and $S \subset \mathbb{R}^n$ be an arbitrary subset. Then for all choices of real λ_x $(x \in S)$,*

$$P\Big(\bigcap_{x\in S} [\|G(\omega)(x)\|_2 + \|x\|_2\, g \ge \lambda_x]\Big) \ge P\Big(\bigcap_{x\in S} [\|x\|_2 (\sum_1^k g_j^2)^{1/2} + \sum_1^n x_i h_i \ge \lambda_x]\Big) .$$

Proof: For $x \in S$ and $y \in S_2^{k-1}$ we define the two Gaussian processes,

$$X_{x,y} = \langle G(\omega)(x), y\rangle + \|x\|_2\, g \quad \text{and} \quad Y_{x,y} = \|x\|_2 \sum_1^k g_j y_j + \sum_1^n x_i h_i ,$$

where h_i, g_j, g denote orthonormal Gaussian r.v.'s. It is easy to check that $\mathbb{E}(X_{x,y}^2) = \mathbb{E}(Y_{x,y}^2)$ and

$$\begin{aligned}\mathbb{E}(X_{x,y}X_{x',y'}) - \mathbb{E}(Y_{x,y}Y_{x',y'}) &= \langle x, x'\rangle\langle y, y'\rangle + \|x\|_2\|x'\|_2 - \|x\|_2\|x'\|_2\langle y, y'\rangle - \langle x, x'\rangle \\ &= \big(\|x\|_2\|x'\|_2 - \langle x, x'\rangle\big)\big(1 - \langle y, y'\rangle\big)\end{aligned}$$

which is always nonnegative and equal to zero if $x = x'$.

For each $x \in S$ the set $\bigcup_{y\in S_2^{k-1}} [X_{x,y} \ge \lambda_x] = [\|G(x)\|_2 + g\|x\|_2 \ge \lambda_x]$ is closed in the probability space $\{\mathbb{R}^{nk+1}, P\}$ where P is the canonical Gaussian measure of $\mathbb{R}^{nk+1}$. Hence $\bigcap_{x\in S} \bigcup_{y\in S_2^{k-1}} [X_{x,y} \ge \lambda_x]$ is closed. The same can be said for the corresponding expression with the $Y_{x,y}$. By Theorem B, for each finite set $\{x_i\}_1^N \subset S$ we have

$$P\Big(\bigcap_{i=1}^{N} \bigcup_{y\in S_2^{k-1}} [X_{x_i,y} \ge \lambda_{x_i}]\Big) \ge P\Big(\bigcap_{i=1}^{N} \bigcup_{y\in S_2^{k-1}} [Y_{x_i,y} \ge \lambda_{x_i}]\Big)$$

and so, ordering the collection of finite subsets of S, F, by inclusion, we obtain that the limits exist and satisfy the inequality

$$\lim_F P\Big(\bigcap_{i=1}^{N} \bigcup_{y\in S_2^{k-1}} [X_{x_i,y} \ge \lambda_{x_i}]\Big) \ge \lim_F P\Big(\bigcap_{i=1}^{N} \bigcup_{y\in S_2^{k-1}} [Y_{x_i,y} \ge \lambda_{x_i}]\Big)$$

But as the sets $\bigcap_{x\in S}\bigcup_{y\in S_2^{k-1}}[X_{x,y}\geq\lambda_x]$ and $\bigcap_{x\in S}\bigcup_{y\in S_2^{k-1}}[Y_{x,y}\geq\lambda_x]$ are closed and P is a regular measure, it follows easily that the two respective limits over F are equal to and satisfy the inequality

$$P\Big(\bigcap_{x\in S}\bigcup_{y\in S_2^{k-1}}[X_{x,y}\geq\lambda_x]\Big)\geq P\Big(\bigcap_{x\in S}\bigcup_{y\in S_2^{k-1}}[Y_{x,y}\geq\lambda_x]\Big)\,. \qquad \square$$

We shall also use the following theorem due to Pisier on the normal tail distribution of a Lipschitz function [P3]. Pisier proved in [P3] a more general result with the constant $c=2\pi^{-2}$.

Theorem 3.2. *Let $f:\mathbb{R}^n\to\mathbb{R}^1$ be a Lipschitz function satisfying $|f(x)-f(y)|\leq\sigma\|x-y\|_2$, let $c=\frac{1}{2}$ and P be the canonical Gaussian measure on $\mathbb{R}^n$. Then for all $\lambda>0$*

$$P(f(x)-\mathbb{E}f>\lambda)<\exp(-c\lambda^2/\sigma^2)\,.$$

Remarks (3.1). Let $\mathcal{G}_{n,n-k}$ denote the Grassman manifold of k-codimensional subspaces of $\mathbb{R}^n$ and $\gamma_{n,n-k}$ the normalized Haar measure on $\mathcal{G}_{n,n-k}$. If $S\subset\mathbb{R}^n$ is a subset, a subspace $\xi\in\mathcal{G}_{n,n-k}$ meets S if and only if it meets S_{rad}, which will denote the symmetric radial projection of $S\cup(-S)$ onto the unit sphere S_2^{n-1}. When $k=n-1$ and S is a measurable subset of $\mathbb{R}^n$, $\gamma_{n,1}(\xi\in\mathcal{G}_{n,1};\xi\cap S\neq\Phi)=m_{n-1}(S_{\text{rad}})$, where m_{n-1} is the normalized surface area of the sphere S_2^{n-1}.

Let Σ denote the set of all symmetric measurable subsets of S_2^{n-1}, in particular $S_{\text{rad}}\in\Sigma$ if S is a measurable subset of $\mathbb{R}^n$. $\gamma_{n,n-k}$ induces the measure $\gamma_{n,n-k}(\xi\in\mathcal{G}_{n,n-k}\,;\,\xi\cap S\neq\Phi)$ on the elements $S\in\Sigma$. This is a probability measure which is invariant under orthogonal transformations $u\in O_n$ applied to S.

Similarly, if $G=G(\omega)=\sum_{i=1}^{n}\sum_{j=1}^{k}g_{i,j}(\omega)e_i\otimes e_j$ is the Gaussian operator from $\mathbb{R}^n$ to $\mathbb{R}^k$, then a.e. $G(\omega)^{-1}(0)$ is a k-codimensional subspace of $\mathbb{R}^n$ and $P(\omega:G(\omega)^{-1}(0)\cap S\neq\Phi)$ is also a probability measure on $S\in\Sigma$, which is invariant under orthogonal transformations $u\in O_n$; because, Gu and G have the same distribution for every $u\in O_n$. By uniqueness of the Haar measure, the two measures are identical on Σ, i.e., $\gamma_{n,n-k}(\xi\in\mathcal{G}_{n,n-k}\,;\,\xi\cap S\neq\Phi)=P(\omega\,;\,G(\omega)^{-1}(0)\cap S\neq\Phi)$ for all $S\in\Sigma$, hence also for all measurable subsets S in $\mathbb{R}^n$.

(3.2) Let $K\subset\mathbb{R}^n$ be a closed convex subset which contains the origin in its interior. Let $\|G\|$ be the "norm" associated with G as a map from $(\mathbb{R}^n,\|\bullet\|_K)$ to $\ell_2^k=(R^k,\|\bullet\|_2)$. By Corollary 1.2(2), taking $S=K$, we have

$$\mathbb{E}\|G\|\leq a_k e_2(K^\circ)+\mathbb{E}(K^\circ)$$

and conversely, if $x\in K$, then

$$\mathbb{E}\|G\| \geq \mathbb{E}\|G(x)\|_2 = \Big(\sum_1^n x_i^2\Big)^{1/2}\mathbb{E}\Big(\sum_1^k g_j^2\Big)^{1/2}$$

implies

$$\mathbb{E}\|G\| \geq a_k\varepsilon_2(K^\circ) ;$$

moreover,

$$\mathbb{E}\|G\| \geq \mathbb{E}\|G^*(e_1)\|_{K^\circ} = \mathbb{E}\Big\|\sum_1^n g_i e_i\Big\|_{K^\circ} = \mathbb{E}(K^\circ) ,$$

that is,

$$\max\big\{a_k\varepsilon_2(K^\circ), \mathbb{E}(K^\circ)\big\} \leq \mathbb{E}\|G\| \leq a_k\varepsilon_2(K^\circ) + \mathbb{E}(K^\circ) .$$

The following theorem estimates from below the measure of the set of k-codimensional subspaces $\xi \in \mathcal{G}_{n,n-k}$ which miss a given closed subset S in S_2^{n-1} by a distance, with respect to the $\|\bullet\|_K$ "norm", greater than 1:

Theorem 3.3. *Let $S \subset S_2^{n-1}$ be a closed subset, $K \subset \mathbb{R}^n$ a closed convex set with $0 \in \operatorname{int} K$ and $s = \mathbb{E}\big(\max_{x\in S}\sum_1^n x_i h_i\big)$. If $a_k\big(1-\varepsilon_2(K^\circ)\big) - \mathbb{E}(K^\circ) > s$, $1 \leq k < n$, then*

$$\gamma_{n,n-k}\big(\xi \in \mathcal{G}_{n,n-k}; \xi\cap(S+K) = \Phi\big) \geq 1 - \frac{7}{2}\exp\left(-\frac{1}{2}\left[\frac{a_k\big(1-\varepsilon_2(K^\circ)\big) - \mathbb{E}(K^\circ) - s}{3+\varepsilon_2(K^\circ) + \mathbb{E}(K^\circ)/a_k}\right]^2\right) .$$

Proof: Let $T = \sum_{i=1}^n\sum_{j=1}^k t_{ij} e_i \otimes e_j$ and $\|T\|$ be its norm as a map from $(\mathbb{R}^n, K)$ to ℓ_2^k. It is easily seen that $\big|\|T\| - \|T'\|\big| \leq \varepsilon_2(K^\circ)\big(\sum_{i,j}(t_{ij} - t'_{ij})^2\big)^{1/2}$, hence $\|T\|$ has Lipschitz constant $\varepsilon_2(K^\circ)$, so by Theorem 3.2

$$P\big(\|G\| \geq (1+\varepsilon)\mathbb{E}\|G\|\big) \leq \exp\big(-\tfrac{1}{2}(\varepsilon\mathbb{E}\|G\|/\varepsilon_2(K^\circ))^2\big) \leq \varepsilon(-\varepsilon^2 a_k^2/2) ,$$

where $\varepsilon > 0$ is to be chosen later.

Let $\lambda = (1+\varepsilon)\mathbb{E}\|G\|$ and $P = P\big(\omega\ ;\ G(\omega)^{-1}(0)\cap(S+K) = \Phi\big)$.

Setting $Q = P(\omega\ ;\ \min_{x\in S}\|G(x)\|_2 \geq \lambda)$, we have

$$\begin{aligned} Q &\leq P\big(\omega\ ;\ \min_{x\in G^{-1}(0), y\in S}\|x-y\|_K\|G\| \geq \lambda\big) \\ &\leq P\big(\|G\| \geq (1+\varepsilon)\mathbb{E}\|G\|\big) + P\big(\min_{x\in G^{-1}(0), y\in S}\|x-y\|_K > 1\big) \\ &= P\big(\|G\| \geq (1+\varepsilon)\mathbb{E}\|G\|\big) + P \leq \exp(-\varepsilon^2 a_k^2/2) + P . \end{aligned}$$

On the other hand,

$$Q + \tfrac{1}{2}\exp(-\varepsilon^2 a_k^2/2) \geq Q + P(g \geq \varepsilon a_k) \geq P\Big(\bigcap_{x\in S}\big[\|G(x)\|_2 + g \geq \lambda + \varepsilon a_k\big]\Big)$$

and by Lemma 3.1

$$\geq P\Big(\bigcap_{x\in S}\Big[\Big(\sum_1^k g_j^2\Big)^{1/2} + \sum_1^n x_i h_i \geq \lambda + \varepsilon a_k\Big]\Big) \equiv R\,.$$

Moreover, since the Lipschitz constant of the function $\big(\sum_1^k g_j^2\big)^{1/2}$ is 1, it follows by Theorem 3.2 that

$$\begin{aligned}
1 - R &= P\Big(\bigcup_{x\in S}\Big[\Big(\sum_1^k g_j^2\Big)^{1/2} + \sum_1^n x_i h_i < \lambda + \varepsilon a_k\Big]\Big)\\
&\leq P\Big(\Big(\sum_1^k g_j^2\Big)^{1/2} < (1-\varepsilon)a_k\Big) + P\Big(\bigcup_{x\in S}\Big[\sum_1^n x_i h_i \leq \lambda - (1-2\varepsilon)a_k\Big]\Big)\\
&\leq \exp(-\varepsilon^2 a_k^2/2) + P\Big(\max_{x\in S}\sum_1^n x_i h_i \geq (1-2\varepsilon)a_k - \lambda\Big)\,.
\end{aligned}$$

Now, the function $f(h) = \max_{x\in S}\sum_1^n x_i h_i$ defined for $h \in \mathbb{R}^n$, has Lipshitz constant equal to 1, $\mathbb{E}(f) = s$ and since for the proper choice of $\varepsilon > 0$

$$a_k(1-2\varepsilon) - \lambda - s \geq a_k(1-2\varepsilon) - s - (1+\varepsilon)\big[a_k\varepsilon_2(K^*) + \mathbb{E}(K^*)\big] > 0\,,$$

we obtain by Theorem 3.2,

$$P\big(f(h) \geq a_k(1-2\varepsilon) - \lambda\big) \leq \exp\big(-\tfrac{1}{2}\{a_k(1-2\varepsilon) - \lambda - s\}^2\big)\,.$$

Combining the inequalities we have

$$\begin{aligned}
P &= \gamma_{n,n-k}\big(\xi \in \mathcal{G}_{n,n-k}\,;\ \xi \cap (K+S) = \Phi\big)\\
&\geq 1 - \frac{5}{2}\exp(-a_k^2\varepsilon^2/2) - \exp(-\sigma^2/2)\,,
\end{aligned}$$

where $\sigma = a_k(1-2\varepsilon) - s - (1+\varepsilon)\big[a_k\varepsilon_2(K^*) + \mathbb{E}(K^*)\big]$. We now choose ε so that $\varepsilon a_k = \sigma$, i.e.,

$$\varepsilon a_k = \frac{a_k(1-\varepsilon_2(K^*)) - \mathbb{E}(K^*) - s}{3 + \varepsilon_2(K^*) + \mathbb{E}(K^*)/a_k} > 0\,,$$

then the required estimate $P \geq 1 - \frac{7}{2}\exp(-\varepsilon^2 a_k^2/2)$ follows. □

Remark (3.3). By continuity Theorem 3.3 remains true if K has dimension lower than n, i.e., when K is a closed convex set which contains the origin (not necessarily in its interior) and $\dim\operatorname{span}(K) < n$. The values $\varepsilon_2(K^*)$ and $\mathbb{E}(K^*)$ are then understood as

$$\varepsilon_2(K^*) = \max\{\|x\|_2 \ ; \ x \in K\} \ , \text{ and}$$

$$\mathbb{E}(K^*) = \mathbb{E}\big(\max_{x\in K} \sum_1^n x_i g_i\big) \ .$$

Taking for example $K = rB_2^\ell = \{x = (x_1, \ldots, x_\ell, 0, \ldots, 0) \ ; \ \sum_1^\ell x_i^2 \le r^2\}$, we obtain $\varepsilon_2(K^*) = r$, $\mathbb{E}(K^*) = ra_\ell$ and so, if $1 \le k, \ell \le n$ and r satisfies $a_k(1-r) - ra_\ell > s$, then the measure of the k-codimensional subspaces $\xi \in \mathcal{G}_{n,n-k}$ which miss the set $S + K$ is greater than

$$1 - \frac{7}{2}\exp\left(-\frac{1}{2}\left[\frac{a_k(1-r) - ra_\ell - s}{3 + r + ra_\ell/a_k}\right]^2\right) \ .$$

In particular, taking $r = 0$ we obtain:

Corollary 3.4. *Let $S \subset S_2^{n-1}$ be a closed subset, $s = \mathbb{E}\big(\max_{x\in S}\sum_1^n x_i g_i\big)$, $1 \le k < n$ and assume $a_k > s$. Then*

$$\gamma_{n,n-k}(\xi \in \mathcal{G}_{n,n-k} \ ; \ \xi \cap S = \Phi) \ge 1 - \frac{7}{2}\exp\left(-\frac{1}{18}(a_k - s)^2\right) \ .$$

Corollary 3.5. *Let $S \subset S_2^{n-1}$ be a closed set and $s = \mathbb{E}\big(\max_{x\in S}\sum_1^n x_i h_i\big)$. If $1 \le k < n$, $\alpha > 0$ and $s < \sqrt{k} - 3\alpha$, then*

$$\gamma_{n,n-k}(\xi \in \mathcal{G}_{n,n-k} \ ; \ \xi \cap S = \Phi) \ge 1 - \frac{7}{2}\exp(-\alpha^2/2) + 0(k^{-1/2}) \ .$$

In particular, if $s < \sqrt{n} - 3\alpha$, then $S \cup (-S)$ has surface area smaller then $\frac{7}{2}\exp(-\alpha^2/2) + 0(n^{-1/2})$.

Proof: Apply Corollary 3.4 and use the fact that the surface area of $S \cup (-S)$ is

$$\gamma_{n,1}(\xi \in \mathcal{G}_{n,1} \ , \ \xi \cap S \ne \Phi) \ . \qquad \square$$

Remark (3.4.). If $0 < \lambda < 1$ and $0 < r < \frac{1}{2}$ are fixed and $s < \lambda(1-2r)\sqrt{n}$, then taking in Theorem 3.3, $k = n-1$ and $K = rB_2^n$ we obtain

$$\gamma_{n,1}(\xi \in \mathcal{G}_{n,1} \ , \ \xi \cap (S + rB_2^n) \ne \Phi) \le c_1 \exp\big(-f(\lambda, r)n\big) \ ,$$

where c_1 is a constant and $f(\lambda, r)$ a positive function. This says that the surface area $m_{n-1}\big((S + rB_2^n)_{\text{rad}}\big)$ of the radial projection of $(S + rB_2^n) \cup (-S + rB_2^n)$ onto S_2^{n-1} tends to 0 exponentially fast with n. A more careful analysis of this situation is as follows:

Corollary 3.6. *Let $S \subset S_2^{n-1}$ be an arbitrary closed set, $s = \mathbb{E}\big(\max_{x\in S}\sum_1^n x_i h_i\big)$ and t be any number such that $s < t < \sqrt{n}$. Let $K_t = \frac{1}{2}(1 - tn^{-1/2})B_2^n$. Then, there are universal constants $c_1, c_2 > 0$ such that, $m_{n-1}\big((S + K_t)_{\rm rad}\big) \le c_1 \exp\big(-c_2(x-s)^2\big)$.*

Proof: We apply Theorem 3.3 with $k = n - 1$. Since $a_n = \sqrt{n}\,(1 - \frac{1}{4n} + 0(n^{-2}))$, $a_{n-1} = \sqrt{n-1}\,(1 - \frac{1}{4n} + 0(n^{-2})) = \sqrt{n}\,(1 - \frac{3}{4n} + 0(n^{-2}))$. We also have $\varepsilon_2(K_t^*) = \frac{1}{2}(1 - tn^{-1/2})$ and $\mathbb{E}(K_t^*) = \frac{1}{2}(1 - tn^{-1/2})a_n$ and the inequality in Theorem 3.3 yields the result. □

Combining the estimates of Corollary 3.4 and Theorem 2.1 we obtain a quantitative version of Theorem 2.1.

Corollary 3.7. *Let $0 < \varepsilon \le 1$ and $\ell + 1 \le \frac{1}{2}n\varepsilon^2 m_{n-1}(A)$, where $A \subset S_2^{n-1}$ is an arbitrary closed subset. Then*

$$\gamma_{n,\ell}(\xi \in \mathcal{G}_{n,\ell}\,;\, \xi \cap S_2^{n-1} \subset A_\varepsilon) \ge 1 - c_1 \exp\Big\{-c_2 n^{-1}\big(\tfrac{1}{2}n\varepsilon^2 m_{n-1}(A) - \ell - 1\big)^2\Big\}\,,$$

where c_1 and c_2 are positive universal constants.

Proof: Let $k = n - \ell$. By inequality 2.2 with $S = (A_\varepsilon)^C$ we have

$$\begin{aligned} a_k - s \ge \sqrt{k-1} - s &\ge \sqrt{k-1} - \sqrt{n}\sqrt{1 - \tfrac{1}{2}n\varepsilon^2 m_{n-1}(A)} \\ &\ge \{k - 1 - n(1 - \tfrac{1}{2}\varepsilon^2 m_{n-1}(A))\}/2\sqrt{n} \\ &\ge \big\{\tfrac{1}{2}n\varepsilon^2 m_{n-1}(A) - \ell - 1\big\}/2\sqrt{n} > 0 \end{aligned}$$

and Corollary 3.4 concludes the proof. □

Remark(3.5). Corollary 3.7 and Lévy's isoperimetric inequality imply that if $n\varepsilon^2$ is big, then the set of subspaces E of dimension $\sim n\varepsilon^2$ for which (2.3) is satisfied has probability close to 1.

4. Missing Convex Sets by Random Subspaces

Let $\{K_i\}_1^N$ be a sequence of N closed convex sets in $\mathbb{R}^n$ which contain the origin 0 in their relative interiors and let $\{z_i\}_1^N$ be arbitrary points in $\mathbb{R}^n$. We denote by t_i (resp., T_i) the smallest (resp., largest) Euclidean distance of a point in $z_i + K_i$ to the origin and set $\mathbb{E}(K_i^*) = \mathbb{E}\big(\max_{x\in K_i}\sum_{j=1}^n x_j h_j\big)$. In the next theorem we estimate from below the measure of the set of all k-codimensional subspaces ξ in $\mathcal{G}_{n,n-k}$ which are distant d_i at least from $z_i + K_i$ for all $i = 1, 2, \ldots, N$.

Theorem 4.1. $\gamma_{n,n-k}\{\xi \in \mathcal{G}_{n,k-k}\ ;\ d(\xi, z_i + K_i) \geq d_i \text{ for all } i = 1,2,\ldots,N\}$

$$\geq 1 - \exp(-\delta^2 a_n^2/2) - (3/2)\exp(-\varepsilon^2 a_k^2/2)$$

$$-\sum_{i=1}^{N} \exp\left(-\left\{(1-2\varepsilon)a_k t_i - \mathbb{E}(K_i^*) - (1+\delta)d_i(a_n + a_k)\right\}^2/2T_i^2\right),$$

where $\varepsilon, \delta > 0$ *and where we assume* $(1-2\varepsilon)a_k t_i > \mathbb{E}(K_i^*) + d_i(1+\delta)(a_n + a_k)$ *for all* $i = 1, \ldots, N$.

Proof: Let $\|G\|$ be the norm of $G = \sum_{i=1}^{n}\sum_{j=1}^{k} g_{ij}(\omega)e_i \otimes e_j$ as a map of ℓ_2^n to ℓ_2^k. We have $a_n \leq \mathbb{E}\|G\| \leq a_n + a_k$. Let $\lambda_i = d_i(1+\delta)\mathbb{E}\|G\|$ and

$$\begin{aligned} P &:= \gamma_{n,n-k}\big(\xi \in \mathcal{G}_{n,n-k}\ ;\ d(\xi, z_i + K_i) \geq d_i\ ,\ 1 = 1,2,\ldots,N\big) \\ &= P\big(\omega\ ;\ d\big(G^{-1}(0), z_i + K_i\big) \geq d_i\ ,\ i = 1,2,\ldots,N\big)\ . \end{aligned}$$

Observe that

$$\begin{aligned} Q &:= P\big(\forall i\ ,\ \min_{x \in z_i + K_i} \|G(x)\|_2 \geq \lambda_i\big) \\ &\leq P\big(\forall i\ ,\ \|G\| d\big(G^{-1}(0), z_i + K_i\big) \geq \lambda_i\big) \\ &\leq P\big(\|G\| \geq (1+\delta)\mathbb{E}\|G\|\big) + P\big(\forall i, d\big(G^{-1}(0), z_i + K_i\big) \geq d_i\big) \\ &\leq \exp\left(-\frac{\delta^2}{2}(\mathbb{E}\|G\|)^2\right) + P\ . \end{aligned}$$

Now, apply Lemma 3.1 with $\lambda_x = \lambda_i + \varepsilon a_k\|x\|_2$ whenever $x \in z_i + K_i$ and with $S = \bigcup_1^N (z_i + K_i)$, to get a lower estimate for Q

$$\begin{aligned} Q + \tfrac{1}{2}\exp(-\varepsilon^2 a_k^2/2) &\geq Q + P(g > a_k\varepsilon) \\ &\geq P\Big(\bigcap_{x \in S} [\|G(x)\|_2 + \|x\|_2 g \geq \lambda_x]\Big) \\ &\geq P\Big(\bigcap_{x \in S} [\|x\|_2\big(\sum_1^k g_j^2\big)^{1/2} + \langle x, h\rangle \geq \lambda_x]\Big) \equiv R\ . \end{aligned}$$

But

$$\begin{aligned} 1 - R &= P\Big(\bigcup_{x \in S} [\|x\|_2\big(\sum_1^k g_j^2\big)^{1/2} + \langle x, h\rangle \leq \lambda_x]\Big) \\ &\leq P\Big(\big(\sum_1^k g_j^2\big)^{1/2} \leq (1-\varepsilon)a_k\Big) + P\Big(\bigcup_{i=1}^{N}\ \bigcup_{x \in z_i + K_i} [\langle x, h\rangle \leq \lambda_i - (1-2\varepsilon)a_k\|x\|_2]\Big) \\ &\leq \exp(-\varepsilon^2 a_k^2/2) + \sum_{i=1}^{N} P\Big(\bigcup_{x \in z_i + K^i} [\langle x, h\rangle \geq (1-2\varepsilon)a_k\|x\|_2 - \lambda_i]\Big) \\ &\leq \exp(-\varepsilon^2 a_k^2/2) + \sum_{i=1}^{N} P\Big(\bigcup_{x \in z_i + K_i} [\langle x, h\rangle \geq (1-2\varepsilon)a_k t_i - \lambda_i]\Big)\ . \end{aligned}$$

Since the function $f(h) = \max_{x \in z_i + K_i} \langle x, h \rangle$, defined for $h \in \mathbb{R}^n$, has Lipschitz constant T_i and, since $\mathbb{E}(f) = \mathbb{E}(K_i^*)$ and, given the assumption $(1-2\varepsilon)a_k t_i > \lambda_i + \mathbb{E}(K_i^*)$, it follows from Theorem 3.2 that $P(f \geq (1-2\varepsilon)a_k t_i - \lambda_i) \leq \exp\big(-((1-2\varepsilon)a_k t_i - \lambda_i - \mathbb{E}(K_i^*))^2/2T_i^2\big)$, concludes the proof. □

Taking $\delta \to \infty$, $(1+\delta)d_i \downarrow 0$ for each i, we obtain the measure of all k-codimensional subspaces which miss $\bigcup_1^N (z_i + K_i)$.

Corollary 4.2. *Under the assumptions $\varepsilon > 0$ and $(1-2\varepsilon)a_k t_i > \mathbb{E}(K_i^*)$ for all $i = 1, \dots, N$, we have*

$$\gamma_{n,n-k}\big(\xi \in \mathcal{G}_{n,n-k}\ ;\ \xi \cap (z_i + K_i) = \Phi \quad \text{for all} \quad i = 1, \dots, N\big)$$
$$\geq 1 - \frac{3}{2}\exp(-\varepsilon^2 a_k^2/2) - \sum_{i=1}^N \exp\big(-\{(1-2\varepsilon)a_k t_i - \mathbb{E}(K_i^*)\}^2/2T_i^2\big)\ .$$

5. Random Subspaces which Escape through at least One Mesh

Theorem 1.4 gave a sufficient condition for the existence of a subspace of given codimension k which misses a given subset of the sphere. However, it may happen that every subspace of codimension k hits the set and yet there might be a subspace which misses a "piece" of this set. To study such cases we consider a finite collection A_ℓ ($\ell = 1, 2, \dots, L$) of closed subsets of S_2^{n-1} and ask for a condition which implies that there exists a subspace of codimension k which misses at least one of the sets A_ℓ. To obtain this condition we shall use the following theorem proved in [G3], of which Theorem A is a special case.

Theorem C. *Let $f(x) = \min_{i_1} \max_{i_2} \min_{i_3} \max \dots x_{i_1, i_2, \dots, i_k}$ where for each $\ell = 1, 3, \dots, k$, the index i_ℓ ranges over some finite non-empty set $C(i_1, i_2, \dots, i_{\ell-1})$ (which, as indicated, shows that this set may depend on the previous choice of $i_1, i_2, \dots, i_{\ell-1}$). Given two distinct vector indices $i = (i_1, \dots, i_k)$ and $j = (j_1, \dots, j_k)$ let $m = m(i,j)$ be the first coordinate such that $i_m \neq j_m$. If $\{X_i\}$ and $\{Y_i\}$ are two Gaussian processes such that*

(i) $\mathbb{E}|X_i - X_j|^2 \leq \mathbb{E}|Y_i - Y_j|^2$ if m is even

(ii) $\mathbb{E}|X_i - X_j|^2 \geq \mathbb{E}|Y_i - Y_j|^2$ if m is odd,

then $\mathbb{E}f(X) \leq \mathbb{E}f(Y)$.

Theorem A is obtained from Theorem C by taking $k = 2$ and $1 \leq i_1 \leq n$, $1 \leq i_2 \leq m$.

Corollary 5.1. *Let $\{A_\ell\}_{\ell=1}^{L}$ be a finite collection of closed subsets of the sphere S_2^{n-1} and $G(\omega) = \sum_{i=1}^{n}\sum_{j=1}^{k} g_{ij}(\omega)e_i \otimes e_j$ a Gaussian operator from $\mathbb{R}^n$ to $\mathbb{R}^k$. Then*

$$\mathbb{E}\big(\max_\ell \{ \min_{x\in A_\ell} \|G(\omega)(x)\|_2 + 2g_\ell\}\big) \geq a_k - a_n \int_{S_2^{n-1}} \min_\ell \max_{x\in A_\ell} \langle x,u\rangle m_{n-1}(du) .$$

Proof: We apply Theorem C and for each $\ell \in \{1,2,\ldots,L\}$, $x \in A_\ell$ and $y \in S_2^{k-1}$ we define the two triple indexed Gaussian processes

$$Y_{\ell,x,y} = \sum_{j=1}^{k} g_j y_j + \sum_{i=1}^{n} h_i x_i \quad \text{and} \quad X_{\ell,x,y} = \langle G(\omega)x, y\rangle + 2g_\ell .$$

For any two triples $\alpha = (\ell,x,y)$ and $\beta = (\ell',x',y')$ we have

$$\mathbb{E}|Y_\alpha - Y_\beta|^2 = \sum_{j=1}^{k}(y_j - y_j')^2 + \sum_{i=1}^{n}(x_i - x_i')^2 = 4 - 2\langle x,x'\rangle - 2\langle y,y'\rangle$$

and

$$\mathbb{E}|X_\alpha - X_\beta|^2 = \sum_{i=1}^{n}\sum_{j=1}^{k}(x_iy_j - x_i'y_j')^2 + 8(1-\delta_{\ell,\ell'}) = 10 - 8\delta_{\ell,\ell'} - 2\langle x,x'\rangle\langle y,y'\rangle .$$

Therefore if $\ell = \ell'$ and $x \neq x'$, i.e., when $m(\alpha,\beta) = 2$, then

$$\mathbb{E}|Y_\alpha - Y_\beta|^2 - \mathbb{E}|X_\alpha - X_\beta|^2 = 2(1 - \langle x,x'\rangle)(1 - \langle y,y'\rangle) \geq 0 .$$

To check condition (ii) for Theorem C, notice that if $\ell = \ell'$ and $x = x'$ then $\mathbb{E}|X_\alpha - X_\beta|^2 = \mathbb{E}|Y_\alpha - Y_\beta|^2$, and if $\ell \neq \ell'$ then

$$\mathbb{E}|Y_\alpha - Y_\beta|^2 - \mathbb{E}|X_\alpha - X_\beta|^2 = -2(3 + \langle x,x'\rangle + \langle y,y'\rangle - \langle x,x'\rangle\langle y,y'\rangle) \leq 0 ,$$

hence, by Theorem C, $\mathbb{E}\min_\ell \max_{x\in A_\ell} \min_{y\in S_2^{k-1}} X_{\ell,x,y} \leq \mathbb{E}\min_\ell \max_{x\in A_\ell} \min_{y\in S_2^{k-1}} Y_{\ell,x,y}$ therefore, replacing X by $-X$ and Y by $-Y$ we obtain

$$\begin{aligned}
\mathbb{E}\max_\ell \min_x \max_y X_{\ell,x,y} &= \mathbb{E}\max_\ell\{ \min_{x\in A_\ell} \|G(\omega)(x)\|_2 + 2g_\ell\} \\
&\geq \mathbb{E}\max_\ell \min_x \max_y Y_{\ell,x,y} = \mathbb{E}\Big(\sum_1^k g_j^2\Big)^{1/2} + \mathbb{E}\max_\ell \min_x \sum_1^n h_i x_i \\
&- a_k - \mathbb{E}\min_\ell \max_x \sum_1^n h_i x_i - a_k - a_n \int_{S_2^{n-1}} \min_\ell \max_{x\in A_\ell} \langle x,u\rangle m_{n-1}(du) .
\end{aligned}$$

□

As a conclusion we obtain the following generalization of Theorem 1.4.

Theorem 5.2. *Let $\{A_\ell\}_{\ell=1}^L$ be a finite collection of closed subsets of the sphere S_2^{n-1} and $1 \le k \le n$. If*

$$a_k > a_n \int_{S_2^{n-1}} \min_\ell \max_{x \in A_\ell} \langle x, u\rangle m_{n-1}(du) + c\sqrt{\log L} \tag{5.1}$$

then there is a subspace of codimension k which misses at least one of the sets A_ℓ.

Proof: Since $\mathbb{E} \max_\ell 2g_\ell \le c\sqrt{\log L}$ we obtain from Corollary 5.1

$$\mathbb{E} \max_\ell \min_{x \in A_\ell} \|G(\omega)(x)\|_2 \ge a_k - a_n \int_{S_2^{n-1}} \min_\ell \max_{x \in A_\ell} \langle x, u\rangle m_{n-1}(du) - c\sqrt{\log L} > 0\,.$$

But, since the subspace $E_\omega = G(\omega)^{-1}(0)$ has codimension k a.e., there is a subspace E_{ω_0} of codimension k which misses at least one of the sets A_ℓ. □

We now give two different sufficient conditions on the sets A_ℓ which imply inequality (5.1) and hence the conclusion of Theorem 5.2.

Corollary 5.3. *Let $\{A_\ell\}_{\ell=1}^L$ be closed subsets of S_2^{n-1}, $A = \bigcup_1^L A_\ell$ and $1 < k < n$. Assume one of the following two conditions are satisfied:*

(a) there exists $0 < \gamma < \pi$ such that $\bigcap_{\ell=1}^L (A_\ell)_\gamma = \phi$ and

$$a_k > a_n \cos\gamma + c\sqrt{\log L}\,. \tag{5.2}$$

(b) Let α, β $(0 \le \alpha,\ \beta \le \pi)$ be defined by

$$\alpha = \max_\ell \operatorname{diam} A_\ell \quad \text{and} \quad \beta = \sup\Big\{t \ge 0\ ;\ (-A^c) \bigcap_{\ell=1}^L (A_\ell)_t = \phi\Big\}$$

(with $\beta = 0$ if $(-A^C) \bigcap_{\ell=1}^L A_\ell \ne \phi$) and assume

$$a_k > a_n\big\{ - m_{n-1}(A)\cos\alpha + m_{n-1}(A^C)\cos\beta\big\} + c\sqrt{\log L}\,. \tag{5.3}$$

Then inequality (5.1) is satisfied.

Proof: We shall estimate the integral which appears in (5.1). If condition (a) holds, then for every $u \in S_2^{n-1}$, $u \notin (A_{\ell_0})_\gamma$ for some ℓ_0, hence for every $x \in A_{\ell_0}$ the geodesic distance $\rho(x,u) \ge \gamma$, therefore

$$\min_\ell \max_{x \in A_\ell} \langle x, u\rangle \le \max_{x \in A_{\ell_0}} \langle x, u\rangle \le \cos\gamma$$

which shows that (5.2) implies (5.1). If condition (b) holds, let $u \in S_2^{n-1}$. If $u \in -A$, then $-u \in A$, so $-u \in A_{\ell_0}$ for some ℓ_0. But then for all $x \in A_{\ell_0}$ we have

$$\rho(x,u) \geq \rho(-u,u) - \rho(-u,x) \geq \pi - \alpha$$

hence

$$\min_{\ell} \max_{x\in A_\ell}\langle x,u\rangle \leq \max_{x\in A_{\ell_0}} \langle x,u\rangle \leq \cos(\pi-\alpha) = -\cos\alpha \,.$$

On the other hand, let $0 \leq t < \beta$ (take $t = 0$ if $\beta = 0$). If $u \in -A^C$, then $u \notin (A_{\ell_0})_t$ for some ℓ_0, so $\rho(u,x) \geq t$ for every $x \in A_{\ell_0}$, hence

$$\min_{\ell} \max_{x\in A_\ell}\langle x,u\rangle \leq \max_{x\in A_{\ell_0}} \langle x,u\rangle \leq \cos t \underset{t\to\beta}{\longrightarrow} \cos\beta \,.$$

Therefore,

$$\int_{S_2^{n-1}} = \int_{-A} + \int_{-A^C} \leq -m_{n-1}(-A)\cos\alpha + m_{n-1}(-A^C)\cos\beta$$

$$= -m_{n-1}(A)\cos\alpha + m_{n-1}(A^c)\cos\beta \,.$$

This proves that (5.3) implies (5.1). □

Remark (5.1). Corollary 5.3(a) is illustrated by the following example. *Let $0 < \varepsilon,\delta < 1$, $\gamma = \sqrt{2\delta}n^{-(1-\varepsilon)/4}$ and $L \leq \exp(c_1\delta^2 n^\varepsilon)$ be an integer. If $\{A_\ell\}_{\ell=1}^L$ is a collection of closed subsets of S_2^{n-1} such that $\bigcap_{\ell=1}^L (A_\ell)_\gamma = \phi$, then there exists a subspace of dimension $m \geq \delta n^{(\varepsilon+1)/2}$ in $\mathbb{R}^n$ which misses at least one A_ℓ, provided $n \geq N(\varepsilon)$.* (c_1 denotes a positive absolute constant.) The proof of this is done by estimating the quantities which appear in inequality 5.2(a). Take $k \sim n - \delta n^{(\varepsilon+1)/2}$, so $a_k \sim k^{1/2} \sim n^{1/2}(1 - \frac{1}{2}\delta n^{(\varepsilon-1)/2})$, $a_n \sim n^{1/2}$, $\cos\gamma \sim 1 - \gamma^2/2 = 1 - \delta n^{-(1-\varepsilon)/2}$ (since $\gamma \underset{n\to\infty}{\longrightarrow} 0$) and choose c_1 to be a suitable positive constant.

As done in section 3, we can assign the probability measure $\gamma_{n,n-k}$ to the set of all subspaces $\xi \in \mathcal{G}_{n,n-k}$ which miss at least one of the sets A_ℓ ($1 \leq \ell \leq 1$). For this estimate we need the following theorem of [G3] which extends Theorem B above.

Theorem D. *Let $f(x)$ and the processes $\{X_i\},\{Y_i\}$ be as in Theorem C and in addition to conditions (i) and (ii) we assume also that $\mathbb{E}X_i^2 = \mathbb{E}Y_i^2$ for all i. Then for all choices of the real sequence $\{\lambda_i\}$,*

$$P\Big(\bigcap_{i_1}\bigcup_{i_2}\bigcap_{i_3}\bigcup\cdots[X_i \geq \lambda_i]\Big) \leq P\Big(\bigcap_{i_1}\bigcup_{i_2}\bigcap_{i_3}\bigcup\cdots[Y_i \geq \lambda_i]\Big) \,.$$

Applying this together with Theorem 3.2 we obtain

Theorem 5.4. *Let $\{A_\ell\}_{\ell=1}^L$ be a finite collection of closed subsets of S_2^{n-1} and $1 \le k < n$. If inequality (5.1) holds then $\gamma_{n,n-k}(\xi \in \mathcal{G}_{n,n-k}$; there exists ℓ such that $\xi \cap A_\ell = \phi) \ge 1 - \frac{7}{2}\exp\big(-(a_k - s - c\sqrt{\log L})^2/2(4+\sqrt{3})^2\big)$, where $s = a_n \int_{S_2^{n-1}} \min_\ell \max_{x\in A_\ell} \langle x,u\rangle m_{n-1}(du)$.*

Proof: Let $X_{\ell,x,y}$ and $Y_{\ell,x,y}$ be defined as in the proof of Corollary 5.1 and $Z_{\ell,x,y} = Y_{\ell,x,y} + \sqrt{3}g$ where g is an $N(0,1)$ normal variable. The two processes $\{-X_{\ell,x,y}\}$ and $\{-Z_{\ell,x,y}\}$ satisfy the conditions of Theorem D and therefore

$$P\Big(\bigcap_\ell\bigcup_x\bigcap_y[-X_{\ell,x,y} \ge -a]\Big) \le P\Big(\bigcap_\ell\bigcup_x\bigcap_y[-Z_{\ell,x,y} \ge -a]\Big)$$

for all $a \in \mathbb{R}^1$, hence

$$P\Big(\bigcup_\ell\bigcap_x\bigcup_y[X_{\ell,x,y} \ge a]\Big) \ge P\Big(\bigcup_\ell\bigcap_x\bigcup_y[Z_{\ell,x,y} \ge a]\Big)\ .$$

We set $a = c\sqrt{\log L} + 2\varepsilon a_k$ where $\mathbb{E}\max 2g_\ell \le c\sqrt{\log L}$ and $\varepsilon > 0$ is to be chosen later. Then setting $P = P\big(\max_\ell \min_{x\in A_\ell} \|G(\omega)(x)\|_2 > 0\big)$ we have

$$\begin{aligned} P\Big(\bigcup_\ell\bigcap_x\bigcup_y[X_{\ell,x,y} \ge a]\Big) &\le P\big(\max_\ell \min_{x\in A_\ell} \|G(\omega)(x)\|_2 > 0\big) + \\ &\quad + P(\max_\ell 2g_\ell - \mathbb{E}\max_\ell 2g_\ell \ge 2\varepsilon a_k) \le P + \exp(-\varepsilon^2 a_k^2/2)\ . \end{aligned}$$

The last inequality is due to Theorem 3.2 and the fact that the function $f(x) = \max_\ell x_\ell$ has Lipschitz constant 1. By Theorem D and Theorem 3.2 we get

$$P + \exp(-\varepsilon^2 a_k^2/2) \ge P\Big[\Big(\sum_1^k g_j^2\Big)^{1/2} + \max_\ell \min_{x\in A_\ell} \sum_1^n h_i x_i + \sqrt{3}g \ge a\Big] = R\ .$$

$$\begin{aligned} 1 - R &\le P\Big(\Big(\sum_1^k g_j^2\Big)^{1/2} \le (1-\varepsilon)a_k\Big) + P(\sqrt{3}g \le -\sqrt{3}\varepsilon a_k) \\ &\quad + P\Big(\max_\ell \min_{x\in A_\ell} \sum_1^n h_i x_i \le a + \sqrt{3}\varepsilon a_k - (1-\varepsilon)a_k\Big) \\ &\le \exp(-\varepsilon^2 a_k^2/2) + \tfrac{1}{2}\exp(-\varepsilon^2 a_k^2/2) \\ &\quad + P\Big(\min_\ell \max_{x\in A_\ell} \sum_1^n h_i x_i \ge (1-\varepsilon)a_k - a - \sqrt{3}\varepsilon a_k\Big)\ . \end{aligned}$$

Now, the function $f(h) = \min_\ell \max_{x\in A_\ell} \sum_1^n x_i h_i$ has Lipschitz constant 1 and $\mathbb{E}(f) = s$, hence we can apply Theorem 3.2 again and obtain

$$\begin{aligned} P\big(f(h) \ge (1-\varepsilon)a_k - a - \sqrt{3}\varepsilon a_k\big) &= P\big(f(h) - s \ge (1-\varepsilon)a_k - a - \sqrt{3}\varepsilon a_k - s\big) \\ &\le \exp(-\varepsilon^2 a_k^2/2) \end{aligned}$$

where we select ε so that $\varepsilon a_k = (1-\varepsilon)a_k - a - \sqrt{3}\varepsilon a_k - s$, i.e., $\varepsilon a_k = \left(a_k - c\sqrt{\log L} - s\right)/(4+\sqrt{3})$.

□

Acknowledgement. I wish to thank Joel Zinn for some stimulating and valuable discussions.

References

[BG] Y. Benyamini and Y. Gordon, Random factorization of operators between Banach spaces, J. d'Analyse Math. 39 (1981),45-74.

[D] R.M. Dudley, The size of compact subsets of Hilbert spaces and continuity of Gaussian processes, J. Funct. Analy. 1 (1967), 290-330.

[DS] S. Dilworth and S. Szarek, The cotype constant and almost Euclidean decomposition of finite dimensional normed spaces, preprint.

[F1] X.M. Fernique, Des resultats nouveaux sur les processus Gaussiens, C.R. Acad. Sci., Paris. Ser. A-B 278 (1974), A363-A365.

[F2] X.M. Fernique, Régularité des trajectoires des fonctions aléatoires Gaussiens, Springer Lecture notes 480 (1975), 1-96.

[FT] T. Figiel and N. Tomczak-Jaegermann, Projection onto Hilbertian subspaces of Banach spaces, Israel J. Math 33 (1979), 155-171.

[G1] Y. Gordon, Some inequalities for Gaussian processes and applications, Israel J. Math. 50 (1985), 265-289.

[G2] Y. Gordon, Gaussian processes and almost spherical sections of convex bodies, The Annals of Probability 16 (1987), to appear.

[G3] Y. Gordon, Elliptically contoured distributions, Probability Theory and Related Fields, to appear.

[K] J.P. Kahane, Une inequalité du type de Slepian et Gordon sur les processus Gaussiens, Israel J. Math. 55 (1986), 109-110.

[L] D.R. Lewis, Ellipsoids defined by Banach ideal norms, Mathematica 26 (1979), 18-29.

[M1] V.D. Milman, Random subspaces of proportional dimension of finite dimensional normed spaces; approach through the isoperimetric inequality, Banach Spaces, Proc. Missouri Conference, 1986, Springer Lecture Notes #1166, 106-115.

[M2] V.D. Milman, Almost Euclidean quotient spaces of subspaces of finite dimensional normed spaces. Proc. AM.S. 94 (1985), 445-449.

[M3] V.D. Milman, Volume approach and iteration procedures in local theory of Banach spaces, Proc. Missouri Conf. 1984, Springer Lecture Notes 1166, 1985.

[M4] V.D. Milman, The concentration phenomenon and linear structure of finite-dimensional normed spaces, to appear.

[M5] V.D. Milman, A new proof of the theorem of A. Dvoretzky on sections of convex bodies, Func. Anal. Appl. 5 (1971), 28-37.

[MS] V.D. Milman and G. Schechtman, Asymptotic theory of finite dimensional normed spaces, Springer Lecture Notes 1200, 1986.

[P1] G. Pisier, Sur les éspaces de Banach K-convexes, Sém. D'Analyse Fonctionelle, exposé XI, 1979-80.

[P2] G. Pisier, Holomorphic semi-groups and the geometry of Banach spaces, Ann. Math. 115 (1982), 375-392.

[P3] G. Pisier, Probabilistic methods in the geometry of Banach spaces, Springer Lecture Notes 1206 (1986), 167-241.

[PT1] A. Pajor and N. Tomczak-Jaegermann, Subspaces of small codimension of finite-dimensional Banach spaces, Proc. A.M.S. 97 (1986), 637-642.

[PT2] A. Pajor and N. Tomczak-Jaegermann, Gelfand numbers and Euclidean sections of large dimensions, Springer Lecture Notes, Proc. Probability Conf., Aarhus, Denmark 1986, to appear.

ISOMORPHIC SYMMETRIZATION AND GEOMETRIC INEQUALITIES

V.D. Milman

Tel Aviv University
Tel Aviv, Israel

1. Introduction

In this paper, we describe a method of obtaining inequalities involving volumes of convex bodies in $\mathbb{R}^n$. The classic approach to such inequalities usually deals with different types of symmetrizations and, as a result, it derives inequalities in which extremal cases coincide with euclidean balls. However, in many important cases, extremal bodies are clearly not euclidean and natural inequalities involve some universal constants, independent of dimension or the bodies involved. It is difficult to compute these constants; however, we usually don't use their exact values in application anyway.

One method of obtaining such inequalities, which we present in this paper, is based on the study of the linear structure of subspaces and quotient spaces of normed finite dimensional linear spaces $X_\alpha = (\mathbb{R}^n, \|\cdot\|_\alpha)$.

The method has two different presentations. In the introduction we describe briefly one which has already been reflected in the literature [M1], [BM], [M2], [M3].

Consider $\mathbb{R}^n$ equipped with a euclidean norm $|\cdot|$ and let $D = \{x \in \mathbb{R}^n \mid |x| \le 1\}$ be the euclidean ball. Denote sX a subspace of X, qY a quotient space of a space Y and $d_X \equiv d(X, \ell_2^{\dim X})$ to be the Banach-Mazur (multiplicative) distance between X and $\ell_2^{\dim X}$:

$$d_X = \inf\{\|T\| \cdot \|T^{-1}\| \mid T : X \longrightarrow \ell_2^{\dim X} \text{ be a linear isomorphism}\} .$$

1.1 Theorem. [M1], [M4] *For any space $X = (\mathbb{R}^n, \|\cdot\|)$ and any $1/2 < \lambda < 1$, there exists qsX - a quotient of a subspace of X - such that $\dim qsX \ge \lambda n$ and $d_{qsX} \le C(1-\lambda)^{-1} \log(1-\lambda)^{-1}$.*

Supported in part by the Fund for Basis Research of the Israel Academy of Sciences and Humanities

1.2. Theorem 1.1 opens a new opportunity to study different kinds of volume inequalities. An important observation in this approach is a *"weak dependence" of volume* on dimension in the following sense. Let K be a symmetric convex compact body in the euclidean space $(\mathbb{R}^n, |\cdot|)$. Denote $sK = K \cap (s\mathbb{R}^n)$ (as before $s\mathbb{R}^n$ is a subspace of $\mathbb{R}^n$), $qK = \operatorname{Proj} K$ (the orthogonal projection of K on a space $q\mathbb{R}^n$) and

$$d_K \equiv d(K, D) \stackrel{\text{def}}{=} \inf\{a \cdot b \mid D \subset bK \subset abD\}$$

Note that if K is the unit ball of the normed space $X = (\mathbb{R}^n, \|\cdot\|)$, then the Banach-Mazur distance, defined above, $d_X = \inf\{d_{TK} \mid T$ runs through all linear isomorphisms of $\mathbb{R}^n\}$. Consider

$$\text{v.r. } K/D \stackrel{\text{def}}{\equiv} (\operatorname{Vol} K / \operatorname{Vol} D)^{1/n}$$

(the volume ratio of K by D). Then ([BM]) v.r.(sK/sD) =v.r.$(K/D)(1+o(1))$ (where $o(1) \to 0$ if $n \to \infty$) if $(n - \dim(s\mathbb{R}^n))/n \simeq o(1/\log d_K)$ and a similar statement is true for qK.

1.3 Isomorphic Symmetrization – "descending"-procedure. By 1.2, Theorem 1.1 allows us, with no essential change of volume and with a small decrease of dimension, to replace the symmetric convex body K (viewed as the unit ball of a space X) by another convex body K_1 (the unit ball of $X_1 = qsX$) which is much closer to a euclidean ball than the original K (say $d_{X_1} \leq C(\log(d_X)^2)$, and so on. Finally, after a finite number of steps, we pass to a space Y in some fixed neighbourhood of the euclidean space (again with no essential change of v.r. of the unit ball of Y with respect to K). Therefore we have a symmetrization type of procedure, but in an *isomorphic* sense (and *not the isometric one*), because Theorem 1.1 does not allow us to approach the euclidean space isometrically. We come to some (fixed, independent of n) neighbourhood of an ellipsoid but cannot come (as some examples indeed show) too close to it. We call this procedure a "descending" symmetrization because, at every step, we decrease the dimension of our bodies. This method was first used by [BM] in solving the old question of Mahler [Ma]. Let $K \subset \mathbb{R}^n$ be a convex compact body with the origin 0 belonging to the interior $\overset{\circ}{K}$ for K. Then $K^\circ = \{x \in \mathbb{R}^n \mid (x, y) \leq 1$ for any $y \in K\}$ is the dual body. Define

$$s(K) \stackrel{\text{def}}{=} \left(\frac{\operatorname{Vol} K \cdot \operatorname{Vol} K^\circ}{\operatorname{Vol} D \cdot \operatorname{Vol} D}\right)^{1/n} = \text{v.r.} K/D \cdot \text{v.r.} K^\circ/D \,. \tag{1.1}$$

Theorem 1.4. [BM]. *There exists a numerical constant $c > 0$ such that for any integer n and any convex body $K \subset \mathbb{R}^n$, $0 \in \overset{\circ}{K}$,*

$$c \leq s(K) \,.$$

We call this inequality an inverse Santalo inequality because Santalo [S] has proved that

$$s(K) < 1$$

(i.e., $\max_K(\operatorname{Vol} K \cdot \operatorname{Vol} K^\circ) = (\operatorname{Vol} D)^2$*).*

The following result gives a more advanced example of an application of a descending symmetrization.

Theorem 1.5. ([M3], the inverse form of the Brunn-Minkowski inequality.) *For every convex compact body* $K \subset \mathbb{R}^n$ *there is a corresponding affine transform* $u_K : \mathbb{R}^n \to \mathbb{R}^n$ ($\det u_K = 1$) *such that for any two such bodies* K_1 *and* K_2 *and any* $\varepsilon > 0$

$$\left\{ \operatorname{Vol}(u_{K_1} K_1 + \varepsilon u_{K_2} K_2)\right\}^{1/n} \le C\left\{(\operatorname{Vol} K_1)^{1/n} + \varepsilon(\operatorname{Vol} K_2)^{1/n}\right\}$$

for some numerical constant C *independent of* n *and the bodies* K_i. *At the same time, this inequality is also satisfied for the polar bodies* $(u_{K_1} K_1)^\circ$ *and* $(u_{K_2} K_2)^\circ$.

This theorem has a number of interesting consequences (see [KM], [M3], [M2] or [P2]). However, its proof by descending symmetrization is quite a heavy one.

So, in this paper, we will present, in detail, a different version of an isomorphic symmetrization with no change of dimension. We call this symmetrization a "convex surgery", because, at every step we will either cut the body, replacing K by $K \cap \mathcal{E}$ for a suitable ellipsoid $\mathcal{E}$, or use a dual operation to a cutting off operation, which is a convex hull $\operatorname{Conv} K \cup \mathcal{E}$. After a few steps, we will come to a body T which is C-isomorphic to an ellipsoid (and, as usual, C is a universal constant independent of dimension or the body K). In this approach, we do not use the Quotient of Subspace Theorem 1.1, although we use all the main ingredients of the proof of this theorem (including an iteration procedure). As we will see, all proofs are simplified significantly by combining volume computations with entropy estimations. (G. Pisier has noted that the use of entropy simplifies some elements in the original proof of [M3].) Such a direct way is very short. We have already demonstrated a part of this technique ("cutting surgery") in [M5] where a direct and short proof is given for the theorem [BM] that a cotype 2 space has a finite volume ratio.

In sections 3 and 4, we develop a convex surgery symmetrization and prove Theorems 1.4 and 1.5 using it. An entropy point of view on the Quotient of Subspace Theorem 1.1 is given in section 5. In section 6, we consider inequalities involving mixed volumes. These results are still in a preliminary form. However, I think that the technique of convex surgery, applied to the study of mixed volumes and some other observations in this section, is already interesting in its present form.

In the next section we give an account of notions and results which we use in the proofs.

2. Notations and Some Known Results

Let $X = (\mathbb{R}^n, \|\cdot\|)$ be an n-dimensional normed space, $K(X)(= K) = \{x \in \mathbb{R}^n \mid \|x\| < 1\}$ its unit ball. We always consider $\mathbb{R}^n$ equipped with the standard euclidean norm $|\cdot|$ (and, i.e., the inner product $(\cdot,\cdot)$) and the Lebesgue measure Vol. Then the polar body $K^\circ = \{x \in \mathbb{R}^n \mid |(x,y)| \le 1 \text{ for } \forall y \in K\}$ is the unit ball of the dual space $X^* = (\mathbb{R}^n, \|\cdot\|^*)$, i.e., $K(X^*) = K^\circ$. Also, we denote $D = K(\mathbb{R}^n; |\cdot|)$ to be the unit euclidean ball. We use some standard quantities in the theory of finite dimensional normed spaces (the so called Local Theory of Normed Spaces). Let

$$M_X = \int_{S^{n-1}} \|x\| d\mu(x)$$

where $S^{n-1} = \partial D$ is the unit euclidean sphere and $\mu(x)$ is the probability rotation invariant measure on S^{n-1}. We also write M_K instead of M_X, where K is the unit ball of X, or just M, if it is clear from the text which norm we mean. We often denote $M^* = M_{X^*}$. Note that we may consider different euclidean structures in $\mathbb{R}^n$ and, consequently, derive different numbers M and M^* for the same space X. We often write $A(n) \lesssim B(n)$ instead of saying that there exists a numerical constant $C > 0$ such that $A(n) \le CN(n)$; e.g., $1 \lesssim f(n)$ means a uniform positive low bound for $f(n)$, and $\lambda_t \simeq 1$ means existence of universal constants $0 < c$ and C such that $c \le \lambda_t \le C$.

One of the important results which we will use below states:

Theorem 2.1. *There exists a euclidean structure such that* $M_X \cdot M_{X^*} \lesssim \log 2d_X$.

We call this structure an ℓ-structure and the unit ball of this euclidean norm the ℓ-ellipsoid of X. The theorem is a combination of a few results (Lewis [L], Figiel and Tomczak [FT] and Pisier [P1]). We refer to the books and surveys [MSch], [P3], [BM], Appendix, for proofs and more details.

Let $N(K_1, K_2)$ be the covering number of K_1 by K_2, i.e.,

$$N(K_1, K_2) = \min\{N \in \mathbb{N} \mid \exists x_1, \dots, x_N \in \mathbb{R}^n, \text{ and } K_1 \subset \bigcup_1^N (x_i + K_2)\},$$

(K_1 and K_2 are compact convex symmetric bodies in $\mathbb{R}^n$).

Then the well known Sudakov inequality [Su] states that for some universal constant $c > 0$

$$N(K, \lambda D) \leq e^{cn(M^*/\lambda)^2} . \tag{2.1}$$

Its dual version (see [PT]) says

$$N(D, \lambda K) \leq e^{cn(M/\lambda)^2} \tag{2.2}$$

(again for a universal constant $c > 0$).

3. Isomorphic Symmetrization Through Convex Surgery; Proof of Theorem 1.4

Define $K_\lambda \overset{\text{def}}{=} K \cap \lambda D$ where, as before, K is the centrally symmetric convex compact body in $\mathbb{R}^n$ and D is the euclidean ball. The following two simple lemmas will be our main geometrical tool.

Lemma 3.1. *Let $N(K, \lambda D)$ be the covering number of K by λD. Then*

$$\operatorname{Vol} K \leq N(K, \lambda D) \cdot \operatorname{Vol} K_\lambda .$$

Proof: is immediate from the Brunn (concavity) theorem. Indeed, by the definition of $N = N(K, \lambda D)$, there exist $\{x_i\}_1^N$ such that $K \subset \bigcup_1^N (x_i + \lambda D)$. Therefore, $K \subset \bigcup_1^N (x_i + \lambda D) \cap K$. It remains to note that, by Brunn's Theorem, for any $x \in \mathbb{R}^n$

$$\operatorname{Vol}\left[(x + \lambda D) \cap K\right] \leq \operatorname{Vol}(\lambda D \cap K) . \qquad \square$$

Lemma 3.2. *Let $N(D, \lambda K)$ be the covering number of D by λK and let $D \subset bK$ for $b \geq 1$. Then for $\lambda \geq 1$*

$$\operatorname{Vol}\left[\operatorname{Conv}\left(K \cup \frac{1}{\lambda} D\right)\right] \leq enbN(D, \lambda K) \operatorname{Vol} K$$

and for $0 < \lambda < 1$

$$\operatorname{Vol}\left[\operatorname{Conv}\left(K \cup \frac{1}{\lambda} D\right)\right] \leq en\frac{b}{\lambda} N(D, \lambda K) \operatorname{Vol} K .$$

Proof: Note that $N(D, \lambda K) = N\left(\frac{1}{\lambda} D, K\right) \overset{\text{def}}{=} N$. If

$$\frac{1}{\lambda} D \subset \bigcup_1^N (x_i + K)$$

then $x_i \in K + \frac{1}{\lambda}D \subset \left(1+\frac{b}{\lambda}\right)K$. Also clearly $\alpha\left(\frac{1}{\lambda}D\right) \subset \bigcup_1^N(\alpha x_i + \alpha K)$ for $0 \le \alpha \le 1$ and, for $\alpha + \beta = 1$

$$\alpha\left(\frac{1}{\lambda}D\right) + \beta K \subset \bigcup_1^N(\alpha x_i + K) .$$

So

$$\text{(3.1)} \qquad \operatorname{Conv}\left(\frac{1}{\lambda}D \cup K\right) \subset \bigcup_{i=1}^N \bigcup_{0\le\alpha\le 1} (\alpha x_i + K) .$$

Now take $[b \cdot T]$ equally distributed on $[0,1]$ numbers α_j, $j = 1, \dots, [bT]$, where $T = 2n$ in the case $\lambda \ge 1$ and $= 2n/\lambda$ for $0 < \lambda < 1$. Then, by (3.1), for any $z \in \operatorname{Conv}\left(\frac{1}{\lambda}D \cup K\right)$, we find α_j such that

$$z \in \alpha x_i + K = \alpha_j x_i + (\alpha - \alpha_j)x_i + K \in \alpha_j x_i + \left(\frac{1 - b/\lambda}{bT} + 1\right) K \subset \alpha_j x_i + \left(1 + \frac{1}{n}\right) K .$$

Finally, we have from (3.1)

$$\operatorname{Conv}\left(\frac{1}{\lambda}D \cup K\right) \subset \bigcup_{i=1}^N \bigcup_{j=1}^{[bT]} \left\{\alpha_j x_i + \left(1 + \frac{1}{N}\right) K\right\}$$

and

$$\operatorname{Vol}\left[\operatorname{Conv}\left(\frac{1}{\lambda}D \cup K\right)\right] \le N \cdot bT \left(1 - \frac{1}{n}\right)^n \operatorname{Vol} K . \qquad \square$$

Proof of Theorem 1.4. With K, we associate a normed space $X = (\mathbb{R}^n, \|\cdot\|_K)$ such that K is the unit ball in this norm. Note that $s(K)$ (defined by (1.1)) is an affine invariant: $s(TK) = s(K)$ for any linear invertible map $T : \mathbb{R}^n \to \mathbb{R}^n$. Therefore, we may assume a euclidean structure for $\mathbb{R}^n$ to be the structure defined by Theorem 2.1 for the space X. Choose also the homotetical normalization such that $M(= M_X) = 1$. It is well known that, with this normalization, b defined in Lemma 3.2, is $\le n$ (indeed, even $\le \sqrt{n}\log n$ – see [MSch]). Now, for $a > 1$, we consider

$$\lambda^{\text{up}} = M^* a \qquad \text{and} \qquad \lambda^{\text{down}} = Ma$$

and we replace the body K by a new body K_1

$$K_1 = \operatorname{Conv}\left[(K \cap \lambda^{\text{up}} D) \bigcup \frac{1}{\lambda^{\text{down}}} D\right] .$$

By Lemmas 3.1 and 3.2, equipped with the estimates (2.1) and (2.2) for the covering numbers $N(K, \lambda D)$ and $(D, \lambda K)$, we obtain

$$\text{(3.2)} \qquad \exp(-cn/a^2) \le \operatorname{Vol} K_1 / \operatorname{Vol} K \le \exp(cn/a^2)$$

for a universal constant $c > 0$. (Note that $\lambda^{\text{down}} > 1$ because $M = 1$ and we ignore $n \cdot b \le n^2$ in Lemma 3.2 because we only use $a < \log n$ and therefore a small change of $c > 0$ will take care of a factor of order n^2.) The dual body K_1° is constructed, in exactly the same way, from the dual body K°:

$$K_1^\circ = \lambda^{\text{down}} D \bigcap \text{Conv}\left(K^\circ \cup \frac{1}{\lambda^{\text{up}}} D\right) .$$

Therefore, the same estimate (3.2) is also valued for the volume ratio $\text{Vol}\, K_1^\circ / \text{Vol}\, K^\circ$. Finally, it is clear that

$$d_{K_1} \le \lambda^{\text{up}} \cdot \lambda^{\text{down}} = MM^* a^2 \lesssim a^2 \log 2 d_X$$

and, as is well known, $d_X \le \sqrt{n}$ for any X.

So, we have realized the first step of symmetrization, replacing the original body K by another body K_1, which is, in general, much closer to the euclidean ball and which has (assuming $a \sim \log n$) almost the same volume ratio as K; similarly, its polar K_1° has almost the same volume ratio as K°. In exact terms

$$\exp(-c/a^2) \le s(K)/s(K_1) \le \exp(c/a^2)$$

for some universal (numerical) constant c.

Now we organize an iteration of the above procedure.

Let $a_1 = \log n$, $a_2 = \log\log n, \dots, a_t = \log^{(t)} n$ - the t-iterated logarithm of n, and let t be the smallest integer such that $\log^{(t)} n < 2$. In the first step, we take $\lambda_1^{\text{up}} = M^* a_1$ and $\lambda_1^{\text{down}} = M a_1$. Then we derive the body K_1. Let X_1 be a normed space $(\mathbb{R}^n, \|\cdot\|_{K_1})$ - i.e. K_1 is the unit ball of X_1. We have

$$d_{X_1} \le d_{K_1} \lesssim (\log n)^3 \quad \text{and} \quad e^{-c/(\log n)^2} \le \frac{s(K)}{s(K_1)} \le e^{c/(\log n)^2} .$$

Similarly, we derive body K_2 starting with body K_1 and choosing $\lambda_2^{\text{up}} = M_{K_1^\circ} a_2$, $\lambda_2^{\text{down}} = M_{K_1} a_2$, where averages M_{K_1} and $M_{K_1^\circ}$ are taken, with respect to a new euclidean structure - the ℓ-structure for the space X_1 (see Theorem 2.1). Therefore, by the way, the euclidean ball D_1 is now the unit ball of the new euclidean norm. Consequently, we have

$$d_{X_2} \le d_{K_2} \lesssim (\log\log n)^3 \quad \text{and} \quad e^{-c\left[\frac{1}{(\log n)^2} + \frac{1}{(\log\log n)^2}\right]} \le \frac{s(K)}{s(K_2)} \le e^{c\left[\frac{1}{(\log n)^2} + \frac{1}{(\log\log n)^2}\right]} .$$

Continuing in such a way we derive the body K_t on the step t. Clearly, $d_{X_t} \le C$ for some universal constant C, i.e., K_t is C-isomorphic to an ellipsoid and then

(3.3) $$\frac{1}{C} \le s(K_t) \le C .$$

Because, obviously, $\sum_1^t a_j^{-2} < C_2$ (is uniformly bounded), we obtain

$$\frac{1}{C_1} \le s(K) \le C_1 \tag{3.4}$$

for some numerical constant C_1 (independent of dimension n and body $K \subset \mathbb{R}^n$). So, we have proved Theorem 1.4 together with a weak version of Santalo inequality $s(K) \le$ Const. instead of $s(K) \le 1$.

4. Isomorphic Symmetrization Through Convex Surgery; Proof of Theorem 1.5

We start with the following lemma.

Lemma 4.1. *Let D, K and P be centrally symmetric convex compact bodies in $\mathbb{R}^n$. Then*

$$\max_{x \in \mathbb{R}^n} \operatorname{Vol}\left[(x + D) \cap K + P\right] = \operatorname{Vol}\left[(D \cap K) + P\right]$$

(i.e., the maximum is attained for $x = 0$).

Proof: follows, of course, by a right application of the Brunn-Minkowski inequality. Define $T_x = \left[(x + D) \cap K\right] + P$. Note that $-T_x = T_{-x}$ and therefore

$$\operatorname{Vol} T_x = \operatorname{Vol} T_{-x} \, .$$

Also

$$\frac{T_x + T_{-x}}{2} = \frac{(x + D) \cap K + (-x + D) \cap K}{2} + P \subset (K \cap D) + P = T_0 \, .$$

So, by Brunn-Minkowski inequality

$$\operatorname{Vol} T_x \le \operatorname{Vol} \frac{T_x + T_{-x}}{2} \le \operatorname{Vol} T_0 \, . \qquad \square$$

An immediate consequence of Lemma 4.1 is the following generalization of Lemma 3.1:

Corollary 4.2. *Let $N = N(K, \lambda D)$ be the covering number of K by λD. Then for any (centrally symmetric) convex body P*

$$\operatorname{Vol}(K + P) \le N(K, \lambda D) \operatorname{Vol}\left[(\lambda D \cap K) + P\right] .$$

Indeed, $K \subset \bigcup_1^N (x_i + \lambda D) \cap K$ for some $x_i \in \mathbb{R}^n$ and $K + P \subset \bigcup_1^N \left[(x_i + \lambda D) \cap K + P\right]$. It remains to apply Lemma 4.1. $\square$

The next lemma gives an improved, entropy-type version of Lemma 3.2.

Lemma 4.3. *Let K and L be symmetric convex bodies in $\mathbb{R}^n$ and $L \subset bK$ for some $b \geq 1$. Then*

$$N\left(\mathrm{Conv}(K \cup L), K\left(1+\frac{1}{n}\right)\right) \leq 2bnN(L,K)\ .$$

Proof: By the definition of $N = N(L,K)$ there exist x_i, $i = 1,\dots,N$, such that $L \subset \bigcup_1^N (x_i + K)$. Clearly $x_i \in L + K \subset (b+1)K$.

Let $\alpha_j = j/2bn$, $j = 1,\dots$. As in the proof of Lemma 3.2, we have

$$\mathrm{Conv}(L \cup K) \subset \bigcup_{0\leq\alpha\leq 1}\bigcup_{i=1}^{N}(\alpha x_i + K) \subset \bigcup_{j=1}^{[2bn]}\bigcup_{i=1}^{N}\left(\alpha_j x_i + \left(1+\frac{b+1}{2bn}\right)K\right) \qquad \square$$

Using this lemma for $L = \frac{1}{\lambda}D$ we obtain:

Corollary 4.4. *Under the above conditions on K and D (i.e., $D \subset \lambda bK$) for any convex body P*

$$\mathrm{Vol}\left(\left(\mathrm{Conv}\, K \cup \frac{1}{\lambda}D\right) + P\right) \leq 2ebn\, N(D,\lambda K)\,\mathrm{Vol}(K+P)\ .$$

The main preliminary result on our way to Theorem 1.5 is the following statement.

Theorem 4.5. *For every convex compact body $K \subset \mathbb{R}^n$ there exists an ellipsoid $\mathcal{E}_K$ such that*

$$\mathrm{Vol}\,\mathcal{E}_K = \mathrm{Vol}\,K$$

and for any convex compact body $P \subset \mathbb{R}^n$

$$\mathrm{Vol}(K+P) \leq C^n\,\mathrm{Vol}(\mathcal{E}_K + P)$$

where C is a numerical constant (i.e., a universal number, independent of n and the bodies involved).

Proof: Note first that it is enough to consider a case of centrally symmetric bodies K and P. Indeed, define for any K the symmetrized body $\widehat{K} = \frac{1}{2}(K - K)$. By Rogers-Shephard [RSh], we know that

$$\mathrm{Vol}\,K \leq \mathrm{Vol}\,\widehat{K} < 2^n\,\mathrm{Vol}\,K\ .$$

So, if Theorem 4.5 is correct for centrally symmetric convex bodies, then

$$\mathrm{Vol}(K+P) \leq \mathrm{Vol}(\widehat{K}+\widehat{P}) \leq C^n\,\mathrm{Vol}(\mathcal{E}_{\widehat{K}} + \widehat{P}) \tag{4.1}$$

where $\operatorname{Vol} \mathcal{E}_{\widehat{K}} = \operatorname{Vol} \widehat{K} \le 2^n \operatorname{Vol} K$. Take α, $\frac{1}{2} \le \alpha \le 1$, such that $\operatorname{Vol}(\alpha \mathcal{E}_{\widehat{K}}) = \operatorname{Vol} K$ and define $\mathcal{E}_K = \alpha \mathcal{E}_{\widehat{K}}$. Then we may continue (4.1)

$$\operatorname{Vol}(K+P) \le \left(\frac{C}{\alpha}\right)^n \operatorname{Vol}(\mathcal{E}_K + \alpha \widehat{P}) \le \left(\frac{C}{\alpha}\right)^n \operatorname{Vol}(\mathcal{E}_K + \widehat{P}) =$$

$$= \left(\frac{C}{\alpha}\right)^n \operatorname{Vol}\left(\tfrac{1}{2}(\mathcal{E}_K + P) - \tfrac{1}{2}(\mathcal{E} + P)\right) \le (2C)^n \operatorname{Vol}(\mathcal{E}_K + P) .$$

Therefore, next we assume K and P to be central symmetric convex compact bodies in $\mathbb{R}^n$.

Our construction of the ellipsoid $\mathcal{E}_K$ follows exactly the steps of the proof of Theorem 1.4 in the previous section. We derive there the body K_t which was C-isomorphic to an ellipsoid $\mathcal{E}$ and

$$(\operatorname{Vol} K_t / \operatorname{Vol} K)^{1/n} \simeq 1 . \tag{4.2}$$

Our $\mathcal{E}_K = \lambda \mathcal{E}$ for such a $\lambda > 0$ that $\operatorname{Vol} \mathcal{E}_K = \operatorname{Vol} K$. Clearly, from (4.2), $\lambda \simeq 1$.

Let us now go through the proof of Theorem 1.4 in section 3 and check one step of the iteration. Choosing λ^{up} and λ^{down}, we construct a body K_1 satisfying (3.2). Using Corollaries 4.2 and 4.4, we may add to this information that for any symmetric convex body P

$$\operatorname{Vol}(K+P) \le e^{cn/a^2} \operatorname{Vol}(K_1 + P) . \tag{4.3}$$

(As usual, c is a universal constant; a has the same sense as in (3.2).) The estimates (3.2) and (4.3) allow us to use the same iteration procedure which we organized in section 3, i.e., choosing the same a_i, $i = 1, \dots, t$, and constructing the same bodies K_i, $i =, \dots, t$. On the last step we derive a body K_t, which is C-isomorphic to the ellipsoid $\mathcal{E}$, and satisfies (4.2) and

$$\operatorname{Vol}(K+P) \le C_1^n \operatorname{Vol}(K_t + P) . \tag{4.4}$$

This immediately implies the theorem.

Remarks. 1. Obviously, for any $t > 0$, we may choose $\mathcal{E}_{tK} = t\mathcal{E}_K$.

2. Let $u : \mathbb{R}^n \to \mathbb{R}^n$ be a linear isomorphism. The normed space $X = (\mathbb{R}^n \| \cdot \|_{uK})$ associated with uK is isometric to the normed space $X = (\mathbb{R}^n, \| \cdot \|_K)$ and therefore the ℓ-euclidean structure, which we considered at the start of the proof of Theorem 1.4, is the same. Geometrically, this means that instead of the pair: body K – the ℓ-ellipsoid D, we have the pair: the body uK – the ℓ-ellipsoid uD. So, at the end of our construction we will have

the body uK_t which is C-isomorphic to the ellipsoid $u\mathcal{E}$. We see that the construction gives $\mathcal{E}_{uK} = u\mathcal{E}_K$. Of course, it is straightforward that, in Theorem 4.5, we may choose $\mathcal{E}_{uK} = u\mathcal{E}_K$.

3. Let K be a centrally symmetric body and let K° denote its polar. Then we can construct K_i and K_i° in parallel, as in the proof of Theorem 1.4 (see section 3). On the final step t, besides (4.4), we obtain that

$$\mathrm{Vol}(K^\circ + P) \leq C_1^n \,\mathrm{Vol}\left((K_t)^\circ + P\right) .$$

Therefore, we may choose $\mathcal{E}_{K^\circ} = \lambda \mathcal{E}_K^\circ$ where $\lambda > 0$ is such a normalization that $\mathrm{Vol}\,\mathcal{E}_{K^\circ} = \mathrm{Vol}\,K_\circ$. By (3.3) and (3.4), i.e., indeed, by Theorem 1.4, we have $\lambda \simeq 1$.

The proof of Theorem 1.5 is an easy exercise in a use of Theorem 4.5. Indeed, it is again enough to deal with centrally symmetric bodies K_1 and K_2. Then, using Theorem 4.5 twice, one first replaces K_1 by $\mathcal{E}_{K_1}$ and then $P = K_2$ by $\mathcal{E}_{K_2}$. So

$$\mathrm{Vol}(K_1 + tK_2)^{1/n} \lesssim \mathrm{Vol}(\mathcal{E}_{K_1} + t\mathcal{E}_{K_2})^{1/n}$$

for any $t > 0$ (we also use Remark 1 above). If it has happened that $\mathcal{E}_{K_1} = \lambda \mathcal{E}_{K_2}$, then $\mathrm{Vol}(\mathcal{E}_{K_1} + t\mathcal{E}_{K_2})^{1/n} = (\mathrm{Vol}\,\mathcal{E}_{K_1})^{1/n} + t(\mathrm{Vol}\,\mathcal{E}_{K_2})^{1/n} = (\mathrm{Vol}\,K_1)^{1/n} + t(\mathrm{Vol}\,K_2)^{1/n}$ and we derive Theorem 1.5.

However, generally, one cannot expect the ellipsoid $\mathcal{E}_{K_1}$ to be proportional to the ellipsoid $\mathcal{E}_{K_2}$. Therefore, we consider linear maps $u_{K_i} : \mathbb{R}^n \to \mathbb{R}^n$, $\det u_{K_i} = 1$, $i = 1, 2$, such that $u_{K_i}\mathcal{E}_{K_i} = \lambda_i D$ where D is the standard euclidean ball and λ_i are defined by a condition $\mathrm{Vol}\,\lambda_i D = \mathrm{Vol}\,K_i$. Then, starting with the bodies $u_{K_1}K_1$ and $u_{K_2}K_2$, and by Remark 2, one derives the ellipsoids proportional to D and this has completed the proof of the main part of Theorem 1.5. Using Remark 3, one obtains the addition involving the polar bodies $(u_{K_i}K_i)^\circ$.

Remark 4. The original proof of Theorem 1.5, as already noted, was heavy and complicated. Its main points were outlined in the note [M3]. After my talk on this proof, in February 1986 (at the Gromov Seminar, IHES), G. Pisier noted that the theorem is equivalent to Theorem 4.5, which logically preceded Theorem 1.5. He later gave his version of the proof of this theorem [P4] which is also extended and non-trivial (see much simplified new version [P2]).

5. Entropy Point of View

The inequalities for the volume of convex bodies, which we have studied in the previous sections, can be reformulated in the language of covering numbers $N(K_1, K_2)$ or entropy numbers which we introduce next. Let X and Y be the normed spaces and K_X, K_Y their unit balls. Then the entropy numbers (in Pietch terminology [Pi]) $e_k(u)$ of an operator $u : X \to Y$ are defined as follows

$$e_k(u) = \inf\{\varepsilon > 0 : N(uK_X, \varepsilon K_Y) \le 2^{k-1}\} .$$

Clearly, $\|u\| = e_1(u) > e_2(u) \ge \cdots$. Roughly speaking, the entropy numbers (as the function on k) are the inverse function for the logarithm of the covering numbers (as the function on ε). In our applications we mostly consider u as being the identity map $(\mathbb{R}^n, \|\cdot\|_X) \to (\mathbb{R}^n, \|\cdot\|_Y)$ and then we write $e_k(K_X, K_Y)$. Also in the typical case of $K_Y = D$ being the euclidean ball, we often use the simplified notation

$$e_k(K, D) = e_k(K) .$$

Remark 5.1. Theorem 4.5 is equivalent to the following statement:

(5.1) for every convex symmetric body $K \subset \mathbb{R}^n$ there exists an ellipsoid $\mathcal{E}$ such that $\mathrm{Vol}\,\mathcal{E} = \mathrm{Vol}\,K$ and $e_{\delta n}(K, \mathcal{E}) \le \lambda$ for some $\delta > 0$ and $\lambda > 1$.

Indeed, let (5.1) be known. Then $e_{\delta n}(K, \mathcal{E}) \le \lambda$ means $N(K, \lambda\mathcal{E}) \le 2^{\delta n}$. Using Lemma 4.1 we have

$$\mathrm{Vol}(K + P) \le N(K, \lambda\mathcal{E}) \cdot \mathrm{Vol}(\lambda\mathcal{E} + P) \le (\lambda \cdot 2^{\delta})^n \,\mathrm{Vol}(\mathcal{E} + P)$$

and, therefore, we have derived Theorem 4.5 with $C = \lambda \cdot 2^{\delta}$ (this direction was pointed out to me by G. Pisier - see Remark 4 of the previous section).

In the opposite direction, it is well known that

$$N(K, \lambda\mathcal{E}) < \frac{\mathrm{Vol}\,(K + \frac{\lambda}{2}\mathcal{E})}{\mathrm{Vol}\,\frac{\lambda}{2}\mathcal{E}} \tag{5.2}$$

(to prove the above, consider the maximal number M of disjoint balls $\frac{\lambda}{2}\mathcal{E}$ with centers in K; clearly $N(K, \lambda\mathcal{E}) < M$ and we derive the above inequality).

Using Theorem 4.5, we may continue, assuming $\mathcal{E} = \mathcal{E}_K$

$$N(K, \lambda\mathcal{E}) \le C^n \frac{\mathrm{Vol}\,(\mathcal{E} + \frac{\lambda}{2}\mathcal{E})}{\mathrm{Vol}\,\frac{\lambda}{2}\mathcal{E}} = \left[C\left(1 + \frac{2}{\lambda}\right)\right]^n .$$

Therefore, $e_{\delta n} \le \lambda$ for some $\delta \le \log\left[C\left(1 + \frac{2}{\lambda}\right)\right]$. □

Applying Remark 5.1 twice for the body K and its polar K° (in this case, using Remark 3 of section 4) we obtain

Proposition 5.2. *Let K be a centrally symmetric compact body in $\mathbb{R}^n$. There exists a euclidean structure $(\mathbb{R}^n, |\cdot|)$ such that for some universal number $\delta > 0$ (independent of n and K)*

$$e_{\delta n}(K) \cdot e_{\delta n}(K^\circ) \leq 1$$

where $e_k(K)$ is $e_k(K, D)$, $D = \{x \in \mathbb{R}^n \mid |x| \leq 1\}$ is the euclidean ball of the defined euclidean structure, and K° is the polar body with respect to D.

Note, that for $\delta < 1$, Proposition 5.2 gives, in a sense, the best possible estimate because:

For every $1 \leq k \leq n$

$$1 \lesssim e_k(K) \cdot e_k(K^\circ) .$$

Indeed, by the definition of the entropy numbers, $K \subset \bigcup_{i \geq 1}^{N} (x_i + e_k D)$ for $N \leq 2^k$; therefore, $\mathrm{Vol}\, K \leq 2^k \cdot e_k^n \,\mathrm{Vol}\, D$. So, by the inverse Santalo inequality [BM], there exists a universal $c > 0$ such that

$$c \leq e_k(K) \cdot e_k(K^\circ) .$$

We use Proposition 5.2 together with the following statement proved in the Appendix of [MP].

Proposition 5.3. *Let $v : X \to \ell_2^n$ be a linear isomorphism. There exists a universal number $\theta > 0$ such that if $e_k(v) \leq 1$ for some $k < \theta n$, then there exists a set $\mathcal{A} \subset G_{n,n-m}$ of subspaces of codimension $m \simeq k$ (i.e., $m = [ck]$ for a universal constant c) such that for every subspace $S \in \mathcal{A}$*

$$\|v|_S\| \lesssim \sqrt{n/k}$$

and $\mathcal{A}$ has an "almost full" measure; precisely, for μ be the probabilistic Haar measure on the Grassmann manifold $G_{n,n-m}$

$$\mu(\mathcal{A}) \geq 1 - e^{-ck} \tag{5.3}$$

(as usual, $c > 0$ is some universal constant).

We will use this Proposition in the following form (using different parameters):

Proposition 5.3$'$. *If $e_{\delta n}(v) \leq \lambda$ for some $\delta < \theta$, then there exists a set of almost full measure $\mathcal{A} \subset G_{n,n-m}$, $m \simeq \delta n$, such that for every subspace $S \in \mathcal{A}$*

$$\|v|_S\| \lesssim \lambda/\sqrt{\delta} \qquad \text{(i.e., } \frac{\sqrt{\delta}}{\lambda}|x| \lesssim \|x\| \quad \text{for every } x \in S) .$$

This proposition was formulated in [MP] in a slightly different form with $S(v) = \sup_{k\geq 1} \sqrt{k} e_k(v)$, substituting $\sqrt{n} e_k(v)$. However, the proof given in that Appendix corresponds exactly to Proposition 5.3 above. In that proof, we do not write the estimate (5.3) which can be taken from the proof of the Lemma from the same appendix.

We note in passing the following question parallel to Proposition 5.2.

Problem. Does there exist a universal constant C such that for any $n \in \mathbb{N}$ and any normed space $X = (\mathbb{R}^n, \|\cdot\|)$ there exists a linear isomorphism $v : X \to \ell_2^n$ (i.e., in our terminology, a euclidean structure) such that

$$S(v) \cdot S\big((v^{-1})^*\big) \leq Cn$$

where $S(v) = \sup_{k\geq 1} \sqrt{k} e_k(v)$?

Note a few simple observations involving entropy numbers.

Remark 5.4. a) Let $E \hookrightarrow X$ be a subspace of X; then $e_k(v|_E) < e_k(v)$ (i.e., $e_k(K \cap E) \leq e_k(K)$).

b) Define $q_E K \equiv \mathrm{Proj}_E K$ being the orthogonal projection of a body K onto a subspace E. Then $e_k(q_E K) \leq e_k(K)$.

We are now ready to put forward an entropy point of view on the quotient of a subspace Theorem 1.1.

Proposition 5.5. *Let $e_{\delta n}(K) \cdot e_{\delta n}(K^\circ) \leq \lambda$ for some $\delta < \theta < 1$ (θ is a universal constant). Then there exists a set $\mathcal{A} \subset G_{n,n-m}$, $m \simeq \delta n$, of an "almost full" measure (in the sense of Proposition 5.3) such that for every $E \in \mathcal{A}$ and some subspace $S \hookleftarrow E$ the body $\mathrm{Proj}_E(K \cap S)$ is $\sim \lambda/\delta$ close to a euclidean ball, i.e.,*

$$d_{q_E(K\cap S)} \lesssim \lambda/\delta$$

(recall, that the codimension of $q_E(K \cap S)$ is $\sim \delta n$).

In other words, if K is the unit ball of a space X, then there exists a subset of $G_{n,p}$ of a large measure (in the sense of the euclidean structure induced by D) of quotients of subspaces qsX such that

$$p = \dim qsX \geq (1 - c\delta)n \quad \text{and} \quad d_{qsX} \lesssim \lambda/\delta\ .$$

Proof: Let $v : X = (\mathbb{R}^n, \|\cdot\|_K) \to \ell_2^n = (\mathbb{R}^n, \|\cdot\|_D)$ be a proportion of the identity map such that $e_{\delta n}(v) = \lambda$ and $e_{\delta n}(K^\circ) \leq 1$. Then, by Proposition 5.3′, we find a subspace S, codim $S \lesssim \delta n$ (indeed, we have a set of large measure of such subspaces) and

$$\frac{\sqrt{\delta}}{\lambda}|x| \lesssim \|x\| \equiv \|v^{-1}x\|$$

for $x \in S$. Therefore, for the dual space S^*

$$\|(v^{-1})^* x\| \lesssim \frac{\lambda}{\sqrt{\delta}}|x|$$

for $x \in S^*$. The unit ball of S^* is $\mathrm{Proj}_S K^\circ (= q_S K^\circ)$ and, by Remark 5.4 $e_{\delta n}(q_S K^\circ) \leq 1$. Therefore, we may again apply Proposition 5.3′ for the space S^* and find a subspace $E \hookrightarrow S^*$, codim $E \lesssim \delta n$, such that

$$\sqrt{\delta}|x| \lesssim \|(v^{-1})^* x\| \lesssim \frac{\lambda}{\sqrt{\delta}}|x|$$

for any $x \in E$. So $d_{E^*} = d_E \lesssim \lambda/\delta$. Clearly, the unit ball of E^* is $q_E(K \cap S)$.

Unfortunately, at this stage, Proposition 5.5 does not yet imply Theorem 1.1 (even in a rough sense $d_{q_s X} \leq f(\lambda)$, i.e., ignoring how the function $f(\lambda)$ depends on λ) because we cannot apply Proposition 5.2; the number δ in this proposition can be larger than θ in Proposition 5.5. So, next, we will show that by a suitable increase of λ we can put δ in the inequality

$$e_{\delta n}(K, D) e_{\delta n}(K^\circ, D) \leq \lambda$$

below any given bound. Moreover, we are interested in the dependence $\delta(\lambda)$ in this inequality. A good dependence $\delta(\lambda)$ of λ is not achieved for the same D for all the range of λ, $\lambda \in (0, 1)$. However, for every λ_0 we construct an ellipsoid $\mathcal{E}_{\lambda_0}$ (which plays the role of D) which, for our purpose, also works in a large range of λ around λ_0.

Indeed, starting with K, in section 3, we have already constructed the ellipsoid which we are interested in. We go through the same iteration procedure as in section 3 to construct bodies K_1, K_2 and so on. However, we introduce the last step p differently. First, we recall the situation which we have after p steps of the iteration (see Proof of Theorem 1.4 in section 3). We derived body K_p and an ellipsoid $\mathcal{E}_p$ such that $K_p \subset C_p \mathcal{E}_p$, $c_p \lesssim a_p^2(\log^{(p)} n)$, $\mathrm{Vol}\, \mathcal{E}_p = \mathrm{Vol}\, K$ and

$$\mathrm{Vol}(K + \lambda \mathcal{E}_p) \leq e^{o_p(n)} \mathrm{Vol}(K_p + \lambda \mathcal{E}_p) \tag{5.4}$$

where $o_p(n) \lesssim n/a_p^2$. (Indeed, we have elaborated this with some remarks from the proof of Theorem 4.5 which we use for $P = \lambda \mathcal{E}_p$.) So, we may continue (5.4)

$$\mathrm{Vol}(K + \lambda \mathcal{E}_p) \leq e^{o_p(n)} \mathrm{Vol}(C_p \mathcal{E}_p + \lambda \mathcal{E}_p) = e^{o_p(n)} (C_p + \lambda)^n \mathrm{Vol}\, K\ .$$

Now, we can estimate $N(K,\lambda\mathcal{E}_p)$ by (5.2)

$$N(K,\lambda\mathcal{E}_p) \le \frac{\mathrm{Vol}\left(K+\frac{\lambda}{2}\mathcal{E}_p\right)}{\mathrm{Vol}\left(\frac{\lambda}{2}\mathcal{E}_p\right)} \le e^{o_p(n)}\left(1+\frac{2C_p}{\lambda}\right)^n \simeq 2^{n\left\{\log\left(1+\frac{2C_p}{\lambda}\right)+\frac{c}{a_p^2}\right\}}.$$

Therefore, the entropy number

$$e_{\delta_p(\lambda)n}(K,\mathcal{E}_p) \le \lambda$$

for some

$$\delta_p(\lambda) \lesssim \frac{1}{a_p^2} + \log\left(1+\frac{ca_p^2\log^{(p)} n}{\lambda}\right) \simeq \frac{1}{a_p^2} + \frac{a_p^2\log^{(p)} n}{\lambda} \simeq \frac{a_p^2\log^{(p)} n}{\lambda}$$

for $a_p^4\log^{(p)} n \gtrsim \lambda \gtrsim a_p^2\log^{(p)} n$. Note, that in section 3 we chose $a_i = \log^{(i)} n$. We are doing the same in this construction for $i \le p-1$. However, a choice of a_p and p will be defined later. At this stage, we note only that

$$1 < a_p < \log^{(p-1)} n .$$

Summing up, we consider the euclidean structure defined by the ellipsoid $\mathcal{E}_p$ (i.e., we take $D = \mathcal{E}_p$). Then the same inequalities are satisfies for the polar body K° (with respect to $\mathcal{E}_p = D$). We have already used this fact in sections 3 and 4. Now we discuss the choice of p. Let $T_p = [\log^{(p-1)} n]^4 \log^{(p)} n$. Given λ, we define p as the smallest number such that $\lambda \le T_p$. Then a number a_p is defined by $\lambda = a_p^4(\log^{(p)} n)^2$. So

$$\delta_p(\lambda) \lesssim \frac{1}{a_p^2\log^{(p)} n} = \frac{1}{\sqrt{\lambda}} .$$

Finally, we may choose $a_p = a$ in various ways so that $\lambda' = a^4(\log^{(p)} n)^2$ will range between $T_p \ge \lambda' \gtrsim \log T_p$ and for any such λ' we have

$$e_{\delta(\lambda')n}(K,\mathcal{E}_p) \le \lambda'$$

for some

$$\delta_p(\lambda') \lesssim \frac{1}{\sqrt{\lambda'}}$$

We sum up the described results in the following theorem.

Theorem 5.6. *Let $K \subset \mathbb{R}^n$ be a centrally symmetric convex body in $\mathbb{R}^n$. For any $\lambda > 1$ there exists a euclidean structure (with the unit euclidean ball $D = \mathcal{E}_\lambda(K)$) such that*

1) $\mathrm{Vol}\, K = \mathrm{Vol}\, D$

2) *the entropy numbers* $e_{cn/\sqrt{\lambda}}(K,D) \le \lambda$ *and* $e_{cn/\sqrt{\lambda}}(K^\circ,D) \le \lambda$ *where* K° *is the polar body of* K *with respect to* D, *(*$c > 0$ *is some numerical constant independent of* n, λ *or* K*),*

3) *there exists an interval* $I = [A, e^A]$ *with* $\lambda \in I$ *such that for any* $\lambda' \in I$

$$e_{cn/\sqrt{\lambda'}}(K,D) \le \lambda' \quad \text{and} \quad e_{cn/\sqrt{\lambda'}}(K^\circ,D) \le \lambda' .$$

The probabilistic version of Theorem 1.1 follows now by applying Theorem 5.6 to Proposition 5.5.

Theorem 5.7. *For any space* $X = (\mathbb{R}^n, \|\cdot\|)$ *and any* $1 > \varepsilon > 0$ *there exists a euclidean structure* $(\mathbb{R}^n, |\cdot|)$ *such that for some set of* k*-dimensional subspaces* $\mathcal{A} \subset G_{n,k}$ *of a large measure* $\mu(\mathcal{A}) > 1 - e^{-ck}$ *(*$c > 0$ *is a universal constant,* μ *is the probability measure on the Grassmann manifold* $G_{n,k}$ *rotation invariant with respect to the constructed euclidean structure* $(\mathbb{R}^n, |\cdot|)$*) where* $k = [(1-\varepsilon)n]$, *we have for any* $E \in \mathcal{A}$ *and some* m*-dimensional subspace* F, $F \hookleftarrow E$, $m \sim (n+k)/2$, *the quotient of the subspace* $(X \cap F)/E = q_E s_F X$ *is* C/ε^5*-isomorphic to the euclidean* k*-dimensional space. Moreover, if* K *is the unit ball of* X *and* D *is the unit euclidean ball of* $(\mathbb{R}^n, |\cdot|)$ *then*

$$d_{q_E(K \cap F)} \lesssim 1/\varepsilon^5 .$$

Proof: Use Proposition 5.5 in the situation achieved in Theorem 5.6. Note that $e_{cn/\sqrt{\lambda}}(K) \cdot e_{cn/\sqrt{\lambda}}(K^\circ) \le \lambda^2$. Therefore, $\dim qsX \ge (1 - c/\sqrt{\lambda})n$ and $d_{qsX} \lesssim \lambda^2 \cdot \sqrt{\lambda}$. So define $\varepsilon = c/\sqrt{\lambda}$.
□

Remark 5.8. By minor modifications of the proof of Theorem 5.6, we can show that *any* interval $I = [A, e^A]$ with $\lambda \in I$ may be considered in Theorem 5.6, 3), however, the ellipsoid $\mathcal{E}_{\lambda;A}(K)$ will depend also on which A is chosen. So, we may choose, for example, $I = [\lambda, e^\lambda]$ or $I = [\log\lambda, \lambda]$, but the constructed ellipsoids may differ in these two cases.

6. Inequalities Involving Mixed Volumes

In this section, we extend to the mixed volume case some of the volume inequalities which were previously discussed. Unfortunately, results here are still in a preliminary form. Recall first some well known notions. Let K be a convex compact in $\mathbb{R}^n$. We define

$$V_k(K) = V(\underbrace{K, \dots, K}_{k}, \underbrace{D, \dots, D}_{n-k})$$

the mixed volume (in the standard Minkowski sense) of k-times body K and $(n-k)$-times the euclidean ball $D \subset \mathbb{R}^n$. Then it is well known that

$$\frac{V_k(K)}{\operatorname{Vol} D} = \int_{\xi \in G_{n,k}} \frac{\operatorname{Vol}_k(\operatorname{Proj}_\xi K)}{\operatorname{Vol}_k D_k} d\mu(\xi) .$$

Here Vol_k denotes the k-dimensional volume of k-dimensional subspace $\xi \hookrightarrow \mathbb{R}^n$ (the euclidean structure in $\mathbb{R}^n$ induced by the euclidean unit ball D produces the k-dimensional volume on any k-dimensional subspace), D_k denotes a k-dimensional euclidean ball ($= \operatorname{Proj}_\xi D$) and $\operatorname{Proj}_\xi K$ means the orthogonal projection of K on ξ. We also keep the same letter μ to define the probabilistic rotation invariant measure on the Grassmann manifold $G_{n,k}$ (for any n and $k < n$).

The following generalized Brunn-Minkowski inequality is also well known (as a consequence, say, of Alexandroff-Fenchel inequalities): for every integer $k \le n$, any convex compacts K_1 and $K_2 \subset \mathbb{R}^n$,

$$\left[V_k(K_1 + K_2)\right]^{1/k} \ge \left[V_k(K_1)\right]^{1/k} + \left[V_k(K_2)\right]^{1/k} .$$

In this section, we discuss the reverse type inequality (up to a factor C^k) in the spirit of Theorem 1.5. Indeed, at the present level, the Local Theory of Normed Spaces is also involved in the formulation of a result. We refer the reader to the book [MSch] or the survey [P3] for notions and results, which will be used next. Some of them have already been defined in section 2.

Theorem 6.1. *For every symmetric convex body $K \subset \mathbb{R}^n$ and $T \subset \mathbb{R}^n$ and any integer k, $1 \le k \le n$.*

$$V_k(K+T)^{1/k} \le M_K \cdot M_{K^\circ} V_k(K)^{1/k} + M_T \cdot M_{T^\circ} V_k(T)^{1/k} .$$

Proof: This is an easy consequence from the so-called Alexandroff inequalities which state that for any $1 \le k \le \ell \le n$

$$\left(\frac{V_\ell(K)}{\operatorname{Vol} D}\right)^{1/\ell} < \left(\frac{V_k(K)}{\operatorname{Vol} D}\right)^{1/k} .$$

So

$$\frac{1}{M_K} \le \left(\frac{\operatorname{Vol} K}{\operatorname{Vol} D}\right)^{1/n} < \left(\frac{V_k(K)}{\operatorname{Vol} D}\right)^{1/k} \le \frac{V_1(K)}{\operatorname{Vol} D} = M_{K^\circ} \tag{6.1}$$

(the left side inequality is the immediate consequence of the Hölder inequality). Then

$$\left(\frac{V_k(K+T)}{\operatorname{Vol} D}\right)^{1/k} \le M_{(K+T)^\circ} = M_{K^\circ} + M_{T^\circ} \le$$

$$\le M_K M_{K^\circ} \left(\frac{V_k(K)}{\operatorname{Vol} D}\right)^{1/k} + M_T M_{T^\circ} \left(\frac{V_k(T)}{\operatorname{Vol} D}\right)^{1/k} . \qquad \square$$

Combining Theorem 6.1 with the known results in Local Theory, such as Theorem 2.1 and crucial resuts of Pisier [P1] on K-convexity, we derive the following consequences.

Corollary 6.2. *For every symmetric convex body $K \subset \mathbb{R}^n$ there exists a linear transformation $u_K : \mathbb{R}^n \to \mathbb{R}^n$, $\det u_K = 1$, such that for any two such bodies K and T and any $\lambda > 0$ and an integer k, $1 \le k \le n$,*

(a) $V_k(u_K K + \lambda u_T T)^{1/k} \le C \log 2n \big(V_k(u_K K)^{1/k} + \lambda V_k(u_T T)^{1/k}\big)$ for a universal constant $C \le e + 1$;

(b) for any integer m there exists a constant $C(m)$ depending only on m, such that if the normed spaces $X = (\mathbb{R}^n, \|\cdot\|_K)$ and $Y = (\mathbb{R}^n, \|\cdot\|_T)$ with the unit balls K and T respectively, do not contain 2-isomorphic copies of ℓ_1^m then

$$V_k(u_K K + \lambda u_T T)^{1/k} \le C(m)\big(V_k(u_K K)^{1/k} + \lambda V_k(u_T T)^{1/k}\big) .$$

Note also, that the polar bodies $(u_K K)^\circ$ and $(u_T T)^\circ$ satisfy the same inequalities a) and b).

The next statement demonstrates a convex surgery technique applied to the mixed volumes of a convex body. Note, that the only interesting case here is $k \lesssim o(n)$. We use the following lemma.

Lemma 6.3. *If $k < n/2$ then for a "random" k-dimensional subspace ξ and the orthogonal projection P_ξ onto ξ*

$$\frac{1}{M_K} D \subset C \cdot P_\xi K$$

where C is a numerical (universal) constant. The exact meaning of "random" is , in this Lemma, that the statement is true for any ξ from a set $\mathcal{A} \subset G_{n,k}$ with $\mu(\mathcal{A}) > 1 - \alpha^n$ for a small $\alpha \ll 1$ (α indeed depends on C and by increasing C we may put α below any fixed positive bound).

The lemma was first used in [M2] in the dual form. Its importance for the study of convex bodies in $\mathbb{R}^n$ was explained in [M6] where a number of applications was presented. It has been used regularly during the last three years. For the proof we refer the reader to the papers just mentioned, or to [Go] for a recent probabilistic approach. However, the estimate on $\mu(\mathcal{A})$ as it is written above can only be found in [M2].

The following statement is an easy consequence of Lemma 6.3.

Proposition 6.4. *For some universal constant $C \ge 1$ and $k \le n/2$*

$$C^k \cdot V_k(K) \ge V_k\left(\operatorname{Conv}\left(K \cup \frac{1}{M_K} D\right)\right) \ge V_k(K) .$$

The following discussion uses "cutting" operations.

Theorem 6.5. *For every symmetric convex body $K \subset \mathbb{R}^n$, any symmetric convex set $T \subset C_1 n^{10} D \subset \mathbb{R}^n$ and any integer k, $1 \le k < n$,*

$$\left(V_k(K+T)\right)^{1/k} \le C_2 \sqrt{\log n}\, M_K \cdot M_{K^\circ} \left(V_k(rD+T)\right)^{1/k}$$

where $r = (V_k(K)/\operatorname{Vol} D)^{1/k}$ (the "mixed" volume ratio of K) and C_i are numerical universal constants.

Remark. The bound $T \subset C_1 n^{10} D$ is clearly a technical condition. Of course, we could put n^{100} instead of n^{10} by changing the constant C_2. However, it has to be remembered that, usually, we have a distance $d(T, D)$ at most n, i.e., $T \subset C \cdot nD$.

Proof: Our purpose is to compare $V_k(K+T)$ and $V_k(K \cap M_{K^\circ} D + T)$. By the definition of V_k, we have to estimate $\operatorname{Vol}_k(P_\xi K + P_\xi T)$ where ξ is a k-dimensional subspace. Note that $(P_\xi K)^\circ = K^\circ \cap \xi$ and therefore $M_{(P_\xi K)^\circ} = M_{K^\circ \cap \xi}$. Also, clearly

$$\int_{\xi \in G_{n,k}} M_{K^\circ \cap \xi}\, d\mu(\xi) = M_{K^\circ} ,$$

but we have to estimate the deviation of the function $f(\xi) = M_{K^\circ \cap \xi}$ from its average M_{K°. This is done in the next lemma, which we present in a slightly more general form than we will employ, using a notion of type. Indeed, only the Corollary 6.7 is applied later and it does not contain the notion of type.

Lemma 6.6. *Let $X = (\mathbb{R}^n, \|\cdot\|)$ be a normed space equipped with a euclidean structure $(\mathbb{R}^n, |\cdot|)$ and let $b = \|Id : (\mathbb{R}^n, |\cdot|) \to X\|$, i.e., $\|x\| \le b|x|$ for any $x \in X$. Define $M = M_X$ and $M(\xi) = M_{X \cap \xi}$ for a k-dimensional subspace $\xi \in G_{n,k}$. Let $T_2(k)$ denote the type 2 constant of X, computed on k elements. Then for some universal constant $c > 0$*

$$\mu\{\xi \in G_{n,k} \mid |M(\xi) - M| > \delta M\} < e^{-c\frac{M^2 k}{b^2 T_2(k)^2} n\delta^2} \le e^{-c(M/b)^2 n\delta^2} .$$

Proof: We use the standard "concentration phenomenon" argument on $G_{n,k}$ (see [GrM] or [MSch], Chapter 6). Let ξ and $\eta \in G_{n,k}$; define a distance $\rho(\xi, \eta) = \inf\left\{ \sqrt{\sum_1^k |e_i - a_i|^2} \mid \{e_i\}_1^k \right.$ is an orthonormal basis in ξ and $\{a_i\}_1^k$ is an orthonormal basis on η; $|\cdot|$ is a euclidean norm $\left.\right\}$. Given $\rho(\xi, \eta) \le \varepsilon$ we have to estimate $|M(\xi) - M(\eta)|$. Let $g_i(\omega)$, $i = 1, \ldots, k$, be independent Gaussian standard variables, $\bar{e} = \{e_i\}_1^k$ - some orthonormal basis in ξ and $\bar{a} = \{a_i\}_1^k$ - some orthonormal basis in η. Then

$$\int_\Omega \Big\|\sum_1^k g_i(\omega)(e_i - a_i)\Big\| d\omega \ge \left| \int \Big\|\sum_1^k g_i(\omega) e_i\Big\| d\omega - \int \Big\|\sum_1^k g_i(\omega) a_i\Big\| d\omega \right| \simeq |\sqrt{k}(M(\xi) - M(\eta))| .$$

Using a type 2 condition and bases $\bar{e}$ and $\bar{a}$, which produce the distance $\rho(\xi,\eta)$, we obtain

$$|M(\xi) - M(\eta)| \lesssim \frac{1}{\sqrt{k}} T_2(k) \sqrt{\sum_1^k \|e_i - a_i\|^2} \leq \frac{1}{\sqrt{k}} T_2(k) \cdot b\rho(\xi,\eta) .$$

So, by a concentration property of Grassmann manifolds, we have

$$\mu\{\xi \in G_{n,k} \mid |M(\xi) - M| \geq \frac{1}{\sqrt{k}} T_2(k) b\varepsilon\} \leq e^{-c\varepsilon^2 n}$$

(indeed, this inequality is first established for the median of the function $M(\xi)$ substituting for M; however, as shown in [MSch], App. V, we may replace the median by the expectation M just by changing a universal constant $c > 0$). Now put $\delta = T_2(k)b\varepsilon/\sqrt{k}M$. To finish the second inequality of the lemma, note that $T_2(k) \leq \sqrt{k}$ always.

Corollary 6.7. *Under the notation of Lemma 6.4 and for any $t \geq 2$*

$$\mu\{\xi \in G_{n,k} \mid M(\xi) > tM\} < e^{-c(M/b)^2 nt^2} .$$

Remark. It is well known (see, e.g., [MSch]) that for any k-dimensional subspace ξ

$$M(\xi) \lesssim \sqrt{\frac{n}{k}} M .$$

We use Corollary 6.5 for X with the unit ball K°. Then M is M_{K° and $M(\xi) = M_{K^\circ \cap \xi}$.

Note that the same maximum principle for mixed volume holds as we had for volume in Lemma 4.1:

Lemma 6.8. *Let K, T and P be centrally symmetric compact bodies in $\mathbb{R}^n$. Then*

$$\max_x V_k[K \cap (x+T) + P] = \operatorname{Vol}_k(K \cap T + P) .$$

Proof is the same as for Lemma 4.4. Of course, we have to use the generalized Brunn-Minkowski inequality for mixed volumes instead of the usual Brunn-Minkowski inequality for volumes.

Using Lemma 6.8 with estimate (2.1) we obtain

Corollary 6.9. *For any integer $k \leq n$, any symmetric bodies K and $T \subset \mathbb{R}^n$ and any $\lambda_0 > 0$*

$$V_k(K+T) \leq N(K, \lambda_0 D) V_k(K \cap \lambda_0 D + T) \leq$$
$$\leq e^{cn(M_{K^\circ}/\lambda_0)^2} V_k(K \cap \lambda_0 D + T) .$$

Take $\lambda_0 \simeq \sqrt{n/k}M_{K^\circ}$ *and introduce the new body* $K_1 = K \cap \lambda_0 D$.

Now we start to estimate the integral

$$I = \int\limits_{\xi \in G_{n,k}} \mathrm{Vol}_k P_\xi(K_1 + T) d\mu(\xi) ,$$

using Corollary 4.2 together with estimate (2.1). Take $\lambda_1 \simeq a_1 M_{K^\circ}$ for $a_1 = \sqrt{\log n}$. Then

$$I \le \int\limits_\xi e^{ck(M_{K^\circ \cap \xi}/\lambda_1)^2} \mathrm{Vol}_k(P_\xi K_1 \cap \lambda_1 D + P_\xi T) d\mu(\xi) \tag{6.1}$$

$$\le \int\limits_{\xi \in G_{n,k}} \exp\left[c\frac{k}{\log n} \cdot \left(\frac{M_{K_1^\circ \cap \xi}}{M_{K^\circ}}\right)^2\right] \cdot \mathrm{Vol}_k(P_\xi K_1 \cap \lambda_1 D + P_\xi T) d\mu(\xi)$$

Use next Corollary 6.7 for body K_1°. The number "b" for the body K_1° is at most $\sqrt{n/k}M_{K^\circ}$ (by the choice of λ_0). Therefore, $\mu\{\xi \mid M_{K_1^\circ \cap \xi} > tM_{K_1^\circ}\} < \exp\left[-c(M_{K_1^\circ}/M_{K^\circ})^2 kt^2\right]$, and for $q = \log n$ the $L_q(\mu)$ norm

$$\left(\int\limits_\xi \left\{\exp\left[c\frac{k}{\log n}\left(\frac{M_{K_1^\circ} \cap \xi}{M_{K^\circ}}\right)^2\right]\right\}^q d\mu(\xi)\right)^{1/q} < C^k . \tag{6.2}$$

So, by applying the Hölder inequality with $p = 1 + \frac{1}{\log n}$ and $q = 1 + \log n$, we continue (6.1)

$$I \le C^k \left(\int \left[\mathrm{Vol}_k(P_\xi K_1 \cap \lambda_1 D + P_\xi T)\right]^{1+\frac{1}{\log n}} d\mu(\xi)\right)^{1/p} .$$

Now we use the condition on $T : T \subset cn^{10}D$. Then

$$\left[\mathrm{Vol}_k(P_\xi K_1 \cap \lambda_1 D + P_\xi T)\right]^{1/\log n} \le (Cn^{10})^{k/\log n} < C^k .$$

Finally, we have, for some universal constant C,

$$I \le C^k \int\limits_{\xi \in G_{n,k}} \mathrm{Vol}_k(P_\xi K_1 \cap \lambda_1 D - P_\xi T) d\mu(\xi) \le C^k \int \mathrm{Vol}_k\left(P_\xi(\lambda_1 D + T)\right) d\mu(\xi) =$$

$$= C^k V_k(\lambda_1 D + T) \cdot \frac{\mathrm{Vol}\, D}{\mathrm{Vol}_k D_k} .$$

This inequality and Corollary 6.9 imply

$$V_k(K + T) \le C^k V_k(\lambda_1 D + T)$$

with $\lambda_1 \simeq \sqrt{\log n}M_{K^\circ}$. Use (6.1) to finish the proof of Theorem 6.5. □

In the next statement, we improve Theorem 6.5 for a very broad family of convex bodies. In the excepted terminology of Local Theory, we restrict our attention to the so-called K-convex bodies. We have already used this restriction in Corollary 6.2.(b).

Theorem 6.10. *For any integer m and numbers $C_1 < \infty, c_2 > 0$ there exist constants $\alpha = \alpha(m) > 0$ depending only on m and $C = C(m, C_1, c_2)$ depending only on m, C_1 and c_2, such that, if a normed space $X = (\mathbb{R}^n, \|\cdot\|_K)$ with the unit ball K does not contain a 2-isomorphic copy of ℓ_1^m, then, for any symmetric convex set $T \subset C_1 n^{10} D \subset \mathbb{R}^n$ and any integer k, $c_2(\log n)^{1/\alpha} \leq k \leq n$,*

$$[V_k(K+T)]^{1/k} \leq C(m, C_1, c_2) M_K \cdot M_{K^\circ} [V_k(rD+T)]^{1/k}$$

where $r = (V_k(K)/\operatorname{Vol} D)^{1/k}$.

(So, under the above condition on K, we do not need the factor $\sqrt{\log n}$ in Theorem 6.5.)

Proof: We have to use Lemma 6.6 in its full strength. Then we have an estimate on a measure of "large" $M(\xi)$ which depends (in the exponent) on the factor $k/T_2(k)^2$. Then it is known (by Maurey and Pisier [MaPi]) that our condition on K implies an existence of $\alpha(= \alpha(m)) > 0$ such that

$$k/T_2(k)^2 \geq k^\alpha \ . \tag{6.3}$$

(By the way, this condition is symmetric for K and K°; we use it in the proof for the polar body K°.) Keeping this remark in mind, we repeat the proof of Theorem 6.5 but choose $\lambda_1 \simeq aM_{K^\circ}$ for $a \simeq$ const (this constant is dependent on C_1 and c_2). The formula (6.2) will be rewritten as follows:

$$\left(\int_\xi \exp\left[c\frac{k}{a^2} \log n \left(\frac{M_{K_1^\circ \cap \xi}}{M_{K^\circ}} \right)^2 \right] d\mu(\xi) \right)^{1/\log n} < C$$

because of (6.3) and the condition on k. The end of the proof is the same as for Theorem 6.5.□

We leave it to the reader to formulate a Corollary to Theorem 6.10 in the spirit of Corollary 6.2(b).

References

[BM] J. Bourgain and V.D. Milman, Sections euclidiennes et volume des corps symetriques convexes dans $\mathbb{R}^n$, C.R. Acad. Sc. Paris, t.300 Serie 1, N. 13 (1985) 435-438. (Also: New volume ratio properties for convex symmetric bodies in $\mathbb{R}^n$, Invent. Math. 88, 319-340 (1987).

[FT] T. Figiel, N. Tomzcak-Jaegermann, Projections onto Hilbertion subspaces of Banach spaces. Israel J. Math. 53, 155-171 (1979).

[Go] Y. Gordon, On Milman's inequality and random subspaces which escape through a mesh on $\mathbb{R}^n$, Gafa-Seminar 86-87, Lecture Notes in Mathematics, Springer-Verlag,

[KM] H. König and V.D. Milman, On the covering numbers of convex bodies, GAFA-Israel Seminar, Springer-Verlag Lecture Notes in Math., 1267, 82-95 (1987).

[L] D.R. Lewis, Ellipsoides defined by Banach ideal norms. Mathematica 26, 18-29 (1979).

[Ma] K. Mahler, Ein Übertragungsprincip für konvexe Körper. Casopis Pest. Mat. Fys. 68, 93-102 (1939).

[MaPi] B. Maurey and G. Pisier, Series de variables aléatoires vectorielles indépendantes et propriétés géométriques des espaces de Banach, Studia Math. 58, 45-90 (1976).

[M1] V.D. Milman, Almost Euclidean quotient spaces of subspaces of finite dimensional normed spaces, Proc. AMS 94, 445-449 (1985).

[M2] V.D. Milman, Geometrical inequalities and mixed volumes in Local Theory of Banach Spaces, Asterisque 131, 373-400 (1985).

[M3] V.D. Milman, An inverse form of the Brunn-Minkowski inequality with applications to Local Theory of Normed Spaces, C.R. Acad. Sc. Paris, t. 302, Ser. I, No. 1, 25-28 (1986).

[M4] V.D. Milman, Concentration phenomenon and linear structure of finite dimensional normed spaces, Proceedings ICM, Berkeley 1986.

[M5] V.D. Milman, Some remarks on Urysohn's inequality and volume ratio of cotype 2-spaces, GAFA-Israel Seminar Notes, Springer-Verlag Lecture Notes in Math. 1267, 75-81 (1987).

[M6] V.D. Milman, Random subspaces of proportional dimension of finite dimensional normed spaces: Approach through the isoperimetric inequality, Banach Spaces, Proc. Missouri Conference 1984, Springer Lecture Notes # 1166, 106-115.

[MSch] V.D. Milman and G. Schechtman, Asymptotic theory of finite dimensional normed spaces, Springer Lecture Notes #1200 (1986).

[P1] G. Pisier, Holomorphic semi-groups and the geometry of Banach spaces, Ann. of Math. 15, 375-392 (1982).

[P2] G. Pisier, Volume of convex bodies and geometry of Banach spaces (book in preparation).

[P3] G. Pisier, Probabilistic methods in the geometry of Banach spaces, Probability and Analysis, Springer-Verlag, Lecture Notes in Math. 1206, 167-241.

[P4] G. Pisier, A simpler proof of several results of Vitali Milman, Manuscript, March 86.

[Pi] A. Pietsch, Operators Ideals, North Holland, 1987.

[PT] A. Pajor, N. Tomczak-Jaegermann, Remarques sur les nombres d'entropie d'un opérateur et de son transposé. C.R. Acad. Sci., Paris

[RSh] C.A. Rogers, G.C. Shepard, The difference body of a convex body, Arch. Math. 8, 220-233 (1957).

[S] L.A. Santalo, Un invariante afin pasa los cuerpos convexos del espacio de n dimensiones, Portugal Math. 8 (1949) 155-161.

[Su] V.N. Sudakov, Gaussian random processes an measures of solid angles in Hilbert spaces, Soviet Math. Dokl. 12, 412-415 (1971).

DIMENSION, NON-LINEAR SPECTRA AND WIDTH

M. Gromov

IHES
France

Abstract

This talk presents a Morse-theoretic overview of some well known results and less known problems in spectral geometry and approximation theory.

§0 Motivation: Various Descriptions of the linear spectrum

The main object of the classical spectral theory is a *linear* operator Δ on a Hilbert space X. We assume Δ is a self-adjoint possibly unbounded (e.g., differential) operator and then consider the normalized *energy*

$$E(x) = \langle \Delta x, x\rangle / \langle x, x\rangle \ ,$$

which is defined for all non-zero x in the domain of Δ. Since the energy E is homogeneous,

$$E(ax) = E(x) \quad \text{for all} \quad a \in \mathbb{R}^{\times} \ ,$$

it defines a function on *the projective space* P consisting of the lines in the domain $X_\Delta \subset X$ of Δ,

$$P = P(X_\Delta) = X_\Delta \backslash \{0\} / \mathbb{R}^{\times} \ .$$

This function on P is also called the *energy* and denoted by $E : P \to \mathbb{R}$. Notice that since Δ is a *linear* operator the function E on P is *quadratic*, that is the ratio of two quadratic functions on the underlying linear space.

Now, the spectrum of Δ can be defined in terms of the energy E on P. To simplify the matter we assume below that Δ is a *positive* operator with *discrete* spectrum and then we have the following three ways to characterise the spectrum of Δ, that is the set of the eigenvalues $\lambda_0 \leq \lambda_1 \leq \ldots$ of Δ appearing with due multiplicities.

0.1 The Morse-theoretic description of the spectrum. Denote by $\Sigma = \Sigma(E) \subset P$ the *critical* set of E where the differential (or gradient) of E on P vanishes. A trivial (and well known) argument identifies Σ with the union of the 1-dimensional eigenspaces of Δ. In other words, if $x \in X$ is a non-zero vector from the line in X representing a point $p \in P$, then $p \in \Sigma$ if and only if $\Delta x = \lambda x$ for some real λ. Then

$$E(x) = \langle \Delta x, x \rangle / \langle x, x \rangle = \lambda$$

and so $E(p) = \lambda$ as well. It follows that the *spectrum of Δ equals the set of critical values of the energy $E : P \to \mathbb{R}$.* It is equally clear that the critical point of E corresponding to a *simple* eigenvalue λ_i is nondegenerate and has *Morse* index i. More generally, the multiplicity of λ_i equals $\dim \Sigma_i + 1$ for the component $\Sigma_i \subset \Sigma$ on which E equals λ_i, since Σ_i consists of the lines in the eigenspace $L_i \subset X$ associated to λ_i.

Notice that the definition of critical values of E is purely topological and applies to not necessarily quadratic functions on P. In fact, the set of critical values serves as a nice substitute for the spectrum for some non-quadratic energy functions (e.g., for the energy on the loop space in a compact *symmetric* space). But the essentially *local* nature of the critical values and non-stability of these under small perturbations (every point can be made critical by an arbitrary small C^0-perturbation of the energy function) forces us to look for another candidate for the non-linear spectrum.

0.2 Characterization of the spectrum by linear subspaces contained in the level sets $X_\lambda = \{x \in X \mid E(x) \leq \lambda\}$. Denote by $L_i \subset X$ the linear subspace spanned by the eigenvectors corresponding to the first $i+1$ eigenvalues $\lambda_0, \lambda_1, \ldots, \lambda_i$ of Δ and observe that

$$L_i \subset X_{\lambda_i} .$$

This signifies the inequality

$$\langle \Delta x, x \rangle \leq \lambda_i \langle x, x \rangle$$

for all $x \in L_i$, as $E(x) = \langle \Delta x, x \rangle / \langle x, x \rangle$.

The following *extremal property* of X_{λ_i} is more interesting. *If $\lambda < \lambda_i$, then X_λ contains no linear subspace of dimension $i+1$.* In fact, let $L \subset X$ be a linear subspace of dimension $i+1$. Then there is a non-zero vector $x \in L$ which is orthogonal to the subspace $L_{i-1} \subset X$. That is $\langle x, x_j \rangle = 0$ for the first i eigenvectors $x_0, \ldots, x_j, \ldots, x_{i-1}$ of Δ. It is trivial to prove that this x satisfies

$$\langle \Delta x, x \rangle \geq \lambda_i \langle x, x \rangle ,$$

which shows $L \not\subset X_\lambda$ for $\lambda < \lambda_i$.

Let us summarize this discussion in terms of the projective space $P = P(X_\Delta)$ and the energy of E on P.

The eigenvalue λ_i is the minimal number, such that the level

$$P_\lambda = E^{-1}[0,\lambda] = \{x \in P \mid E(x) \leq \lambda\} \subset P$$

contains a projective subspace of dimension i.

Remark. (a) The above characterization of λ_i is geometrical rather then topological as it makes use of the projective (linear) structure of P. On the other hand this projective definition of the spectrum obviously generalizes to *non-quadratic* energies E on P.

(b) An advantage of the projective definition of λ_i over the Morse-theoretic one (see 0.1) is the stability under small perturbations of the energy. Besides, the above existence proof of a "λ_i-hot" vector x in an arbitrary subspace $L \subset X$ of (asymptotically large) dimension $i+1$ gives a glimpse of general methods used for obtaining *lower bounds* for λ_i.

(c) An interesting generalization of the projective view on λ_i consists in replacing P by another geometrically signficant (homogeneous) space with a distinguished class of subspaces. The most obvious candidate for such a space is the Grassmann manifold $G = G_k(X)$ of the k-dimensional subspaces on X. Distinguished subspaces in G are Grassman manifolds $G_k(L) \subset G = G_k(X)$ for all linear subspaces $L \subset X$. (If $k = 1$, then $G = P$.) Now "the lower bound for λ_i" (see the above (b)) takes the following shape: any linear subspace $L \subset X$ contains "an interestingly hot" k-dimensional subspace $K \subset L$, where K becomes hotter and hotter as $\dim L \to \infty$ for $k = \dim K$ being kept fixed. (Compare Dvoretzky's theorem discussed in 1.2.)

0.3 Topological characterization of the eigenlevels $P_{\lambda_i} \subset P$. If we denote $\operatorname{pro} P_\lambda$ the maximal dimension of projective subspaces contained in P_λ, then we can say that the spectrum points λ_i are exactly those (see 0.2) where the function $\operatorname{pro} P_\lambda$ is *strictly* increasing in λ. In fact if λ_i is an eigenvalue of multiplicity m_i, then $\operatorname{pro} P_\lambda$ jumps up at λ_i by m_i.

Now we want to replace $\operatorname{pro} P_\lambda$ by a purely topological invariant of P_λ.

0.3.A Essential dimension. Consider a subset A in a topological space P and define the *essential dimension of A in P*,

$$\operatorname{ess} A = \operatorname{ess}_P A$$

as the smallest integer i, such that A is contractible in P onto an i-dimensional subset $A' \subset P$. This means there exists a continuous map (homotopy) $h : A : [0,1] \to P$, such that h on A at

$t = 0$ is the identity map,

$$h \mid A \times 0 : A \underset{\mathrm{Id}}{\subset} P$$

and such that

$$\dim h(A \times 1) \leq i ,$$

that is the image $h(A \times 1) \subset P$ admits arbitrarily fine coverings by open subsets where no $i+2$ among these subsets intersect.

0.3.B Basic example. If P is a projective space and A is a *projective* subspace, then

$$\operatorname{ess} A = \dim A . \qquad (*)$$

Notice that the inequality $\operatorname{ess} A \leq \dim A$ is trivial while the opposite inequality $\operatorname{ess} A \geq \dim A$ amounts to the following (simple but not totally trivial).

0.3.C Topological fact. *The dimension of a projective subspace $A \subset P$ cannot be decreased by a homotopy of A in P.* (See 4.1 for the proof and further discussion.)

0.3.D. Now we return to our positive quadratic energy function E on P and observe that the level $P_\lambda = \{x \in P \mid E(x) \leq \lambda\}$ can be contracted in P onto the projective subspace corresponding to the linear span of the eigenvectors belonging to the eigenvalues $\lambda_i \leq \lambda$. (This is more or less obvious.) This property combined with 0.3.C and the discussion in 0.2 implies that

$$\operatorname{ess} P_\lambda = \operatorname{pro} P_\lambda$$

for all λ. Therefore the definition of λ_i for *quadratic functions* on P can be formulated purely topologically, *the eigenvalue λ_i is the minimal number λ, such that the level $P_\lambda \subset P$ has* $\operatorname{ess} P_\lambda \geq i$, *which means P_λ cannot be contracted onto an $(i-1)$-dimensional subset in P.*

0.3.D Remarks. (a) The notion of ess makes sense for subsets in an *arbitrary topological space Q* and therefore one can speak of the ess-spectrum of an energy E on Q.

(b) If an energy E on Q is amenable to Morse theory, then the number $M(\lambda)$ of *λ-cold eigenpoints* of E, that are critical points q of E where $E(q) \leq \lambda$, can be bounded from below in terms of the ess-spectrum by

$$M(\lambda) > N(\lambda) = \operatorname{ess} E^{-1}[0, \lambda] .$$

(See $[\mathrm{Gr}]_1$ for another estimate of this nature for spaces of closed curves in Riemannian manifolds.)

0.4 Definitions of "dim"-spectrum for any "dimension". Let "dim" be a monotone increasing function on subsets A of a given space P, that is

$$A_1 \subset A_2 \Longrightarrow \text{"dim"} A_1 \leq \text{"dim"} A_2 \ .$$

If such a "dimension" is originally defined only on a certain class of admissible subsets, we agree to extend "dim" to all subsets A in P by taking all admissible subsets $A' \subset A$ and by setting

$$\text{"dim"} A = \sup_{A'} \text{"dim"} A' \ .$$

For example, the ordinary dimension on linear (or projective) subspaces extends in this way to all subsets of a linear (projective) space.

Now, with a given "dim" we define the "dim"-*spectrum* $\{\lambda_i\}$ of an energy $E : P \to [0, \infty]$, *as follows*, λ_i *is the upper bound of those* $\lambda \in \mathbb{R}$, *for which the level* $E^{-1}[0, \lambda]$ *has "dim"* $< i$.

In "physical" terms, every $A \subset P$ with "dim"$A \geq i$ contains a λ-hot point ($a \in A$, where $E(a) \geq \lambda$) for every $\lambda \leq \lambda_i$ and λ_i is the maximal number with this property.

The spectrum $\{\lambda_i\}$ can be more conveniently defined via the *spectral function* which, roughly speaking, counts the number of eigenvalues (or rather, of energy levels) of E below λ for all $\lambda \geq 0$. More precisely, this number $N(\lambda)$ is defined by

$$N(\lambda) = \text{"dim"} E^{-1}[o, \lambda] \ .$$

0.4.A Remarks on the range of E. (a) We allow infinite values for the energy in order not to bother with the domain of definition of E (and Δ as in §0.1). Namely, if E is originally defined on a dense subset $P_0 \subset P$ we extend E to P by

$$E(p) = \limsup_{u \to p} E \mid U \cap P_0$$

over a fundamental system of neighbourhoods U of p.

(b) There is no reason to restrict oneself to $[0, \infty]$-valued energies. In fact, for an arbitrary map $E : P \to T$ one can define the spectral function on the subsets $S \subset T$ by

$$N(S) = \text{"dim"} E^{-1}(S) \ .$$

(According to the physical terminology such an E should be called *observable*. The standard example of this is the position $P \to \mathbb{R}^3$ of a particle in $\mathbb{R}^3$.)

Example. Let $\|x\|_1, \ldots, \|x\|_m$ be norms on a linear space X. These naturally define a map E of the projective space $P = P(X)$ to the $(m-1)$-simplex $\Delta^m = I\!R_+^m / I\!R_+^\times$. A typical case of interest is $\|x\|_i = \|D_i x\|_{L_{p_i}}$ for some differential operators D_i on a function space X.

0.4.B Dimension-like properties of pro and ess. Let us axiomatize certain common features of the "dimensions" pro and ess by calling a function "dim" on subsets in a projective space P *dimension-like* if it has the following six properties.

(i) INTEGRALITY AND POSITIVITY. If $A \subset P$ is a non-empty subset, then "dim"A may assume values $0, 1, 2, \ldots, \infty$. If A is empty then "dim"$= -\infty$.

(ii) MONOTONICITY. If $A \subset B$ then "dim A"$\leq$ "dim"B for all A and B in P.

(iii) PROJECTIVE INVARIANCE. If $f : P \to Q$ is a projective embedding between projective spaces, then

$$\text{"dim"} f(A) - \text{"dim"} A$$

for all $A \subset P$.

(iv) INTERSECTION PROPERTY. *If $P' \subset P$ is a projective subspace of codimension k, then*

$$\dim A \cap P' \geq \dim A - k$$

for all $A \subset P$.

(v) NORMALIZATION PROPERTY. If A is a *projective* subspace in P then "dim A" equals the ordinary dimension $\dim A$.

(vi) THE $*$-ADDITIVITY. Let $A_1 * A_2 \subset P$ denote the union of the projective lines meeting given subsets A_1 and A_2 in P. Then

$$\text{"dim"} A_1 * A_2 = \text{"dim"} A_1 + \text{"dim"} A_2 + 1 \ ,$$

provided A_1 and A_2 are *projectively disjoint*. This means the *projective spans* PA_1 and PA_2 do not intersect, where the *projective span* PA indicates the minimal projective subspace in P containing A. (Notice that this additivity implies the above normalization property, as $P^{m+n+1} = P^m * P^n$.)

Remark. It is obvious that pro satisfies (i)-(vi) and that ess satisfies (i) and (ii). The properties (iii)-(vi) for ess follow from

0.4.B$_1$ Subadditivity of ess. The following property makes the "dimension" ess especially useful,

$$\text{ess}\, A \cup B \leq \text{ess}\, A + \text{ess}\, B + 1$$

for all subsets A and B in P. See 4.1 for the proof.

0.5 Codimension and width. Define the projective *codimension* copro A for A in P as the minimum of the codimensions of projective subspaces P' contained in P. Then define the *coprojective dimension* by

$$\mathrm{pro}^{\perp} A = \mathrm{copro}\, P \backslash A \ .$$

Observe that $\mathrm{pro}^{\perp}$ satisfies the "dimension" properties (i)-(vi) in 0.4.B. In fact (essentially because of (iv)) this $\mathrm{pro}^{\perp}$ is the *maximal* set function on P satisfying (i)-(vi). (Notice that pro is the minimal such function.)

0.5.A Definition of i-width. Let B be a subset in a Banach space X and define the *width function* of B on the dual space X' by

$$\mathrm{Wid}(B, y) = \sup_B y - \inf_B y$$

for all linear functions y on X. Then define *the i-width* of B by

$$\mathrm{Wid}_i\, B = (\lambda'_i)^{-1} ,$$

where λ'_i is the i-th $\mathrm{pro}^{\perp}$-eigenvalue of the energy

$$E' = \| \quad \|' / \mathrm{Wid}(B, \quad) : P(X') \longrightarrow [0, \infty] \ .$$

For example,

$$\mathrm{Wid}_0\, B = (\min E)^{-1} = \max \mathrm{Wid}(B, \quad) / \| \quad \|' = \mathrm{Diam}\, B \ .$$

In the special case, where B is a *centrally symmetric* subset in X our definition is equivalent to the usual one,

$\mathrm{Wid}_i\, B$ *equals the lower bound of those $\delta > 0$ for which there exists an i-dimensional linear subspace L in X whose $(\delta/2)$-neighbourhood contains B, that is*

$$\mathrm{dist}(b, L) \leq \delta/2$$

for all $b \in B$.

0.5.B Coprojective dimension and width. Recall the *duality correspondence* D which maps subsets $Y \subset X'$ to those in X by

$$D(Y) = \bigcup_{y \in Y} D(y) \ ,$$

for

$$D(y) = \{x \in X \mid |y(x)| = \|y\|'\|x\|\}\ .$$

We use the same notation D for the associated correspondence on the projective spaces, $P(X') \rightsquigarrow P(X)$, and call a subset $Q \subset P(X)$ an *i-coplane* if it is the D-image of an i-codimensional projective subspace in $P(X')$. Then we define *copro*$^{\perp} A$ for all $A \subset P(X)$ as the maximal i such that the complement $P(X)\backslash A$ contains no i-coplaine. In other words $\text{copro}^{\perp} A \leq i \iff A$ meets every i-coplane in $P(X)$. One easily sees with Bezout's theorem (compare §4) that

$$\text{ess}\, A \cap Q \geq \text{ess}\, A - i$$

for all $A \subset P(X)$ and all i-coplanes Q. In particular, if ess $A \geq i$, then A meets every i-coplane in $P(X)$, which is equivalent to

$$\text{copro}^{\perp} A \geq \text{ess}\, A$$

for all A. It follows that

$$\text{copro}^{\perp} A \geq \text{pro}\, A. \tag{$*$}$$

Notice that $(*)$ is a reformulation of the following

Tichomirov Ball Theorem. *Let $B^{i+1}(\varepsilon) \subset X$ be the ε-ball in some linear $(i+1)$-dimensional subspace of X and let L be a linear i-dimensional subspace in X. Then there exists a point $b \in B$, such that* $\text{dist}(b, L) = \varepsilon$.

In fact, the projectivization of the subset L^* of non-zero vectors $x \in X$ for which

$$\text{dist}(x, L) = \|x\|\ ,$$

is an i-coplane in $P(X)$, and every i-coplane comes from some L. Now, both $(*)$ and the ball theorem claim that L^* meets every $(i+1)$-dimensional linear subspace in X.

Coming back to the width of B, where B is the unit ball of some (semi)norm $\| \ \|_0$ on X, we see that

$$\text{Wid}_i\, B = 2(\lambda_i^{\perp})^{-1}$$

for the copro$^{\perp}$-spectrum $\{\lambda_i^{\perp}\}$ of $E = \| \ \|/\| \ \|_0$ and the above discussion relates these $\lambda_i^{\perp}$ to the ess and pro$^{\perp}$-spectra by the inequalities

$$\lambda_i^{\perp} < \lambda_i^{\text{ess}} \leq \lambda_i^{\text{pro}} \tag{$**$}$$

Remark. The number $(\lambda_i^{\text{pro}})^{-1}$ is called in [I-T] *the Bernstein i-width* of the unit ball of $\| \ \|_0$ in $(X, \| \ \|)$.

0.6 Complementary dimensions and $\{\lambda_{ij}\}$. Let d be a "dimension" function on subsets $A \subset P$ and take $i = 0, 1, \ldots$. Represent A as the difference of subsets, $A = B \backslash C$, and let

$$d^i A = \sup_{B,C}(i - dC + 1)$$

over all B and C, where $dB - i$. If d is subadditive (as ess, see 0.4.B_1). That is, if

$$dB < dA + dC \cdot 1 ,$$

then $d^i \leq d$ (and usually $d^i A = dA$, for $dA \leq i$) but in general d^i can be greater than d.

Next, for a given energy we define λ_{ij} for all $j \leq i$ as the $(i-j)$-th d^i-eigenvalue of E. In other words λ_{ij} is the upper bound of those λ for which every i-dimensional subset $B \subset P$ contains a λ-hot subset $C \subset B$ of dimension $> j$, where "λ-hot" signifies $E \mid B \geq \lambda$.

0.7 Generalized dimension. There are many interesting situations, where the ordinary (pro or ess) "dimension" of the levels of E is infinite, but there is some additional structure which allows a "renormalization". Here are two examples.

(a) Suppose E is a perturbation of E_0 for

$$E_0 = E_0(x) = \langle \Delta_0 x, x\rangle / \langle x, x\rangle ,$$

where Δ_0 is a selfadjoint operator with discrete spectrum which is *not assumed* positive anymore. If Δ_0 has infinitely many negative eigenvalues (e.g., Δ is the Dirac operator), then $\operatorname{pro} E^{-1}(-\infty, \lambda] = \infty$ for all λ. Yet one can define a *finite difference*

$$\operatorname{pro} E^{-1}(-\infty, \lambda] - \operatorname{pro} E^{-1}(-\infty, \lambda']$$

(representing the number of eigenvalues between λ and λ') as the index of an appropriate *Fredholm correspondence* between maximal linear subspaces in $E^{-1}(-\infty, \lambda)$ and $E^{-1}(-\infty, \lambda')$. This kind of situation arises, for example, in the *symplectic Morse theory*, where E is a perturbation of the action (see [Z], [Fl]) and also in the recent unpublished work by Floer on 3-dimensional gauge theory.

(b) VON-NEUMANN DIMENSION. This is defined, for example, on Γ-invariant linear subspaces of a Hilbert space X, where Γ is a given subgroup of unitary operators acting on X. The classical spectral theory does generalize to the Von-Neumann (algebras) framework but one does not know yet if there are suitable delinearization and de-Hilbertization of this theory.

§1 The spectrum of the ratio (L_p-norm)/(L_q-norm) and the concentration phenomenon for measurable functions

Consider the measure space (V, μ) and let

$$E = E_{p/q}(x) = \|x\|_p / \|x\|_q ,$$

where $\|x\|_p$ is ordinary L_p-norm on functions x on V,

$$\|x\|_p = \Big(\int_V |x|^p \Big)^{1/p} ,$$

and where $1 \leq q < p \leq \infty$. It is well known that every "sufficiently large" space L of functions on V contains a function x "concentrated near a single point" in V, where the concentration is measured by the energy $E(x)$. We shall prove in this section the simplest (and the oldest) result of this kind, and refer to [Pi] for deeper theorems.

We assume below that (V, μ) is a probability space, that is $\mu(V) = 1$. Then we define the *projective eigenvalue* $\lambda_i = \lambda_i(L_p/L_q)$ of $E = E_{p/q}$ as the minimal λ, such that $\operatorname{pro} P_\lambda \geq i$ (compare 0.3). Notice that here the inequality $\operatorname{pro} P_\lambda \geq i$ is equivalent to the following property: there exists on $(i+1)$-dimensional linear space L' of L_p-functions on V, such that $\|x\|_p \leq \lambda \|x\|_q$ for all $x \in L$. Observe that $1 - \lambda_0 < \lambda_1 \leq \ldots \leq \lambda_i \leq \ldots$ and let $\lambda_\infty = \lim_{i \to \infty} \lambda_i$.

1.1 Theorem. *The number $\lambda_\infty = \lambda_\infty(L_p/L_q)$ is bounded from below by*

$$\lambda_\infty \geq \gamma_\infty(p,q) = \pi^{\frac{1}{2q} - \frac{1}{2p}} \left(\Gamma\left(\frac{p+1}{2}\right)\right)^{\frac{1}{p}} \Big/ \left(\Gamma\left(\frac{q+1}{2}\right)\right)^{\frac{1}{q}} , \qquad (*)$$

for the Euler Γ-function. Furthermore, if the measure μ is continuous (i.e., without atoms) then also the opposite inequality holds true,

$$\lambda_\infty(L_p/L_q) \leq \gamma_\infty(p,q) , \qquad (**)$$

and thus $\lambda_\infty = \gamma_\infty$.

Proof: For a finite dimensional linear space L of functions of V we consider its dual L' and interpret functions $\ell \in L$ on V as linear functions on L'.

For a measure ν on L' we denote by $I_p(\ell, \nu)$ the integral

$$I_p(\ell, \nu) = \int_{L'} |\ell|^p d\nu$$

for all $\ell \in L$ and write

$$E_{p/q}(\ell,\nu) = I_p^{\frac{1}{p}}(\ell,\nu)/I_q^{\frac{1}{q}}(\ell,\nu) .$$

for $\ell \in L\backslash\{0\}$.

Then we observe that (almost) every point $v \in V$ defines a linear function ℓ' on L that is $\ell'(\ell) = \ell(v)$ for all $\ell \in L$. This gives us a canonical map $V \to L'$, such that every function $\ell \in L$ on V "extends" to a linear function on L'. We denote by μ' the probability measure on L' which is the push-forward of μ under this map and observe that the L_p-norms in L are recaptured by μ. Namely

$$\int_V |\ell(v)|^p d\mu = I_p(\ell,\mu')$$

for all $\ell \in L$ and all p and accordingly

$$E_{p/q}(\ell) = E_{p/q}(\ell,\mu') .$$

If the measure μ on V is continuous, then obviously, for every i and every probability measure ν on $\mathbb{R}^{i+1}$ there exists an $(i+1)$-dimensional space L of functions on V such that the measure μ' on L' is linearly isomorphic to ν. That is μ' goes to ν by some linear isomorphism between L' and $\mathbb{R}^{i+1}$. In particular, such an L exists for the *normalized Gauss measure*

$$d\nu = dt_0 \dots dt_i \exp \sum_{j=0}^{i} t_j^2 \Big/ \pi^{\frac{i+1}{2}} .$$

A straight forward computation shows for this ν that

$$E_{p/q}(\ell,\nu) = \gamma_\infty(p,q)$$

for all $i = 0,1,\dots$, $1 \le q < p \le \infty$, and all $\ell \in L\backslash\{0\}$. Since $E_{p/q}(\ell) - E_{p/q}(\ell,\nu)$ for $\nu = \mu'$, we obtain with the definition of λ_i the inequality

$$\lambda_i(L_p/L_q) \le \gamma_\infty(p,q) \quad \text{for} \quad 1 = 0,1,\dots,$$

which is equivalent to inequality $(**)$ of the theorem.

Now we turn to the proof of $(*)$ and start with the case where either p or q equals two and where we shall give a *sharp* bound for *each* λ_i. To do this we need the normalized measure ν_ρ on the sphere $S_\rho^i \subset \mathbb{R}^{i+1}$ of radius ρ. In other words ν_ρ is the probability measure on $\mathbb{R}^{i+1}$

which is invariant under the orthogonal group $O(i+1)$ and has support S^i_ρ. The $O(i+1)$-invariance of μ_ρ implies that $E_{p/q}(\ell, \nu_\rho)$ is constant in $\rho > 0$ and in ℓ for all non-zero linear functions ℓ on $\mathbb{R}^{i+1}$, which allows us to define

$$\gamma_i(p,q) = E_{p/q}(\ell, \nu_\rho) .$$

This agrees with our γ_∞ defined earlier as $\gamma_i(p,q) \to \gamma_\infty(;,q)$ for $i \to \infty$ by a straightforward computation.

Now, observe that the proof of $(**)$ also yields the following

1.1.A Trivial Proposition. *If the measure μ is continuous then*

$$\gamma_i(L_p/L_q) \leq \gamma_i(p,q) \tag{++}$$

for all $i = 0, 1, \ldots,$ and $1 \leq q < p \leq \infty$.

Notice that $(++)$ is stronger than $(**)$ as $\gamma_i < \gamma_\infty$ for $i < \infty$.

A more interesting fact is that $(++)$ is sharp if either p or q equals two.

1.1.B Theorem. *If p or q equals two then*

$$\lambda_i(L_p/L_q) \geq \gamma_i(p,q) , \tag{+}$$

for all $i = 0, 1, \ldots$.

Proof: Let L be an arbitrary $(i+1)$-dimensional linear space of functions on V. To prove $(+)$ we must show that

$$E(L) \overset{\text{def}}{=} \sup_{\ell \in L \setminus \{0\}} E_{p/q}(\ell) \geq \gamma_i(p,q) .$$

First we recall (L', μ') and observe that

$$E(L) = \sup_{\ell \in L \setminus \{0\}} E_{p/q}(\ell, \mu') .$$

Then we invoke the group G of linear isometries of L with the L_2-norm (induced from $L_2(V,\nu) \supset L$) and consider the natural action of G on L' and on measures on L'. Notice that the dual L_2-norm on L' turns L' into a Euclidean space and G becomes the orthogonal group $O(i+1)$ acting on $\mathbb{R}^{i+1} = L'$ in the usual way. Then we average μ' over G and set

$$\overline{\mu}' = \int_G g\mu' dg$$

for the normalized Haar measure dg on G. Notice, that $\overline{\mu}'$ is a $O(i+1)$-invariant measure on $\mathbb{R}^{i+1} = L'$ and so the energy $E_{p/q}(\ell, \overline{\mu}')$ is independent on ℓ for all $\ell \in L \setminus \{0\}$.

1.1.B$_1$ Basic Lemma. *If p or q equals 2, then*

$$E(L) \geq E_{p/q}(\ell, \overline{\mu}') \tag{$*$}$$

for $\ell \in L \backslash \{0\}$.

Proof: Recall that $E_{p/q}$ is the ratio

$$E_{p/q}(\ell, \nu) = I_p^{1/p}(\ell, \nu) / I_q^{1/q}(\ell, \nu)$$

for

$$I_p(\ell, \nu) - \int_{L'} |\ell|^p d\nu .$$

To be specific let $p = 2$. Then the integral $I_p(\ell, \nu')$ is invariant under the action of G on μ' that is $I_p(\ell, g\mu')$ is constant in g as follows from the definition of G. Thus

$$E_{p/q}(\ell, g\mu') - C I_q^\alpha(\ell, g\mu')$$

for $\alpha = -\frac{1}{q}$ and some $C > 0$. This implies that

$$\sup_{g \in G} E_{p/q}(\ell, g\mu') \geq C \overline{I}_q^\alpha ,$$

where

$$\overline{I}_q = \int_G I_q(\ell, g\mu') dg .$$

Now, by the linearity of $I_q(\ell, \nu)$ in ν,

$$\overline{I}_q \quad I_q(\ell, \overline{\mu}')$$

and by the transitivity of G on the sphere $S^i \subset \mathbb{R}^{i+1} = L'$,

$$E(L) - \sup_{g \subset G} E_{p/q}(\ell, g\mu') .$$

This all together yields $(*)$ for $p = 2$ and the same argument works for $q = 2$.

Now, the proof of $(+)$ follows from $(*)$ and the following simple lemma applied to the measure $\nu = \overline{\mu}'$,

1.1.B$_2$. *Let ν be a rotationally invariant (i.e., $O(i+1)$-invariant) probability measure in $\mathbb{R}^{i+1}$. Then*

$$E_{p/q}(\ell, \nu) \geq E_{p/q}(\ell, \nu_p) = \gamma_i(p, q)$$

for all non-zero linear functions ℓ on $\mathbb{R}^{i+1}$, all $\rho > 0$ and all $1 \le q < p \le \infty$.

Proof: We shall need the following trivial

1.1.B$_2'$ Calculus lemma. *Let $A_1(t) = a_1t + b_1$ and $A_2(t) = a_2t + b_2$ be linear functions in t whose derivatives A_i' are non-zero of same sign, that is $a_1a_2 > 0$, and let $A_1(t_0)$ and $A_2(t_0)$ be positive at some point t_0. If $0 \le q < p < \infty$, then t_0 is not a local minimum point of the ratio $A_1^{\frac{1}{p}}/A_2^{\frac{1}{q}}$.*

We are going to apply this lemma to $E_{p/q} = I_p^{\frac{1}{p}}(\nu)/I_q^{\frac{1}{q}}(\nu)$ keeping in mind that I_p and I_q are linear in ν. We observe that every extremal point ν in the space of $O(i+1)$-invariant measures on $\mathbb{R}^{i+1}$ is a measure supported on a single sphere $S_\rho^i \subset \mathbb{R}^{i+1}$ for some $\rho > 0$, that is $\nu = \nu_\rho$. We also notice that the derivatives in ρ

$$I_p'(\nu_\rho) \quad \text{and} \quad I_q'(\nu_\rho)$$

are strictly positive. Now, 1.1.B$_2'$ shows that $E_{p/q}$ has no local minimum point apart from $\{\nu_\rho\}$ and so by an obvious compactness argument $E_{p/q}$ assumes the minimum exactly on the set $\{\nu_\rho\}_{\rho>0}$. Q.E.D.

1.1.B$_3$ Example. The best known and most useful case of Theorem 1.1.B is that where $p = \infty$ and $q = 2$. In this case $\gamma_i = \sqrt{i+1}$ and so 1.1.B amounts to the following property.

Let L be an $(i+1)$-dimensional linear space of functions on a probability space (V, μ). Then there exists a non-zero $\ell \in L$, such that

$$\sup_{v \in V} |\ell(v)| \ge \sqrt{i+1}\Big(\int_v |\ell(v)|^2 d\mu\Big)^{\frac{1}{2}} \tag{+}$$

Besides the case where $(V, \mu) = (\mathbb{R}^{i+1}, \mu_\rho)$ the equality is achieved for the finite measure space V consisting of $i+1$ equal atomes. This suggests that the averaging is not indispensible for the proof and the following (standard) argument gives a confirmation.

Let $\ell_0, \dots, \ell_i$ be an L_2-orthonormal basis in L. Then every L^2-unit vector $\ell \in L$ is a linear combination

$$\ell = \sum_{i=0}^{i} a_i \ell_i$$

for $\sum a_i^2 = 1$. Therefore the inequality $|\ell(v)| \le \lambda(v)$ for a given $v \in V$ and all unit vectors $\ell \in L$ is equivalent to the inequality

$$\sum_{i=1}^{i} \ell_i^2(v) \le \lambda^2(v) .$$

Hence,

$$\int_V \lambda^2(v) \geq \int_V \sum \ell_i^2(v) = i + 1 ,$$

which implies the required inequality

$$\sup_{v \in V} |\lambda(v)| \geq \sqrt{i+1} .$$

1.1.B$_3$ The above (+) frequently applies to spaces of solutions x of an elliptic equation $\Delta x = 0$ (see [Ka], [Me], [G-M']). For example, if V is a Riemannian manifold of *bounded (local) geometry* and Δ on V is invariantly related to the geometry of V, then

$$\|x\|_\infty \leq \text{const} \|x\|_2 ,$$

where the constant depends only on the implied bound on the geometry. Then the above (+) applied to the normalized Riemannian volume of V, yields

$$\dim \operatorname{Ker} \Delta < \text{const}^2 \operatorname{Vol} V .$$

If V is complete non-compact of *infinite* volume and L is an *infinite dimension* space of solutions x of $\Delta x = 0$, then one can sometimes make sense of the inequality $\dim L / \operatorname{Vol} V > 0$ and use (+) to prove the existence of a non-zero L_2 -solution x on V. (For example, see [Ka].)

1.1.C The proof of 1.1 for all p and q. The basic averaging argument (see 1.1.B$_1$) applies, in principle, to the linear isometry group of $(L, \| \ \|_p)$ for all p, but for $p \neq 2$ this group is usually two small to be useful. However, by Dvoretzky theorem (see 1.2), there exists a j-dimensional subspace $M \subset L$ whose L_p-norm is ε invariant under the L_2 isometry group $G = O(j)$ of $(M, \| \ \|_2)$,

$$(1 - \varepsilon)\|x\|_p \leq \|gx\|_p < (1 + \varepsilon)\|x\|_p$$

for all $x \in M$ and $g \in G$, where ε admits an universal bound in terms of $j = \dim M$ and $i = \dim L - 1$,

$$\varepsilon < \varepsilon_0(i, j) ,$$

such that for every fixed j,

$$\varepsilon_0(i, j) \longrightarrow 0 \quad \text{for} \quad i \to \infty .$$

Now the L_2-argument applies up to an ε-error to $(M, \| \ \|_p)$ and the error goes to zero for $i \to \infty$.

Q.E.D.

1.1.D Remarks. (a) The above argument using Dvoretzky theorem also applies to the spectrum $\{\lambda_{ij}\}$ (see 0.6) and shows that for every fixed j

$$\lim_{i\to\infty} \lambda_{ij} \geq \gamma_\infty(p,q) .$$

In other words, every i-dimensional subspace $L \subset L_q(V,\mu)$ contains a j-dimensional subspace $M \subset L$, such that

$$E_{p/q} \quad M > (1-\varepsilon_{ij})\gamma_\infty(p,q)$$

where $\varepsilon_{ij} \to 0$ for $i \to \infty$.

(b) To prove 1.1 one actually needs only the *weak* Dvoretzky *theorem* (see 1.2.C) whose proof is obtained by an integration argument similar to that used in 1.1.B$_1$. (See §9.3 of [Gr] for yet another application of this argument.)

(c) Theorems 1.1 and (especially) 1.1.B look a century old but I made no effort to find early references. (The earliest frequently cited papers I know of are [Ru] and [Ste].) A very interesting use of 1.1.B$_3$ appears in [Ka] and the averaging argument of 1.1.B$_1$ can also be found in [G-M].

(d) If the measure space V in question is finite and consists of N atoms, then the i-th eigenvalue λ_i of L_p/L_q is related to the $(N-i)$-width of the unit ball $B_{p'} \subset L_{p'}$ with respect to the $L_{q'}$-norm by

$$(N-i)\text{-width}(B_{p'}, L_{q'}) = 2\lambda_i^{-1} ,$$

where p' and q' are determined by

$$\frac{1}{p'} + \frac{1}{p} = 1 , \qquad \frac{1}{q'} + \frac{1}{q} = 1 .$$

In the case where $N = 2i$ and the atoms of the underlying measure space V have unit mass the width, and hence λ_i, were estimated by Kašin (see [Pi]) as follows

$$\lambda_i \asymp \begin{cases} 1 & \text{for } p > 1 \geq 2 \\ i^{\frac{1}{q}-\frac{1}{2}} & \text{for } q < 2 < p \\ i^{\frac{1}{q}-\frac{1}{p}} & \text{for } q \leq p \leq 2 \end{cases}$$

where $a_i \asymp b_i$ signifies that a_i/b_i is pinched between two positive constants for $i \to \infty$. Similar (but more difficult) estimates for all N are due to Gluskin (see [Pi] and [Kaš]).

(e) **Question.** Let H be a homogeneous function in k variables of degree zero. Then for a given k-tuple $(p_1,\ldots,p_k)$ one defines the energy

$$E(x) = H(\|x\|_{p_1},\ldots,\|x\|_{p_2}) ,$$

and asks what the spectrum of this E is. If $k = 2$, the question reduces to L_p/L_q. If $k = 3$, the simplest energy is $\|x\|_{p_1}\|x\|_{p_2}/\|x\|^2_{p_3}$.

In fact one is interested in the spectrum of the multi-parametric "energy"

$$x \longrightarrow (\|x\|_{p_1}, \ldots, \|x\|_{p_k}) ,$$

as it is defined in 0.4.

1.2 Dvoretzky theorem. We state below for reader's convenience several versions of Dvoretsky thoerem and we refer to [Mi-Sh] for the proofs.

The classical version of the theorem claims that the ratio $E(x) = \|x\|'/\|x\|$ of two norms on a linear space L becomes "nearly constant" when restricted to an "appropriate" subspace $M \subset L$, provided $\dim L$ is sufficiently large. Here the non-constancy of E is measured by *the logarithmic oscillation*

$$\operatorname{los} E = \log(\sup E/\inf E)$$

and the precise statement is as follows.

1.2.A. *For every $j \leq i = \dim L$ there exists a linear subspace $M \subset L$ of dimension j, such that*

$$\operatorname{los} E \mid M \leq \varepsilon(i,j) , \tag{*}$$

where $\varepsilon(i,j)$ is a universal constant depending on i and j, such that for every fixed j, $\varepsilon(i,j) \to 0$ for $i \to \infty$.

1.2.B Remark. The most important special case of 1.2.A is where $L = \mathbb{R}^i$ and $\|\ \|$ is the Euclidean norm on $\mathbb{R}^i$. In this case the theorem applied to $E = \|\ \|'$ restricted to the unit sphere in $\mathbb{R}^i$. Notice that this special case (applied first to L and then to M) yields the general case.

1.2.C Weak Dvoretzky. In this version of the theorem the constant ε is allowed to depend on $C = \operatorname{los} E \mid L$. Namely, one assumes $\operatorname{los} E \mid L = C < \infty$ and only claims the existence of an $M \subset L$, such that

$$\operatorname{los} E \mid M \leq \varepsilon(i,j,C) ,$$

where $\varepsilon \to 0$ for $i \to \infty$ and j and C fixed. Here again the most important case is $(L, \|\ \|) = \mathbb{R}^i$. This Euclidean Dvoretzky is equivalent (this is easy, see [Mi-Sh]) to the following subadditivity of the function $\operatorname{pro} X$ for $X \subset P$, which is, we recall, the maximal dimension of projective subspaces contained in X,

$$\operatorname{pro}(X \cup Y) \leq A(\operatorname{pro} X, \operatorname{pro}(Y + \varepsilon), \varepsilon^{-1}) ,$$

where X and Y are subsets in P, where $Y + \varepsilon \subset P$ denotes the ε-neighbourhood of Y with respect to the standard (Euclidean) metric in P, and where A is some function in three real variables. This is worth comparing with the subadditivity of the essential dimension,

$$\operatorname{ess}(X \cup Y) \leq \operatorname{ess} X + \operatorname{ess} Y + 1 ,$$

(see 0.4.B_1).

1.2.D Non-symmetric Dvoretzky. The Dvoretzky theorem remains true if we drop the symmetry requirement for the norms $\|\ \|$ and $\|\ \|'$. This is possible due to the following version of Bezout (Borsuk-Ulam) theorem (compare §4).

Let $E : \mathbb{R}^i \to \mathbb{R}$ be a continuous function and $x_1, \dots, x_k$ be some vectors in $\mathbb{R}^i$. If $k < i$, then there exist an orthogonal transformation g of $\mathbb{R}^i$, such that

$$Eg(x_\nu) = E(-g(x_\nu))$$

for all $\nu = 1, \dots, k$.

1.2.E Dualization. Dvoretzky theorem can be stated as the existence of an ε-round j-dimensional section of a convex subset K in $\mathbb{R}^i$. This yields, by duality, the existence of ε-round projections of K. Since projections commute with taking convex hulls one can drop the convexity assumption on K and arrive at the following proposition.

Let K be a compact subset in $\mathbb{R}^i$ which linearly spans $\mathbb{R}^i$. Then for every $j \leq i$ there exists a surjective linear map $A : \mathbb{R}^i \to \mathbb{R}^j$, such that K goes into the unit Euclidean ball in $\mathbb{R}^j$,

$$A(K) \subset B_1^j = \{x \in \mathbb{R}^j \mid \|x\| \leq 1\} ,$$

and $A(K)$ is ε-dense in B_1^j, where as earlier, for each j,

$$\varepsilon = \varepsilon(i, j) \longrightarrow 0 \quad \text{as} \quad i \to \infty .$$

(Recall, that a subset of a metric space is called ε-dense in B if its ε-neighbourhood contains B.) *Moreover, one can find the above A of form λp, where $p : \mathbb{R}^i \to \mathbb{R}^j \subset \mathbb{R}^i$ is an orthogonal projection onto a subspace and λ is the multiplication by a scalar $\lambda > 0$.*

1.2.F Projection of measures. With little extra effort the above discussion applies to projection of measure on $\mathbb{R}^i$ rather than of subsets. Namely, let μ_i be a probability measure on $\mathbb{R}^i$, for all $i = 1, 2, \dots$, such that the support of μ_i linearly spans $\mathbb{R}^i$.

Then for every $j = 1, 2, \dots$, there exists an orthogonally invariant measure $\overline{\mu}$ on $\mathbb{R}^j$ and a sequence of linear maps $A_i : \mathbb{R}^i \to \mathbb{R}^j$, such that the push-forward measures $A_*(\mu_i)$ on

$\mathbb{R}^j$ *weakly converge to* $\overline{\mu}$. *Moreover one can choose* $A_i = \lambda_i p_i$ *as in 1.2.E.* (This version of Dvoretzky theorem nicely fits the fixed-point philosophy of Fürstenberg, see [Gr-Mi].)

1.3 On the topological version of the $E_{p/q}$-spectra. If the measure space (V,μ) is infinite then the Ess-spectrum for $E_{p/q}$ collapses to the single point $\lambda_0 = 1$. This is immediate with the following.

Trivial observation. *Let $C_\varepsilon \subset P$ be the subset of (the projective classes of) functions x on V, such that $|x(v)| \leq 1$ for all $v \in V$ and*

$$\mu\{v \in V : |x(v)| = 1\} \geq 1 - \varepsilon .$$

Then $\operatorname{ess} C_\varepsilon = \infty$ *for all* $\varepsilon > 0$. (Notice that $\operatorname{pro} C_\varepsilon = 0$ for all $\varepsilon > 0$.)

Now let us compute the topological spectrum of $E_{p/q}$ on the finite measure space (V,μ) consisting of n equal atoms of mass $1/n$.

1.3.A. *The ess-spectrum of $E_{p/q}$ on V is*

$$\lambda_i = \left(\frac{n-i}{n}\right)^{\frac{1}{p}-\frac{1}{q}} . \tag{$*$}$$

Proof: A trivial computation shows that the critical points of the function E of index i are the baricenters of i-codimensional faces (which are $(n-i-1)$-dimensional simplices) of the L_1-sphere $\{\|x\|_1 = 1\} \subset L_1(V,\mu)$ and $E_{p/q}$ equals the above λ_i (given by $(*)$) at these baricenters. Hence $(*)$ follows by the Morse theory.

Remark. One can avoid using Morse theory by applying the following simple topological facts (A) and (B) to the unit L_1- and L_∞-balls

$$\left\{ \|x\|_1 = \frac{1}{n}\sum_{i=1}^{n} |x_i| \leq 1 \right\} \subset \mathbb{R}^n$$

and

$$\left\{ \|x\|_\infty = \sup_i |x_i| \leq 1 \right\} \subset \mathbb{R}^n .$$

(A) *Let $Q \subset P = P(\mathbb{R}^n)$ satisfy $\operatorname{ess} Q \geq i$ and let $B \subset \mathbb{R}^n$ be a convex centrally symmetric polyhedron with non-empty interior. Then the cone $\widetilde{Q} \subset \mathbb{R}^n$ over Q meets some $(n-i-1)$-dimensional face of B.*

Notice that a similar fact for $\operatorname{pro} Q \geq i$ holds true *without* assuming B is symmetric. In fact the meeting points of an $(i+1)$-plane $L \subset \mathbb{R}^n$ with the $(n-i-1)$-faces of B are exactly the extremal points of $B \cap L$.

(B) *Let $\overline{B}_i \subset P$ be the projection to P of the union of i-faces of B. Then*

$$\operatorname{ess} \overline{B}_i = i$$

Notice that $\operatorname{pro} \overline{B}^i = 0$ for the L_∞-ball B and $i \leq n-2$ (this is the case for our ess-spectral discussion), which explains the sharp discrepency between the ess- and pro-spectra.

§2 Variation, oscillation and ess-spectra for spaces of continuous maps

The measure theoretic conentration phenomenon of the previous section has the following topological counterpart.

A "large" subspace in the space of continuous maps between topological spaces V and W must contain a "topologically complicated" map $x : V \to W$.

If $W = \mathbb{R}$ and V is connected, then the complexity of a function $x : V \to \mathbb{R}$ can be measured by the *variation* of X,

$$\operatorname{Var} x = \int_{\mathbb{R}} b_0\left(x^{-1}(t)\right) dt$$

where b_0 is the zero Betti number, that is the number of connected components of the pull-back $x^{-1}(t)$ for all $t \in \mathbb{R}$.

Notice that every map $x : V \to \mathbb{R}$ can be uniquely factorized as follows, $V \xrightarrow{\bar{x}} \overline{V} \xrightarrow{y} \mathbb{R}$ (with x the composite), where $\overline{V}$ is a 1-dimensional space (graph) and $\bar{x}$ is a *connected* map of V onto $\overline{V}$, that is $\bar{x}^{-1}(\bar{v}) \subset V$ is connected for all $\bar{v} \in \overline{V}$. Then

$$\operatorname{Var} x = \operatorname{Var} y ,$$

where $\operatorname{Var} y$ may be thought of as the "total length" of $\overline{V}$ with the metric induced from $\mathbb{R}$. For example, if $V = [0,1]$, then $V = \overline{V}$, $x = y$ and

$$\operatorname{Var} x = \int_0^1 |x'(v)| dv .$$

Remarks. (a) The variation of $x : V \to \mathbb{R}$ is not an especially good measure of complexity as it is unstable under small perturbations of x. But one can stabilize $\operatorname{Var} x$ by introducing for every $0 < \varepsilon < 1$,

$$\operatorname{Var}_\varepsilon x = \inf_y \operatorname{Var}(x + y)$$

over all continuous functions $y : V \to \mathbb{R}$ satisfying

$$\|y\|_\infty < \varepsilon \|x\|_\infty$$

for the norm

$$\|x\|_\infty = \sup_{v \in V} x(v) \ .$$

(b) We are mainly concerned here with functions on $[0,1]$, but we have presented the definitions keeping an eye on possible generalizations. (Compare §2.1.3.B in [Gr]$_3$).

2.1. The number of oscillations of a function. For a function $x : V \to \mathbb{R}$ we write

$$\operatorname{Osc} x = \sup_{v \in V} x(v) - \inf_{v \in V} x(v)$$

and then for every positive $\gamma \leq 1$ define the number of γ-oscillations of x as follows. First we say that subsets V_1 nd V_2 and V are *x-independent* if there exists no *connected* subset $U \subset V$ on which x is constant and which meets both subsets V_1 and V_2. Then we define $\#_\gamma \operatorname{Osc} x$ as the maximal integer k for which there exists x-independent subsets $V_j \subset V$ for $j = 1, \ldots, k$, such that

$$\operatorname{Osc} x \mid V_j \geq \gamma \operatorname{Osc} x$$

for $j = 1, \ldots, k$. We abbreviate $\# \operatorname{Osc} = \#_1 \operatorname{Osc}$ and call this *the number of full oscillations* of x. If $V = [0,1]$ then $\# \operatorname{Osc} x$ equals the maximal number k such that $[0,1]$ can be partitioned into k subintervals with equal x-images.

Also notice that

$$\operatorname{Var} x \geq (\gamma \#_\gamma \operatorname{Osc} x) \operatorname{Osc} x$$

and that $\#_\gamma \operatorname{Osc} x$ enjoys an obvious kind of stability under perturbations of x.

2.2 Theorem. *Let Q be a subset in the projectivized space P of continuous functions on $[0,1]$. Then there exists a function $x \in Q$, such that*

$$\# \operatorname{osc} x \geq \operatorname{ess} Q \ .$$

Proof: Apply 4.3 A and A_1 to the space T of partitions of $[0,1]$ into $k+1 = \operatorname{ess} Q$ subintervals. This gives us a partition $[0,1] = \bigcup_{i=0}^{k} I_i$, and an $x \in Q$, such that

$$\operatorname{osc} x \mid I_i = \operatorname{osc} x \ ,$$

for $i = 0, \ldots, k$. Q.E.D.

2.2.A Remarks and corollaries. (a) The above theorem applies, in particular to every $(k'+2)$-dimensional *linear* (sub)space L of functions on $[0,1]$ and claims the existence of a non-zero $x \in L$ having

$$\# \operatorname{osc} x \geq k+1 \ .$$

(b) One easily sees with 2.2 that the ess-spectrum (as well as the pro-spectrum) of the energy $E(x) = \operatorname{var} x / \|x\|_\infty$ is $\lambda_i = i$ for all $i = 0, 1, \ldots$.

(c) The divergence $\lambda_i \to \infty$ of the pro-spectrum can also be derived from the Dvoretzky theorem (see 1.2.E) as follows. Given an $(i+1)$-dimensional space L of functions on $[0,1]$, we have a continuous map of $[0,1]$ into the dual L', such that the functions from L appear as the restrictions of linear functions on L' to $[0,1]$ (see the proof of 1.1). As $i \to \infty$, we can find a k-dimensional subspace $L_0 \subset L$, such that $k \to \infty$ and the corresponding image of $[0,1]$ is ε-dense in the unit ball of L_0' for some Euclidean metric in L_0' where $\varepsilon \to 0$ for $i \to \infty$. Then obviously $E(x) \to \infty$ for all $x \in L_0$ and $i \to \infty$.

(d) Theorem 2.2 and its corollaries must be as old as the Bezout-Borsuk-Ulam theorem, but I have not checked the literature.

§3 Asymptotic additivity and homogeneity of Dirichlet energies

3.1 Examples of Dirichlet energies. The classical Dirichlet energy is defined on functions x on a bounded Euclidean domain V by

$$E(x) = \|dx\|_2 / \|x\|_2$$

where d denotes the differential of a function and where the L_2-norms of dx and x are taken with the ordinary Lebesgue measure in V. A more general class of integro-differential energies can be defined as follows. Let X and Y be smooth vector bundles over a manifold V and $D : x \mapsto y$ a linear (or non-linear) differential operator between the sections of X and Y. In order to define what we call the *L_pD/L_q-energy*

$$E(x) = \|Dx\|_p / \|x\|_q \ ,$$

we need the following additional structures (a) and (b).

(a) norms in the vector bundles X and Y. With these we have the point-wise norms $\|x(v)\|$ and $\|y(v)\|$ of sections of X and Y on V.

(b) A measure μ on V which is also denoted dv. With this we have the L_p-norm on sections of X and Y,

$$\|x\|_p = \left(\int_V \|x(v)\|^p dv\right)^{1/p}.$$

Notice that for the $L_\infty D/L_\infty$-energy one only needs the measure class of μ rather than the measure itself.

Remarks. (a) if the operator D has infinite dimensional kernel, then, in order to have an "interesting" spectrum, one should either restrict D to a subspace of sections where the kernel is finite dimensional or to pass to an appropriate quotient space. For example, if D is the exterior differential on *forms* (rather than on functions), then one should work modulo closed (sometimes exact) forms.

(b) It is sometimes interesting to use different measures in defining the norms of x and Dx. For example, one may bring into the picture some measure μ' concentrated on a "subvariety" $V' \subset V$ and then to look at $E = L_p D/L_q(\mu')$.

Let us look more closely at the case were $D = d$ is the exterior differential on functions x on V. Here $X = V \times \mathbb{R} \to V$ is the trivial bundle and $Y = T^*(V)$ is the cotangent bundle. We do not have to worry about a norm on X as we already have one, $\|x(v)\| = |x(v)|$, for the ordinary absolute value on $\mathbb{R}$. On the other hand there is no canonical norm on $T^*(V)$ and so we have to choose one. If V is connected, such a norm defines a metric on V by

$$\operatorname{dist}(v_1, v_2) = \sup_x |x(v_1) - x(v_2)|$$

over all C^1-functions x on V, such that

$$\|dx\|_\infty \underset{\text{def}}{=} \sup_{v \in V} \|dx(v)\| \le 1$$

This distance (and sometimes the norm itself) is called a *Finsler* metric on V. A Finsler metric is called *Riemannian*, if the norm in each fiber $T_v'(V)$, $v \in V$ is Euclidean.

Remark. Usually one starts with a (dual) norm in the tangent bundle and define the distance as the length of the shortest path $p : [0,1] \to V$ between v_1 and v_2. Namely, the norm in $T(V)$ allows one to measure the tangent vectors $\frac{dp(t)}{dt} \in T_{p(t)}(V)$ and thus to define *the maximal stretch of p,*

$$\|Tp\| = \sup_{t \in [0,1]} \left\| \frac{dp(t)}{dt} \right\|.$$

Then one gets $\operatorname{dist}(v_1, v_2)$ as $\inf_p \|Tp\|$ over all paths p with $p(0) = v_1$ and $p(1) = v_2$.

To conclude the definition of the $L_p d/L_q$-energy on a Finsler manifold V we need a measure on V. Usually one uses the Finsler norm on $T^*(V)$ to provide V with a measure as follows. One considers the *determinant bundle* $\Lambda(V)$ that is the top exterior power $\Lambda^n T^*(V)$ for $n = \dim V$. There are many (unfortunately too many) natural ways to define a norm on $\Lambda(V)$ starting from our norm on $T^*(V)$. Since $\Lambda(V)$ is one-dimensional, a norm on $\Lambda(V)$ is |section| of $\Lambda(V)$, that is a density on V which integrates to a measure on V.

3.1.A Dirichlet on metric spaces. For a function x on a metric space V we define the Lipschitz constant $\operatorname{Lip} x$ as the supremum of $|x(v_1) - x(v_2)|/\operatorname{dist}(v_1, v_2)$ over all pairs of distinct points v_1 and v_2 in V. Then for a point $v \in V$ we restrict x to the ε-balls $B_\varepsilon \subset V$ around v and set

$$|dx(v)| = |\operatorname{Lip}_v x| = \limsup_{\varepsilon \to 0} \operatorname{Lip} x \mid B_\varepsilon ,$$

and $\|dx\| = \sup_{v \in V} |dx(v)|$. Notice that $\|dx\| \leq \operatorname{Lip} x$ and state the following

Trivial Lemma. *The following two conditions are equivalent*

(i) $\|dx\| = \operatorname{Lip} x$ *for all functions* x *on* V.

(ii) For every two points v_1 *and* v_2 *with some distance* d *in* V *and every* $\varepsilon > 0$ *there exists a (*ε*-middle) point* $v_\varepsilon \in V$, *such that* $\operatorname{dist}(v_i, v_\varepsilon) \leq \varepsilon + \frac{1}{2} d$ *for* $i = 1, 2$.

Metrics satisfying (ii) are called *geodesic*. (They are also called inner metrics, length metrics and local metrics.) Observe that Finsler metrics are geodesic.

Now, with a measure on V we have the $L_p d/L_q$-energy

$$E(x) = \left(\int_V |dx|^p \right)^{1/p} \Big/ \left(\int_V |x|^q \right)^{1/q} .$$

If V is a Finsler space this agrees with the earlier definition. The same can be said for *Carnot spaces* defined below

3.1.B. Carnot spaces. Consider a first order differential operator D on functions x on V, where the range bundle Y is equipped with a norm. The issuing seminorm on C^1-functions,

$$x \longmapsto \|Dx\|_\infty$$

is called a *Carnot structure* on V, provided $D(\text{const}) = 0$, that is $D = h \circ d$ for some homomorphism $h : T^*(V) \to Y$. If h has a constant rank k, then the Carnot structure is uniquely determined by the image bundle of the adjoint homomorphism $h^* : Y^* \to T(V)$, called $\theta = \operatorname{Im} h^* \subset T(V)$, and a norm on θ.

Define Carnot's (semi)metric on V by

$$\operatorname{dist}(v_1, v_2) = \sup_x |x(v_1) - x(v_2)|,$$

over all x with $\|Dx\|_\infty \le 1$. One can equivalently define this "dist" with paths in V tangent to θ. Thus one sees, in particular, that dist is an honest metric, i.e., everywhere $< \infty$, if and only if every two points v_1 and v_2 in V can be joined by a path in V tangent to θ.

Remark. Carnot metrics are sometimes called *Carnot-Caratheodary* (see [G-L-P]) or *subelliptic* (see [St]). Here we reserve the word "*sub-Riemannian*" for the case where the above norm on θ is Euclidean.

3.1.C Alternative definitions of $\|dx\|_p$. Let us recall that the *coboundary* δx of a function x on V is the function on $V \times V$ defined by

$$\delta x(v_1, v_2) = x(v_1) - x(v_2) .$$

Next consider the following function K_ε on $V \times V$,

$$K_\varepsilon(v_1, v_2) = \begin{cases} 0 & \text{if } \operatorname{dist}(v_1, v_2) > \varepsilon \\ \varepsilon^{-1} & \text{if } \operatorname{dist}(v_1, v_2) \le \varepsilon \end{cases}$$

and let $\delta_\varepsilon x$ be the product $K_\varepsilon \delta x$. In other words we restrict δx to the ε-neighbourhood of the diagonal in $V \times V$ and then divide it by ε. Notice that

$$\limsup_{\varepsilon \to 0} \|\delta_\varepsilon x\|_\infty = \|dx\|_\infty .$$

Denote by μ' the measure $\mu \times \mu$ on $V \times V$ and let μ'_ε denote the measure of the ε-neighbourhood of the diagonal, that is

$$\mu'_\varepsilon = \varepsilon \int_{V \times V} K_\varepsilon d\mu' ,$$

and let

$$\|x\|'_p = \limsup_{\varepsilon \to \infty} \|\delta_\varepsilon x\|_p / \mu'_\varepsilon .$$

Notice that for sufficiently smooth Riemannian (and sub-Riemannian) spaces V, $\|x\|'_p = \text{const}_{n,p} \|x\|_p$, where n is the dimension of V (which should be properly defined in the sub-Riemannian case). An advantage of $\|x\|'_p$ over $\|x\|_p$ for non-smooth spaces is clearly seen for $p = 2$ as the norm $\|x\|'_2$ is always Hilbertian and the deviation of $\|dx\|_2$ from being Hilbertian (as well as non-constancy of the norm ratios $\|dx\|_p / \|dx\|'_p$) measures non-smoothness of V. If V is a Finsler manifold (e.g., a domain in a finite dimensional Banach space) this measures how far V is from a Riemannian space. The picture is less clear for nowhere smoth (e.g., fractal) spaces V.

One can generalize the definition of K_ε by taking any function $e(t)$ and by letting

$$K_\varepsilon = e(\operatorname{dist}(v_1, v_2))$$

A classical choice of e is

$$e(t) = \exp -\varepsilon^{-1} t$$

which for $\varepsilon \to 0$ gives us (after a normalization) a regularized version of the above $\|dx\|'_p$.

Finally observe that the functions $K(v_1, v_2) = e(\operatorname{dist}(v_1, v_2))$ define integral operators on V,

$$x \longmapsto K * x = \int_V K(v_1, v_2) x(v_1) dv_1 \ .$$

Spectra of such operators are similar to those of the energies $\|dx\|'_p / \|x\|_p$.

Example. Let $x \mapsto K^0_\varepsilon * x$ be the averaging of x over the ε-balls $B(v, \varepsilon)$ in V, that is

$$K^0_\varepsilon(v_1, v_2) = \begin{cases} 0 & \text{for } \operatorname{dist}(v_1, v_2) \geq \varepsilon \\ [\mu B(v_2, \varepsilon)]^{-1} & \text{for } \operatorname{dist}(v_1, v_2) < \varepsilon \end{cases}$$

(Notice that this K^0_ε is not of the form e(dist), unless the measure $\mu(B(v, \varepsilon))$ is constant in v). If V is "sufficiently smooth" then the operator

$$A_\varepsilon = \varepsilon^{-2}(Id - K^0_\varepsilon)$$

converges for $\varepsilon \to 0$ to the Laplace operator $\Delta - d^* d$ on V. In particular, the eigenvalues of the operator $|A^*_\varepsilon A_\varepsilon|^{1/4}$ converge to those of the energy $\|dx\|_2 / \|x\|_2$. This suggests the definition of the norms $\|\Delta x\|_p = \limsup_{\varepsilon \to \infty} \|A_\varepsilon x\|_p$ for an arbitrary metric space V. Probably, the existence of sufficiently many x with $\|\Delta x\|_p \leq \infty$ implies certain smoothness of X. Otherwise one may try norms associated to more regular operators, for example $\varepsilon^{-\rho}(Id - K^0_\varepsilon)$ for $\rho < 2$.

3.1.D. The above relation between metrics in V and norms on function spaces is of quite general nature. Namely, every seminorm on the space X of (say, continuous) functions x on V defines a norm in the dual X'. As V is canonically mapped into X' by Dirac's $v \mapsto \delta_v$, we get an induced (Caratheadory) metric on V. More generally, if X is the space of sections of a k-dimensional vector bundle over V, then V is naturally mapped into the Grassmanian of the k-planes in X', which again induces a (Bergman) metric in V from a seminorm in X.

The major problem of the geometric spectral theory is to relate the properties of such metrics on V with the spectra of the (ratios between) norms in question.

Remark. The above metric on V may degenerate. For example if we use the norm $\|L_p d\|$ on an n-dimensional manifold V, then the resulting metric on V is degenerate for $n \geq p$. In such a case it is useful to consider the following distance between subsets (rather than points) in V,

$$\operatorname{dist}(V_1, V_2) = \sup_x \|x\|^{-1} ,$$

where x sums over all functions which are equal zero on V_1 and one on V_2. Notice that dist^{-1} is called the *capacity* (associated to the norm $\| \quad \|$) and it has been extensively studied for the above norm $\|L_p d\|$ (see [M-H]).

3.2 Dirichlet energy under cutting and pasting. Start with the simplest case where V is the disjoint union of V_1 and V_2 and canonically decompose each function x on V into the sum $x_1 + x_2$ where $x_1 \mid V_2 = 0$ and $x_2 \mid V_1 = 0$. One trivially has

3.2.A Lemma. *If $p \leq q$ then the energy $E(x) = \|Dx\|_p / \|x\|_q$ satisfies*

$$E(x) \geq \min\left(E(x_1), E(x_2)\right) . \tag{*}$$

On the contrary, if $p \geq q$, then

$$E(x) \leq \max\left(E(x_1), E(x_2)\right) . \tag{**}$$

In particular, if $p = q$ and say $E(x_1) \leq E(x_2)$, then

$$E(x_1) \leq E(x) \leq E(x_2) .$$

This implies the following sub-additivity of the number $N(\lambda)$ of the eigenvalues $\leq \lambda$, that is

$$N(\lambda) = \text{“dim”} E^{-1}(-\infty, \lambda] + 1$$

for a given "dim" (see 0.4).

3.2.A$_1$. *If $p \geq q$ then $N(\lambda) \geq N_1(\lambda) + N_2(\lambda)$ where $N_i(\lambda) = N(\lambda, E \mid V_i)$ for $i = 1, 2$.*

Proof: The inequality (**) shows that the $*$-product (see 0.4) of $E_1^{-1}(-\infty, \lambda) * E_2^{-1}(-\infty, \lambda)$ is contained in $E^{-1}(-\infty, \lambda)$ for all λ (here $E_i = E_i \mid V_i$), and 3.2.A$_1$ follows.

3.2.A$_2$. Suppose our "dim" is sub-additive,

$$\text{“dim”} A \cup B \leq \text{“dim”} A + \text{“dim”} B + 1 .$$

Then for $p \leq q$,

$$N(\lambda) \leq N_1(\lambda) + N_2(\lambda) ,$$

as

$$E^{-1}[0,\lambda] \subset E_1^{-1}[0,\lambda] \cup E_2^{-1}[0,\lambda] .$$

Remind, that ess and $\mathrm{pro}^{\perp}$ are subadditive which implies the above inequality for the respective $N(\lambda)$.

3.2.A$_3$ Remark. If "dim" is not subadditive one can bound the $\{\lambda_{ij}\}$-spectrum (see 0.6) rather than $\{\lambda_i\}$ as follows. Let $M(\lambda, N)$ be the maximal number, such that every N-"dimensional" subset A in P (where the energy E lives) satisfies

$$\text{"dim"}\left(A \cap E^{-1}[\lambda,\infty)\right) \geq M .$$

(Notice that this M can be obviously expressed in terms of λ_{ij}.) Then for $p \leq q$ one trivially has

$$M(\lambda, N) \geq M_2(\lambda, M_1(\lambda, N))$$

for all N and λ, where M_1 and M_2 refer to $E \mid V_1$ and $E \mid V_2$ correspondingly

Let us summarize the previous discussion for $p = q$ and "dim"$=$ ess.

3.2.B Additivity of the spectrum for the energy $E(x) = \|Dx\|_p / \|x\|_p$. *If* V *is the disjoint union of* V_1 *and* V_2 *then the number*

$$N(\lambda) = \text{ess}\, E^{-1}(-\infty,\lambda] + 1$$

is the sum of those for V_1 *and* V_2,

$$N(\lambda) = N_1(\lambda) + N_2(\lambda) .$$

Remarks. (a) *According to our notation this includes the case* $E(x) = \|dx\|_p/\|x\|_p$ *on an arbitrary metric space* V.

(b) The above additivity property trivially generalizes to the case where the measure μ underlying $E(x)$ is decomposed into a sum of measures, $\mu = \mu_1 + \mu_2$, such that the supports of μ_1 and μ_2 are disjoint.

3.2.C Monotonicity of $E(x)$. Let $f : V' \to V$ be a locally homeomorphic map. Then vector bundles on V induce those on V' and a given operator D on V lifts to D' on V'. Now, if our measure μ on V is the push-forward of some μ' on V', then the pull-back map $x \mapsto x' = f^*(x)$ preserves $E = L_pD/L_q$ for $E'(x') = E(x)$, and this remains valid for $E = L_pd/L_q$ on metric spaces.

3.2.C₁ Corollary. *Let $\{V_j\}$, $j = 1, \ldots, k$ be an open cover of V and functions $p_j : V_j \to \mathbb{R}_+$ form a partition of unity. Then the counting function $N(\lambda)$ for $E = L_pD/L_q$ on (V, μ) is bounded by the functions $N_j(\lambda)$ on $(V_j, p_j\mu)$,*

$$N(\lambda) \leq \sum_{j=1}^{k} N_j(\lambda) \, ,$$

provided $p \leq q$ and "dim" is subadditive (compare 3.1.A₂).

3.2.D Energy and $N(\lambda)$ on V/V_0. Denote by $P_0 \subset P$ the space of functions (or sections) vanishing on a given subset $V_0 \subset V$. An important example is where $V_0 = \infty$ and then P_0 by definition of this ∞ consists of functions with compact supports. The energy E restricted to P_0 is also called E on V/V_0 and the corresponding counting function is denoted $N(\lambda, V/V_0)$ or just $N^0(\lambda)$. If V_0 is not specified then $N^0(\lambda)$ refers to $N(\lambda, V/\infty)$.

It is obvious that

$$N^0(\lambda) \leq N(\lambda)$$

and that

$$N(\lambda, U/\infty) \leq N(\lambda, V/\infty)$$

for all open subsets $U \subset V$. It follows (see (*) in 3.2.A) that for $p \geq q$

$$N^0(\lambda) \geq \sum_{j=1}^{k} N_j^0(\lambda)$$

where $N_j^0 = N^0(V_j)$ for disjoint open subsets $V_1, \ldots, V_j, \ldots, V_k$ in V.

3.2.E A bound on the counting function $N(\lambda)$ on V by those on V/V_0 and V_0. Let $V_\varepsilon \subset V$ denote the ε-neighbourhood of V_0,

$$V_\varepsilon = \{v \in V \mid \mathrm{dist}(v, V_0) \leq \varepsilon\} \, ,$$

and $\|x\|_q^\varepsilon$ denote the L_2-norm of the restriction $x \mid V_\varepsilon$. Let $E_\varepsilon(x) = \|Dx\|_p / \|x\|_q^\varepsilon$ and denote by $N_\varepsilon(\lambda)$ the corresponding counting function. Notice that $E_\varepsilon(x) \geq E_\varepsilon(x \mid V_\varepsilon)$ and $N_\varepsilon(\lambda) \leq N(\lambda, V_\varepsilon)$.

Next we recall $N^0(\lambda) = N(\lambda, V/V_0)$ and we assume that $D = d$ and $p = q$. Thus the functions $N(\lambda)$, $N^0(\lambda)$ and $N_\varepsilon(\lambda)$ count the energy levels for L_pd/L_p.

3.2.E$_1$ Lemma. *If the implied dimension is subadditive then*

$$N(\lambda) \leq N^0(\lambda') + N_\varepsilon(\lambda'')$$

for

$$\lambda = \lambda'\lambda''/(\lambda'' + \lambda' + \varepsilon^{-1})$$

and for all positive λ', λ'' and ε.

Proof: Let $a_\varepsilon(v) = \varepsilon^{-1}\operatorname{dist}(v, V_0)$ for $v \in V_\varepsilon$ and $a_\varepsilon(v) = 1$ outside V_ε. Then

$$\|d(a_\varepsilon x)\|_p \leq \|Dx\|_p + \varepsilon^{-1}\|x\|_p^\varepsilon .$$

Now the inequalities

$$\|d(a_\varepsilon x)\|_p \geq \lambda'\|a_\varepsilon x\|_p ,$$

$$\|dx\|_p \geq \lambda''\|x\|_p^\varepsilon$$

and

$$\|a_\varepsilon x\|_p + \|x\|_p^\varepsilon \geq \|x\|_p$$

imply

$$\|dx\|_p \geq \lambda\|x\|_p$$

for $\lambda = \lambda'\lambda''/(\lambda'' + \lambda' + \varepsilon^{-1})$ and the proof follows.

3.2.F Asymptotic additivity of the function $N(\lambda)$. Call a subset $V_0 \subset V$ *thin* if for every $C \geq 0$ there exist $\varepsilon > 0$ and $\lambda_0 \geq 0$, such that N_ε defined in 3.1.E satisfies for all $\lambda \geq \lambda_0$,

$$CN_\varepsilon(C\lambda) \leq N(\lambda) .$$

We call $N(\lambda)$ asymptotically equivalent to $M(\lambda)$ and write

$$N(\lambda) \sim M(\lambda)$$

if

$$N(C\lambda) \geq M(\lambda) > N(C^{-1}\lambda)$$

for every $C > 1$ and all sufficiently large (depending on C) λ.

3.2.F Weyl additivity theorem. *Let the metric space V be decomposed into the union of closed subsets $V = V_1 \cup V_2$, where the intersection $V_0 = V_1 \cap V_2$ is thin. Then the implied counting function $N(\lambda)$ for the energy $E = L_p d/L_p$ and "dim" = ess satisfies*

$$N(\lambda) \sim N_1(\lambda) + N_2(\lambda)$$

where N_i for $i = 1, 2$ are the corresponding functions for V_1 and V_2.

Proof: This follows from 3.2.E$_1$ and 3.2.B.

Remark. Instead of using the specific cut-off function $a_\varepsilon = \varepsilon^{-1}$ dist, one could postulate the existence of such a function with an appropriate notion of *capacity* of V_0 (compare 3.1.D). Thus one would obtain a more general (and more conceptual) version of the additivity theorem.

3.3 The function $N(\lambda)$ and the covering numbers. For a metric space V we consider the numbers COV(ε), which is the minimal number of ε-balls needed to cover V, and the number IN(ε), which is the maximal number of disjoint ε-balls in V. Notice that

$$\mathrm{COV}(\varepsilon) \geq \mathrm{IN}(\varepsilon) \geq \mathrm{COV}(2\varepsilon)$$

for all $\varepsilon \geq 0$.

Also notice that these numbers asymptotically for $\varepsilon \to 0$ are additive as $N(\lambda)$ and in some cases $N(\lambda)$ can be roughly estimated in terms of COV(λ^{-1}). First we give such estimates in the easiest case $E = L_\infty d/L_\infty$.

3.3.A Observation. *The function $N(\lambda) =$ "dim"$E^{-1}(-\infty, \lambda]+1$ for $E(x) = \|dx\|_\infty/\|x\|_\infty$ on a geodesic (see 3.1) metric space V satisfies for all $\lambda > 0$,*

$$\mathrm{IN}(2\lambda^{-1}) \leq N(\lambda) \leq \mathrm{COV}(\lambda^{-1}) \, .$$

Proof: Given disjoint ε-balls $B_1, \ldots, B_N$ in V we consider the linear space L of functions generated by the constants and the functions dist$(v, V \setminus B_i)$, $i = 1, \ldots, N$. Then the (obvious) inequality

$$2\|x\|_{L_\infty} \geq \varepsilon\|dx\|_\infty$$

for all $x \in L$ yields the lower bound on $N(\lambda)$.

To get the upper bound we observe that every "N-dimensional" subspace in the projective space P of functions on V contains (see 0.4) a function x vanishing on a given subset $S \subset V$ consisting of N-points. Since V is geodesic, such an x is bounded by

$$\|x\|_\infty \leq \|dx\|_\infty \sup_{v \in V} \mathrm{dist}(v, S) \, ,$$

which trivially yields the desired upper bound on $N(\lambda)$.

3.3.B The μ-regularity constant and an upper bound on $N(\lambda)$ for $E = L_p d/L_p$. Denote by $\delta = \delta(V, \mu)$ the minimal number such that every two concentric balls on V of radii R and $2R$ satisfy

$$\mu(B(2R)) \leq 2^\delta \mu(B(R))$$

for the given measure μ on V.

Example. If $V = \mathbb{R}^n$ then $\delta = n$. Moreover, if V is a complete Riemannian manifold with non-negative Ricci curvature then also $\delta = \dim V$.

3.3 B₁ Observation. *The function $N(\lambda)$ for $E = L_p d/L_p$ satisfies*

$$N(\lambda) \geq \mathrm{IN}(C\lambda^{-1})$$

for

$$C = 2^{2+\delta/p} .$$

Proof: Consider the linear space L of functions on V generated by constants and the functions $\operatorname{dist}(v, V \setminus B_i)$ for disjoint ε-balls B_i in V. Every $x \in L$ obviously satisfies

$$\|dx\|_p \leq C\varepsilon^{-1}\|x\|_p$$

which immediately yields what we want.

3.3.C Local and global lower bounds on the spectrum. Let V be μ-partitioned into closed subsets V_j, $j = 1, \ldots, k$, that is $V = \bigcup_j V_j$ and doubly covered points in V have measure zero. *If "dim" is subadditive and $p = q$, then, as we know,*

$$N(\lambda, V) \leq \sum_j N(\lambda, V_j) . \tag{$*$}$$

In particular, if

$$\lambda = \min_j \lambda_1(V_j) \tag{$*$}$$

then $N(\lambda) \leq k + 1$. More generally, if

$$\lambda = \min_j \lambda_{i_j}(V_j) ,$$

then

$$N(\lambda) \leq \sum_j i_j + 1 . \tag{$**$}$$

Remark. The presence of constant functions makes $\lambda_0 = 0$ which forces us to use $\lambda_{i_j}(V_j)$ for $i_j \geq 1$. On the other hand the number $\lambda_1(V_j)$ for "nice" small subsets V_j is expected to be $\sim (\operatorname{diam} V_j)^{-1}$. For example, smooth domains in $\mathbb{R}^n$, and more generally, compact Riemannian manifolds do admit arbitrarily fine "nice" partitions. Unfortunately, the construction of "nice" partitions may be quite difficult (if at all possible) for general spaces X. (A trivial obstruction to the "niceness" is disconnectedness. In fact, a set with $m+1$ connected components have $\lambda_0 = \lambda_1 = \ldots = \lambda_m = 0$.) To alleviate this problem we introduce the following.

3.3.C₁ Mollified spectrum. Take a neighbourhood $U \subset V$ of a subset $V_0 \subset V$ and let $\tilde{x}$ denote extensions to $U \supset V_0$ of functions x on V_0. Then we define

$$\|\widetilde{dx}\|_p = \inf_{\tilde{x}} \|d\tilde{x}\|_p$$

and study the corresponding $\widetilde{E}(x) = \|\widetilde{dx}\|_p / \|x\|_q$ and $\widetilde{N}(\lambda)$ for functions x on V_0.

Remark. This $\widetilde{E}$ is a special case of an energy E where one uses two different measures for the definition of $\|dx\|_p$ and $\|x\|_q$. The properties of such energies are quite similar to those where there is only one measure. In fact one can often reduce two measures to one by modifying the operator D in question.

Now, consider a covering $V = \bigcup_j V_j$ and let $U_j \supset V_j$ be neighbourhoods such that the multiplicity of the covering of V by U_j is at most m. Then the function $N(\lambda, V)$ for $E = L_p d / L_p$ and "dim" = ess satisfies

$$N(m^{-\frac{1}{p}}\lambda) \leq k+1$$

where

$$\lambda = \min_j \widetilde{\lambda}_1(V_j) \tag{$\tilde{*}$}$$

for the mollified $\widetilde{\lambda}_1$ of V_j in U_j. This is proven the same way as above $(*)$ and $(**)$ also generalizes to

$$N(m^{-\frac{1}{p}}\lambda) \leq \sum_j i_j + 1 \tag{$\widetilde{**}$}$$

for $\lambda = \min_j \widetilde{\lambda}_{i_j}(V_j \subset U_j)$.

3.3.C₂ Corollary. *Let the μ-constant $\delta(V) < \infty$ (see 3.3.B) and let for every ε-ball $B(\varepsilon)$ in V the mollified eigenvalue $\widetilde{\lambda}_1(B(\varepsilon) \subset B(\rho\varepsilon))$, for the concentric $\rho\varepsilon$-ball satisfies $\widetilde{\lambda}_1 \geq \tau\varepsilon^{-1}$ for given constants $\rho \geq 1$ and $\tau > 0$ and for all $\varepsilon > 0$. Then*

$$N(\lambda) \leq a\,\mathrm{COV}(\lambda^{-1})$$

for some constant $\underline{a} = a(\delta, \rho, \tau) > 0$.

Proof: The inequality $\delta < \infty$ gives us a control over multiplicities of coverings of V by $\rho\varepsilon$-balls, where V is already covered by the concentric ε-balls.

Besides, δ controls the growth of $\mathrm{COV}(\varepsilon)$ which is sufficient for our purpose. We leave the (trivial) details to the reader.

3.3.C$_3$ Remarks. (a) If V satisfies the assumptions of 3.3.C$_2$, then 3.3.B$_1$ also applies, which shows that $N(\lambda)$ has the same order of magnitude for $\lambda \to \infty$ as $\mathrm{COV}(\lambda^{-1})$. In particular, a subset V_0 is thin (see 3.2.F) if and only if its covering number satisfies

$$\mathrm{COV}(\varepsilon, V_0)/\mathrm{COV}(\varepsilon, V) \to 0 \quad \text{for} \quad \varepsilon \to 0\,.$$

Another consequence of the above discussion is the existence of constants $d = d(V) \geq 0$ and $b_i = b_i(V) > 0$ for $i = 1, 2$, such that

$$b_1\lambda^d \leq N(\lambda) \leq b_2\lambda^d\,.$$

We shall see later on that for $\lambda \to \infty$ one can take $b_1 \to b_2$, provided the space V is "infinitesimally renormalizable" (see 3.4).

(b) The conclusion of 3.3.C$_2$ remains valid if the bound $\tilde{\lambda}_1 \geq \tau\varepsilon^{-1}$ is replaced by $\tilde{\lambda}_j \geq \tau\varepsilon^{-1}$ for a fixed $j \geq 1$ and if one uses $a = a(\delta, \rho, \tau, j)$.

(c) Lower bounds on λ_1 often come under the name of *Poincaré-Sobolev inequalities.* By Cheeger's theorem, the first eigenvalue of $E(x) = \|dx\|_2/\|x\|_2$ on a Riemannian manifold can be bounded from below by the *isoperimetric* constant (see below) and Cheeger's argument (based on the coarea formula) can be generalized to non-Riemannian geodesic spaces. Let us indicate several examples where $\lambda_1 \geq \mathrm{const}\,\mathrm{Diam}\,V$.

(c$_1$) V is the interval with the standard metric and measure. The lower bounds on all λ_i are immediate here.

(c$_2$) V is the Euclidean ball or cube. Then the inequality $\lambda_1 \geq \mathrm{const}_n\,\mathrm{Diam}$ follows from the following multiplicativity of λ_1

$$\lambda_1(V_1 \times V_2) \geq \mathrm{const}\,\min\left(\lambda_1(V_1), \lambda_1(V_2)\right)\,.$$

In fact λ_1 of certain "fibered spaces" V can be bounded from below by those of the base and the fibers. We shall show this in another paper where we shall generalize Kato's inequality to non-linear spectra.

(c$_3$) Recall that a (geodesic) *segment* $[v_1, v_2] \subset V$ for v_1 and v_2 in V is the image of an isometric map $[0, d] \to V$ for $d = \mathrm{dist}(v_1, v_2)$ which sends $-1 \to v_1$ and $1 \to v_2$. A subset $V_0 \subset V$ is called a *d-cone* from $v_0 \in V$ if it is a union of segments of length d issuing from v_0. If V_0 is a cone, one naturally defines *ad-cones* $aV_0 \subset V$ for $a \in [0, 1]$.

For a μ-measureable cone V_0, we consider the function $\mu(\alpha) = \mu(\alpha V_0)$ which is monotone in α and so almost everywhere differentiable. Divide the measure of the complement $V_0 \backslash \alpha V_0$ by the derivative of $\mu(\alpha)$ and let

$$b(V_0) = \sup_{\alpha \geq \frac{1}{2}} \mu(V_0 \backslash \alpha V_0)/\mu'(\alpha) .$$

Then take the supremum over all d-cones V_0 in V,

$$b_d = b_d(V) = \sup_{V_0} b(V_0) .$$

It is shown in [Gr]$_4$, for Riemannian manifolds V, that $\widetilde{\lambda}_1$ of $B(\varepsilon) \subset B(10\varepsilon)$ can be bounded from below by $\lambda_1 > C\varepsilon$ where $C > 0$ depends only on $\sup_{d \leq 20\varepsilon} b_d$. In fact the argument in [Gr]$_4$ extends to all metric spaces and (as we shall prove elsewhere) yields the following more general (and especially useful for Carnot spaces) lower bound on $\widetilde{\lambda}_1$.

(c_4) Instead of joining points by segments we join them by *random paths*. Namely, to each pair of points $(v_1, v_2) \in V \times V$ we assign a probability measure $\widetilde{\mu}_{v_1,v_2}$ in the space of continuous maps $[0,1] \to V$ joining v_1 and v_2. By integrating this measure over $V \times V$ we get a measure on the space of maps $[0,1] \to V$, called $\widetilde{\mu}$. Similarly, for each $v_0 \in V$ we have the integrated measure $\widetilde{\mu}_{v_0}$ in the space P_{v_0} of paths issuing from v_0.

Next consider a "hypersurface" in V that is a subset H whose ε-neighbourhoods H_ε satisfy

$$A(H) \underset{\text{def}}{=} \limsup_{\varepsilon \to 0} \varepsilon^{-1} \mu(H_\varepsilon) < \infty$$

and denote by $P_{v_0}(H) \subset P_{v_0}$ the subset of path $p : [0,1] \to V$, such that $p(t) \in H$ for some $t \geq \frac{1}{2}$. Define

$$\widetilde{b} = \sup_{H, v_0} \widetilde{\mu}(P_{v_0}(H))/A(H) .$$

Notice that this $\widetilde{b}$ (as well as b_d of the previous section) is an essentially local invariant in the space of paths.

It is nearly obvious (compare [Gr]$_4$) that the inequality $\widetilde{b} = \widetilde{b}(\mu) < \infty$ for some $\widetilde{\mu} = \mu_{v_1,v_2}$ gives us the following.

Isoperimetric inequality. *Let V be a compact metric space and let V_1 and V_2 be compact subsets in V separated by a hypersurface H in V (i.e., V_1 and V_2 lie in different components of $V \backslash H$). Then* $\min(\mu(V_1), \mu(V_2)) \leq 4\widetilde{b}A(H)$.

By Cheege's theorem this suffices to bound λ_1 (and $\widetilde{\lambda}_1$) from below.

Notice that the "geodesic cone" set-up (see (C_3)) corresponds to the Dirac δ-mass supported on a geodesic segment between v_1 and v_2 (at least for those v_1 and v_2 where such a segment is unique).

3.3.C_4 Reduction of the (isoperimetric) Sobolev inequality to Poincaré inequality. Let every ball $B \subset V$ satisfy the following two conditions

(1) POINCARÉ PROPERTY. Every hypersurface H in B dividing B into two pieces of equal measure satisfies

$$A(H) \geq C[\mu(B)]^{\alpha}$$

for some constant $C > 0$ and $0 < \alpha < 1$.

(2) UNIFORM COMPACTNESS. There are at most k points in B whose mutual distances are all $\geq$ radius of B.

If V is a geodesic space, then the boundary of every subset W with $\mu(W) \leq \frac{1}{2}\mu(V)$ satisfies

$$A(\partial W) \geq K^{-1}C(\mu(W))^{\alpha}. \tag{$*$}$$

Proof: To simplify the matter, assume that $\mu(W \cap B_v(r))$ is continuous in the radius r of the ball around each point $w \in W$. Then there exists a ball of maximal radius say B_1, such that $\mu(B_1 \cap W) = \frac{1}{2}\mu(W)$. Then we take the second such largest ball B_2 with center outside B_1, then B_3 with center outside the union $B_1 \cup B_2$ and so on. Thus we obtain balls $B_1, \ldots, B_i, \ldots$ covering W. If some of these balls intersect at $w \in W$, then their centers, say $v_1, \ldots, v_\ell$, satisfy for all $1 \leq i < j \leq \ell$

$$\text{dist}(v_i, v_j) \geq \max\left(\text{dist}(v_i, w), \text{dist}(v_j, w)\right).$$

Since V is a *geodesic* space, there exist points $v_i' \in B_i$, such that

$$\text{dist}(v_i', w) = \delta = \min_{1 \leq i \leq \ell} \text{dist}(v_i, w)$$

and

$$\text{dist}(v_i', v_i) = \text{dist}(v_i, w) - \delta\,.$$

Clearly,

$$\text{dist}(v_i', v_j') \geq \delta$$

and so $\ell \leq k$. (This argument reproduces the standard proof of Besicoviĉ covering lemma.)

Now, we apply (1) to $H_i = B_i \cap \partial W$ and obtain

$$A(H_i) > C\mu(B_i \cap W)$$

and then (*) by adding these inequalities over all $i = 1, 2, \ldots$.

Application to λ_1. By Mazia-Cheeger inequality our (*) implies

$$\|x\|_{L_q} < \text{const} \|dx\|_{L_1}$$

for const = const$(k^{-1}C, \alpha)$, for $q = \alpha^{-1}$ and all functions x on V whose both levels V_+ where $x \geq 0$ and V_- have measures $\geq \frac{1}{2}\mu(V)$. It follows, that the first eigenvalue λ_1 of $E = L_1 d/L_q$ for $q = \alpha^{-1}$ is $\geq$ const$^{-1} > 0$. (Notice that the inquality (1) we started with expresses a kind of lower bound on the first eigenvalue of $L_1 d/L_1$ on the ball B.)

3.3.D Spectra of disjoint unions $V = \bigcup_k V_k$ for $p \neq q$. As we have seen earlier, the spectral function $N(\lambda) =$ "dim"$E^{-1}[0, \lambda]$ of V is the sum of the corresponding functions $N_k(\lambda)$ of V_k, provided "dim" is subadditive (e.g., "dim" = ess) and $E = L_p d/L_q$ for $p = q$. If $p \neq q$, then the determination of best bounds on $N(\lambda)$ in terms of $N_k(\lambda)$ is a non-trivial problem which is closely related to the spectrum of L_p/L_q (compare §1.). To see this relation we consider several examples, where we assume for simplicity's sake that all pieces V_k, $k = 1, \ldots, \ell$, have the same measure $\mu(V_k) = \ell^{-1}$.

3.3.D$_1$. *Let $N_k(\alpha) \geq 1$ for some $\alpha > 0$ and all $k = 1, \ldots, \ell$ and let $N'(\lambda)$ be the spectral function for the energy $E'(y) = \|y\|_{L_p}/\|y\|_{L_q}$ on the measure space consisting of ℓ atoms of mass ℓ^{-1}. Then*

$$N(\lambda) \geq N'(\beta\lambda) ,$$

for $\beta = \alpha^{-1}\ell^{\frac{1}{p}-\frac{1}{q}}$.

Proof: Take functions x_k on V_k for $k = 1, \ldots, \ell$, such that $E(x_k) \leq \alpha$, and observe that the restriction of E to the span of these x_k is bounded by $\beta E'$.

3.3.D$_2$. Let us apply the above to the spectrum of $L_p d/L_q$ on a metric space V, which satisfied the following strong regularity assumption. Every two (not necessarily) concentric balls B_1 and B_2 in V of radii R and $2R$ satisfy

$$C^{-1} \leq \mu(B_1)/\mu(B_2) \leq C$$

for all $R > 0$ and a fixed $C = C(V) > 0$. We recall the maximal number IN(ε) of disjoint ε-balls in V and look at linear combinations of standard functions supported in such balls. Then *for the ess-spectral function $N^{\text{ess}}(\lambda)$ we obtain with the following lower bound*

$$N^{\text{ess}}(\lambda) \geq b\,\text{IN}(\lambda^{-1})$$

for some constant $b > 0$ depending only on C.

Remark. If one wants to estimate the *pro-spectrum* of $L_p d/L_q$ one should invoke estimates by Kašin and Gluskin of the pro-spectrum of L_p/L_q (see [Pi]).

3.3.D₃. Suppose that the (mollified if necessary) spectral function of every ε-ball satisfies for given p and q,

$$N^{\text{ess}}(\lambda_0, B_\varepsilon) \leq \text{const}\, \varepsilon^{-1}(\mu(B_\varepsilon))^{\frac{1}{p}-\frac{1}{q}} ,$$

for some fixed $\lambda_0 > 0$ and all $\varepsilon > 0$. *Then for $p \geq q$ the function $N^{\text{ess}}(\lambda)$ of V is bounded by*

$$N^{\text{ess}}(\lambda) \leq c\,\mathrm{IN}(\lambda^{-1}) ,$$

by the earlier additivity argument. Thus

$$N^{\text{ess}}(\lambda) \asymp \mathrm{IN}(\lambda^{-1}) .$$

To grasp the meaning of this asymptotic relation, let ε_i be the maximal number for which there are i disjoint ε_i-balls $B_i, B_2, \ldots, B_i$ in V and let x_i denote the distance function to the complement of these balls,

$$x_i(v) = \text{dist}\left(v, V \setminus \bigcup_{j=1}^{i} B_j\right) .$$

Then the above discussion amounts to saying that x_i approximately equals the i-th "eigenfunction" of the energy $E(x) = \|dx\|_{L_p}/\|x\|_{L_q}$, that is

$$\lambda_i^{\text{ess}} \asymp E(x_i) .$$

3.3.E Pro-spectra for $p > q$. Let us show that pro-spectrum in most cases grows faster than the ess-spectrum for $p > q$. Namely $\lambda_i^{\text{pro}}/\lambda_i^{\text{ess}} \to \infty$ for $i \to \infty$.

Start with the simplest case, where $p = \infty$ and $q = 2$. Assume that V can be covered by i balls of radius $\varepsilon = \varepsilon_i$ and show that

$$\lambda_{2i}^{\text{pro}} > \sqrt{i}\varepsilon_i$$

provided $\mu(V) = 1$. In fact, let L be a $2i$-dimensional linear space of functions on V and $L' \subset L$ an i-dimensional subspace of the functions vanishing at the centers of the covering balls. Then every $x \in L'$ has $\|x\|_{L_\infty} \leq \varepsilon^{-1}\|dx\|_{L_\infty}$ and our claim follows from 1.1.B.

This argument applies to all $q < \infty$ and yields the relation $\lambda_i^{\text{pro}}/\lambda_i^{\text{ess}} \to \infty$ under the regularity assumption on V.

3.3.E$_1$. In order to make the above argument work for $p < \infty$ we must first project our L to some finite dimensional L_p-space, and then apply the results of Kašin and Gluskin cited earlier. Such a projection is customarily constructed either with *spline approximations* (discretization) or with *smoothing operators*. Recall that the set S of functions on V is called an *(ε, d)-spline* if the restriction of S on each ε-ball in V is at most d-dimensional. In what follows we shall only use very primitive piece-wise constant splines which correspond to the smoothing with the kernel K_ε in 3.1.C. (A discussion on deep smoothing of Nash can be found in [Gr]$_3$.)

Let us assume every ε-ball $B_\varepsilon \subset V$ satisfies the following:

Mollified Poincaré L_p-lemma. *If a function x on B_ε has $\int_{B_\varepsilon} x dv = 0$, then the L_p-norm B_ε of x on the concentric ball B_δ is bounded by the L_p-norm of dx on B_ε as follows*

$$\|x \mid B_\delta\|_{L_p} \leq C\varepsilon^{-1}\|dx\|_{L_p},$$

for a fixed $C > 0$ and all δ satisfying

$$\delta \leq C^{-1}\varepsilon.$$

Let us also assume V is regular as earlier and prove the following:

Theorem. *If $q < p$ then*

$$\lambda_i^{\text{pro}} \geq \text{const}\, i^\theta \lambda_i^{\text{ess}}$$

for some positive const and θ, and all $i = 1, 2, \ldots$.

Proof: Let L be a $2i$-dimensional linear space of functions on V. Take the minimal $\varepsilon = \varepsilon_i$, such that some δ-balls for $\delta \leq C^{-1}\varepsilon$, say $B_1(\delta), \ldots, B_i(\delta)$ cover V. Notice that we may assume the covering by the concentric ε-balls has bounded (independent of i) multiplicity. Denote by $L' \subset L$ the i-dimensional subspace defined by the equations

$$\int_{B_j(\varepsilon)} x\, dv = 0, \qquad j = 1, \ldots, i,$$

and let

$$\mu_i' = \sup_{x \in L'} \|dx\|_p / \|x\|_p .$$

Notice that

$$\mu_o' \geq \text{const}\, \varepsilon_i^{-1},$$

by the earlier discussion.

Now we take

$$\varepsilon' = \varepsilon'_i = (C'\mu'_i)^{-1}$$

for large (but independent of i) constant, and consider a covering of V by i' balls of radius $\delta' = C^{-1}\varepsilon^1$. We may assume (slightly changing the covering if necessary), that there exists a partition of V into i' subsets V_j of equal mass $= \mu(V)/i'$, such that each subset is contained in a δ-ball of the covering.

Let $x \mapsto \bar{x}$ be the linear operator, which averages x over each V_j, $j = 1,\dots,i'$. Namely $\bar{x}$ is constant and equal $\int_{V_j} x/\mu(V_j)$ on every V_j. Now we see that

$$\lambda_{2i}^{\text{pro}} \geq \lambda'_i \mu'_i ,$$

where λ'_i is the i-the eigenvalue of $E' = L_p/L_q$ on the i'-dimensional space, and the theorem easily follows from the known bound on λ'_i (see [Kaŝ] and [Pi]).

3.4 Selfsimilarity and asymptotics $N(\lambda) \sim \text{const}\,\lambda^d$. This signifies the existence of the limit,

$$\text{const} = \lim_{\lambda\to\infty} N(\lambda)/\lambda^d ,$$

and one is most happy when $0 < \text{const} < \infty$. Notice that the relation $N(\lambda) \sim \text{const}\,\lambda^d$ is equivalent to the *asymptotic homogeneity* of $N(\lambda)$, that is

$$N(a\lambda) \sim a^d N(\lambda)$$

for every fixed $a > 0$ and $\lambda \to \infty$. We shall see below that in certain cases this asymptotics follows from (infinitesimal) homogeneity of the energy.

3.4.A Example. Let V_ε denote the ε-cube $[0,\varepsilon]^n$ and

$$aV_\varepsilon = V_{a\varepsilon} \quad \text{for} \quad a > 0 .$$

We also denote by $a : V_\varepsilon \to aV_\varepsilon$ the obvious (scaling) map which transforms functions x on V_ε to those on aV_ε. Namely $x(v) \mapsto x(a^{-1}v)$, that is $x \mapsto x \circ a^{-1}$. It is obvious that the energy $E(x) = \|dx\|_p/\|x\|_p$ is homogeneous

$$E(x \circ a^{-1}) = a^{-1}E(x) .$$

Next we observe that for every $k = 1,2,\dots,$ the cube V_ε can be partitioned into k^n cubes $k^{-1}V_\varepsilon$. Then the asymptotic additivity of $N(\lambda)$ (see 3.1.F$_1$) implies for "dim" $=$ ess that $N(k\lambda) \sim k^n N(\lambda)$ for all *integers* $k > 0$.

3.4.B Asymptotic homogeneity of $N^0(\lambda)$ for domain $V \subset \mathbb{R}^n$. Recall that $N^0(\lambda, V) = N(\lambda, V/\infty)$ refers to E on functions with compact supports in V, where V is an open subset in $\mathbb{R}^n$. We denote by $aV \subset \mathbb{R}^n$ the homothety (scaling) of V by $a \in \mathbb{R}$ and write

$$\sum_{i=1}^{k} a_i V \prec W ,$$

if there exist vectors $b_i \in \mathbb{R}^n$, such that the translates $a_i V_i + b_i \subset \mathbb{R}^n$ do not intersect and are all contained in W. Now the homogeneity of $E(x)$ together with the obvious superadditivity of $N^\circ(\lambda)$ imply the following property of $N^0(\lambda) =$ "dim"$E^{-1}(-\infty, \lambda)$ for $E = L_p d/L_q$, and $p \geq q$, and for all "dim" satisfying (i)-(vi) in 0.4.

(*) *The relation*

$$\sum_{i=1}^{k} a_k V \prec W$$

implies the inequality

$$\sum_{i=1}^{k} N^0(a_i\lambda, V) \leq N^0(\lambda, W) ,$$

for all open subsets V and W in $\mathbb{R}^n$ and all strings of real numbers a_i.

Now we recall the following

Trivial Lemma. *Let V be a bounded open supset in $\mathbb{R}^n$ and $N(\lambda)$ a positive function in $\lambda \in (0, \infty)$, such that*

$$\sum_{i=1}^{k} N(a_j\lambda) \leq N(a_0\lambda)$$

for all strings of real numbers a_i satisfying

$$\sum_{i=1}^{k} a_i V \prec a_0 V .$$

Then

$$\limsup_{\lambda \to \infty} \lambda^{-n} N(\lambda) = \liminf_{\lambda \to \infty} \lambda^{-n} N(\lambda) ,$$

that is

$$N(\lambda) \sim C\lambda^n$$

for some $C \in [0, \infty]$, provided the boundary $\partial V \subset \mathbb{R}^n$ has measure zero.

3.4.B$_1$. On Positivity and finiteness of constant C. The above discussion shows that the spectral function $N^0(\lambda) = N^0(\lambda, V/\infty)$ for $E = L_p d/L_q$ and $p \geq q$ satisfies Weil's relation $N^o(\lambda) \sim C\lambda^n$, where $C < \infty$ for all "dim" and $p \geq q$ by Poincaré's Lemma. It is obvious that $C > 0$ for $p = q$ and all "dim". Furthermore, if "dim" = ess, then $C > 0$ for all $p \geq q$, as it follows from 3.3. On the other hand if $p > q$ and "dim" = pro then $C = 0$. In fact

$$N^0(\lambda) \asymp \lambda^{n-\theta}$$

for some $\theta > 0$ which can be explicitly determined by the standard approximation techniques, (see [Kaŝ] and [Pi]). Probably, the $\sim$asymptotics also follows by those techniques.

3.4.B$_2$. Determination of $C_0 = C/\operatorname{Vol} V$. It is clear from the previous discussion that $C = C_0 \operatorname{Vol} V$ where $C_0 = C_0(n,p,q)$ a universal constant. If $p = q = 2$ one known this C_0 from the spectrum of the Laplace operator $\Delta = d^*d$, but apart from this case the exact determination of C_0 (or of the asymptotics for $n \to \infty$) seems to run into the same problem as for the covering constant of $\mathbb{R}^n$ by equal balls.

3.4.C. Asymptotics $N(\lambda) \sim C\lambda^n$ for Riemannian manifolds V. Small balls in V are almost isometric to those in $\mathbb{R}^n$ for $n = \dim V$. It follows that

$$N(\lambda) \sim C_0(\operatorname{Vol} V)\lambda^n$$

for the above C_0 and under the same conditions as p and q as for domains in $\mathbb{R}^n$. Notice, that for $p = q = 2$ one obtains much sharper asymptotics using heat and (or) wave equations. One might try to extend the heat equation method to other p and q by using some functional integral of $\exp -tE(x)$.

3.4.D Homogeneous Lie groups. Let V be a Lie group with a left invariant geodesic metric, such that for every $a > 0$, V admits an a-selfsimilarity, that is a map $a : V \to V$, such that

$$\operatorname{dist}(av_1, av_2) - a \operatorname{dist}(v_1, v_2)$$

for all v_1 and v_2 in V. It is well known that such a V is a nilpotent Lie group of Hausdorff dimension $d \geq n = \dim_{\text{top}} V$, where $d = n$ iff $V = \mathbb{R}^n$. The argument of 3.4.B immediately yields Weyl's relation

$$N(\lambda) \sim C\lambda^d$$

for $p \geq q$. Furthermore, one knows (see [F-S], [Pa], [Var]) that this C behaves as that in 3.4.B$_1$.

3.4.E Smooth metric spaces. For metric spaces V_1 and V_2 one defines the Hausdorff distance, called $|V_1 - V_2|_H$, by the condition:

$|V_1 - V_2|_H \leq \varepsilon \Longleftrightarrow$ their exists a metric on the disjoint union $V_1 \cup V_2$, which extend those on V_1 and on V_2, and such that the ordinary Hausdorff distance between the subsets V_1 and

V_2 in $V_1 \cup V_2$ is $\leq \varepsilon$. A more invariant but somewhat less convenient definition consists of mapping the Cartesian power V^N into $\mathbb{R}^M$ for $M = N(N-1)/2$ by $\{v_i\} \mapsto \operatorname{dist}(v_i, v_j)$ and then by measuring the Hausdorf distances of the images in $\mathbb{R}^M$ of V_1 and V_2 for all N.

If V_1 and V_2 carry some measures, we can incorporate these into the definition of the Hausdorff distance by either looking at the pushforward of the measures to $\mathbb{R}^M$, or with the following additional requirement on the metric in $V_1 \cup V_2$:

Every ε-ball B in $V_1 \cup V_2$ has $\mu_1(B) - \mu_2(B) \leq \varepsilon$, where μ_1 and μ_2 are the measures on V_1 and on V_2 respectively.

Now, for every metric space $V = (V, \operatorname{dist})$ we write

$$aV = (V, a\operatorname{dist}) ,$$

for all $a > 0$, and we call V (uniformly) C^1-*smooth*, if every two balls $B_{\varepsilon_1}(v_1)$ and $B_{\varepsilon_2}(v_2)$ in V satisfy

$$|\varepsilon_1^{-1} B_{\varepsilon_1}(v_1) - \varepsilon_2^{-1} B_{\varepsilon_2}(v_2)|_H \leq \delta \qquad (*)$$

where δ depends only on $\operatorname{dist}(v_1, v_2)$ and $\delta \to 0$ for $\operatorname{dist}(v_1, v_2) \to 0$.

It is easy to show that every smooth geodesic space admits a *tangent cone* $T_v(V)$ at all $v \in V$, that is a homogeneous Lie group as in 3.4.D, such that $\varepsilon^{-1} B_\varepsilon(v)$ Hausdorff converges to the unit ball $B_1 \subset T_v(V)$,

$$|B_1 - \varepsilon^{-1} B_\varepsilon(v)|_H \longrightarrow 0 \quad \text{for} \quad \varepsilon \to 0 .$$

Next we say that V is μ-smooth for a given measure μ on V if $(+)$ incorporates the measure, where the ball $\varepsilon^{-1} B_2$ is given the measure of total mass one obtained by the normalization of $\mu \mid B_\varepsilon$. In this case $\varepsilon B_\varepsilon(v)$ converges to $B_1 \subset T_v$ together with μ and one can see that the spectrum D of $L_p d / L_q$ is semicontinuous that is $N(\lambda, T_v) \leq \liminf_{\varepsilon \to 0} N(\lambda, \varepsilon^{-1} B_\varepsilon(v))$. Furthermore, if the (mollified) first eigenvalue of each ball B_ε in V is bounded from below by $\operatorname{const} \varepsilon^{-1} \mu(B)^{\frac{1}{p} - \frac{1}{q}}$, then the spectrum in continuous. It easily follows (under the same conditions as in 3.4.B) that

$$N(\lambda, V) \sim C \lambda^d$$

where d is the Hausdorff dimension is constant in v) and

$$C = \int_V C_0(T_v(V))\, dv .$$

Remarks. (a) The asymptotics $N(\lambda) \sim C\lambda^d$ remains valid under milder (non-uniform) smoothness condition, where the tangent cone may not exist on some "thin" subset of V. In fact one can even replace the Hausdorff distance by another one which is concerned with the measure-images of V^N in $\mathbb{R}^M$ rather than the set-images. It would be interesting to find meaningful examples to justify such generalizations.

(b) The previous discussion has the following discrete counterpart, where D is a *difference operator* on a discrete set V. For example, we may consider the coboundary operator on 0-cochains on the set V of vertices of some graph. Then we consider an exhaustion of V by finite subsets V_i and study the asymptotics of the spectrum of $D \mid V_i$ for $i \to \infty$. The standard example is that of $V = \mathbb{Z}^n \subset \mathbb{R}^n$ where V_i is a ball of radius i around the origin. The smoothness of V must be now expressed in terms of the *tangent cone at infinity*, (which for metric spaces V refers to the Hausdorff limit of $(\operatorname{diam} V_i)^{-1} V_i$ for $i \to \infty$) and the spectral asymptotics are closely related to the thermodynamics limit in statistical mechanics. The existence of such limits in $\mathbb{R}^n$ is easy by the non-Abelian nilpotent case is non-trivial (see [Pa]$_2$).

3.4.F Remarks on the case $p < q$. If $\dim V = 1$, then the energy $E = L_p d / L_q$ satisfies

$$N(\lambda) \asymp \lambda$$

for *all* p and q as it follows from 2.2. In general, if for example V is a domain in $\mathbb{R}^n$, one asks what happens for p and q in the range of the Sobolev embedding theorem, that is for

$$s = 1 - \frac{n}{p} + \frac{n}{q} \geq 0 .$$

Notice that the energy $E(x) = \|dx\|_p / \|x\|_q$ is *scale homogeneous* of degree s,

$$E(x \circ a) = a^s E(x) ,$$

and so the spectrum of E accumulates at zero for $s < 0$. On the other hand, by the embedding theorem the spectrum is discrete for $s > 0$ but the asymptotics (say for "dim" = ess) seems to be unknown for $p < q$. The most interesting case is that of $p = 1$ and $q = n/n - 1$ where $s = 0$ and the (non-compact) embedding theorem is still valid. This theorem bounds λ_1 away from zero (for all "dim") but I do not know if the spectrum is discrete (i.e., $\lambda_i \to \infty$), say for "dim" = ess.

3.4.G The asymptotics $N^0(\lambda) \sim C\lambda^{rn}$ for operators D of order r. Let D be a differential operator of *pure order* r on $\mathbb{R}^n$ with *constant coefficients*. In other words D is invariant under translations and

$$D(x \circ a) = a^r D(x) .$$

Then the previous argument implies that the corresponding $N^\circ(\lambda)$ for $E = L_p D / L_q$ and $p \geq q$ is asymptotic to $C\lambda^{\frac{n}{r}}$ where $0 < C \leq \infty$. If $D = \partial^r$, where $\partial^r x$ denotes the string of the partial derivatives of x of order r (e.g., $\partial^1 = d$) then $C < \infty$ by Poincaré's lemma.

The inequality $C < \infty$ remains true for all elliptic operators D and $1 < q \leq p < \infty$ as

$$\|\partial^r x\|_p < \text{const} \|Dx\|_p$$

for functions x with compact support in $\mathbb{R}^n$. In fact this is even true for pseudo-differential operators of order r which may be any *real* number, e.g., for $(\sqrt{\Delta})^r$ where Δ is the Laplace operator. On the other hand if one wishes to keep $p = \infty$, one should require that D has finite dimensional kernel on every open subset in $\mathbb{R}^n$ which is much stronger than ellipticity. Properties of such D are identical in most respects to those of ∂^r. (If $r = 1$ then $\partial^1 = d$ essentially is the only example, but for $r \geq 2$ there are plenty of such D. For example

$$D : x \mapsto \left(\frac{\partial^2 x}{\partial u_1^2}, \frac{\partial^2 x}{\partial u_2^2}, \ldots, \frac{\partial^2 x}{\partial u_n^2} \right) .$$

3.4.G$_1$. The above discussion extends to homogeneous (nilpotent) Lie groups in place of $\mathbb{R}^n$. Here we look at left invariant operators of order r such that

$$D(x \circ a) = a^r D(x) .$$

Then the corresponding energy $E = L_p D / L_q$ is a-homogeneous of degree $s = r - \frac{d}{p} + \frac{d}{q}$, where d is the Hausdorff dimension of some (and hence any) left invariant and a-homogeneous geodesic metric on our group. Such homogeneity insures, as earlier, the asymptotics $N(\lambda) \sim \text{const}\, \lambda^{d/r}$. What is less trivial is the bound const $< \infty$ and, more generally, the discreteness of the spectrum for $s > 0$. For this we need some (hypo)-ellipticity of D. Probably, if D everywhere (formally as well as locally) has finite dimensional kernel, then the above spectrum is discrete. In fact this finiteness condition makes any mentioning of the group structure unnecessary but nilpotent groups enter through the back door anyway.

3.4.G$_2$. Another generalization consists of allowing *polylinear* operators on $\mathbb{R}^n$ of *pure degree* r, which means $D(x \circ a) = a^r D(x)$. Instead of the finite kernel condition, one should now postulate the discreteness of the spectrum of $L_\infty D / L_\infty$ (on all domains in $\mathbb{R}^n$). More interesting examples are provided by (elliptic) Monge-Ampere operators and the Yang-Mills operator.

3.4.G₂. Let us indicate some (very) non-elliptic operators D on $\mathbb{R}^n$ with (spectrally) interesting energy $E = L_p D / L_q$. First, let

$$Dx = \frac{\partial^n x}{\partial u_1, \dots, \partial u_n},$$

and restrict E to functions with compact support in a bounded domain in $\mathbb{R}^n$. This is especially attractive for $p = 1$ and $q = \infty$ where the problem is non-trivial even for $D = \partial^r$.

Another example is $D = \frac{\partial^3}{\partial u_1^3} + \alpha \frac{\partial^3}{\partial u_2^3}$ on $\mathbb{R}^2$ for some real α with the periodic boundary conditions. (Which means passing to the torus $\mathbb{R}^2/\mathbb{Z}^2$). Here the spectrum of E is intimately related to arithmetic properties of α. For example the discreteness of the spectrum for $\alpha = 2$ and $E = L_2 D / L_2$ is a non-trivial theorem of Thue.

§4 Bezout intersection theory in P, Δ and $P \times \Delta$

4.1 Cohomological definition of ess. Recall that the $\mathbb{Z}_2$-cohomology of the projective space P^k is multiplicatively generated by a single 1-dimensional element, say α, such that $\alpha^i \neq 0$ for $i \leq k$ (and, of course $\alpha^i = 0$ for $i > k = \dim P^k$). With this one sees that $\operatorname{ess} P^k = k$, since the cohomology is homotopy invariant. In fact, one knows (and the proof is very easy) that for all locally closed (i.e., open $\cap$ closed) subsets $Q \subset P^k$, the "dimension" $\operatorname{ess} Q$ equals the greatest i, such that the class α^i does not vanish on Q. Now the subadditivity of ess follows from the fact that cohomology classes multiply like functions. Namely, *if α vanishes on A and β on B then the cup-product $\alpha \vee \beta$ vanishes on $A \cup B$*, where A and B are locally closed subsets in a topological space and α, and β are some cohomology classes of this space.

Here is another immediate corollary of the cohomological definition of "dimension" ess.

4.1.A. *Let S be a connected topological space with a continuous involution called $s \longleftrightarrow -s$, and let f be a symmetric continuous map of S into the sphere S^k, where symmetric means $f(-s) = -f(s)$. If the $\mathbb{Z}_2$-cohomology of S vanish in the dimensions $1, 2, \dots, i-1$, then the image in $P^k = S^k/\mathbb{Z}_2$ of the induced map $\bar{f} = f/\mathbb{Z}_2$ satisfies*

$$\operatorname{ess} \bar{f}(S/\mathbb{Z}_2) \geq i .$$

Remark. The vanishing assumption is satisfied for example, for the sphere S^j for $j \geq i$, since every $(i-1)$-dimensional subset is contactible in S^j for $j \geq i$. In particular, the existence of a symmetric map $f : S^j \to S^k$ implies that $j \leq k$. This fact is often called the *Borsuk-Ulam theorem*.

4.2 Bezout theorem. The Poincaré-Lefschetz duality between the cup-product and intersection shows that

$$\operatorname{coess} A \cap B \leq \operatorname{coess} A + \operatorname{coess} B \ , \qquad (*)$$

where the coessence of a subset in P^k is $k -$ ess, and where A and B are locally closed subsets in P^k.

Here is another form of Bezout theorem. Let $f : S^i \to S^k$ be a symmetric continuous map and $\overline{f} : P^i \to P^k$ be the induced map. Then

$$\operatorname{coess} \overline{f}^{-1}(A) \geq \operatorname{coess} A \qquad (**)$$

for all $A \subset P^k$.

Example. Let $\varphi : S^i \to \mathbb{R}^j$ be a continuous map and $A \subset P^i$ consists of the pairs $(s, -s)$ such that $\varphi(s) = \varphi(-s)$. Then

$$\operatorname{coess} A \geq i - j \ . \qquad (***)$$

In fact, let $\overline{f} : P^i \to P^{i+j-1}$ be defined by

$$\overline{f} : (s_0, \ldots, s_i) \longmapsto (s_0, \ldots, s_i, \varphi_1(s) - \varphi_1(-s), \ldots, \varphi_j(s) - \varphi_j(-s))$$

where $s_0, \ldots, s_i$ are the coordinates of a (one out of two) point in S^i over $\overline{s} \in P^i$ and where $\varphi_1, \ldots, \varphi_j$ are the components of φ. Then A equals the pullback of the obvious j-codimensional subspace in P^{i+j+1} and Bezout theorem applies because $\overline{f}$ is covered by some f.

Remarks. (a) If $i \geq j$, then $(***)$ says that A is non-empty. This is another formulation of the Borsuk-Ulam theorem.

(b) Let us *define* coess′ of a subset A in a (possibly infinite dimensional) projective space P as the minimal i such that there exists a continuous map of P into another projective space, say $\overline{f} : P \to P'$, such that $\overline{f}$ can be covered by a symmetric map of the (spherical) double-coverings of P and P' and such that A contains the pull-back of a projective subspace in P' of codimension i. This coess′ satisfies $(*)$ and $(**)$ (almost) by definition. Moreover, by the Poincaré Lefschetz duality

$$\operatorname{coess}' \geq \operatorname{coess} = \dim P - \operatorname{ess}$$

if P is finite dimensional.

Example. (a) Every i-coplane (see 0.5.B) obviously has coess′ $= i$. Hence, it meets every i-plane by the above discussion.

(b) Let P be the projective space of continuous functions on V and $U \subset V$ be a measurable subset. Denote by $P_u \subset P$ the subset of functions x *equidividing* U, that is the subsets $x^{-1}(-\infty, 0] \cap U$ and $x^{-1}[0, \infty) \cap U$ have measure at least $\frac{1}{2}\mu(U)$. (If the zero level $x^{-1}(0) \subset V$ of X has measure zero, then these equal $\frac{1}{2}\mu(U)$.) Obviously coess$'\, P_u \leq 1$.

4.2.A Corollary (Borsuk-Ulam again). *Let a subset $A \subset P$ have* ess $A \geq i$ *(e.g., A is projective of dimension i) and $U_1, \ldots, U_i$ are subsets in V. Then there exists a function $x \in A$ equidividing all i subsets.*

4.2.B An archetypical spectral application. Let V be a compact n-dimensional Riemannian manifold and $E(x)$ denotes the $(n-1)$-dimensional volume *of the zero set* $x^{-1}(0) \subset V$. Then *the spectrum* $\{\lambda_i\}$ *of this E satisfies*

$$\lambda_i \asymp i^{\frac{1}{n}} .$$

Proof: To bound λ_i from below partition V into i subsets U_i which are roughly isometric to the Euclidean ball of radius $\varepsilon = i^{-\frac{1}{n}}$. Then the above equidividing function x satisfies according to the (isoperimetric) Poincaré lemma,

$$E(x) \geq \text{const}\, i\varepsilon^{n-1} = \text{const}\, i^{\frac{1}{n}} .$$

Next, for the upper bound, first let V be a domain in $\mathbb{R}^n$. Then the space P_d of polynomials of degree $\leq d$ has $i = \dim P_d \asymp d^n$. Since the zero set Σ of a polynomial of degree $\leq d$ meets every line at no more than d points,

$$\mathrm{Vol}_{n-1}(\Sigma \cap V) \leq d(\mathrm{Diam}\, V)^n ,$$

which provides the required upper bound on λ_i for $V \subset \mathbb{R}^n$. In the general case, one may apply a similar argument to an algebraic realization (due to J. Nash) of V in some Euclidean space $\mathbb{R}^N$.

Question. Let P be the space of maps $x : V \to \mathbb{R}^m$ for some $m < n$ and

$$E(x) = \mathrm{Vol}_{n-m} E^{-1}(0) .$$

Then the above polynomial example shows that

$$\lambda_i \leq \text{const}\, i^{\frac{m}{n}} .$$

But I do not even know how to prove that $\lambda_i \underset{i \to \infty}{\longrightarrow} \infty$ for $m \geq 2$.

4.3 $\mathbb{Z}_2$-simplices. . Consider a topological space S and a continuous map π of S into a (finite or infinite dimensional) simplex. A *k-face*, say $S_k \subset S$ by definition is the pull-back of a k-face Δ_k in Δ and the boundary ∂S_k is the pull-back of the boundary of Δ_k. Recall that the $\mathbb{Z}_2$-cohomology of the pair $(\Delta_k, \partial\Delta_k)$ equals $\mathbb{Z}_2$ in dimension k and say that S is a *$\mathbb{Z}_2$-simplex* if the generator of this cohomology group, say $h(\Delta_k)$, goes by π^* to a *non-zero* element in $H^k(S_k, \partial S_k; \mathbb{Z}_2)$, say to $h(S_k)$, for all finite dimensional faces S_k of S.

Examples. (a) If S contains a subset S', such that $\pi : S' \to \Delta$ is a *homeomorphism*, then S is a $\mathbb{Z}_2$-simplex.

(b) Let $\pi : \Delta \to \Delta$, where π sends every face of Δ *into* itself. Such a map is homotopic to the identity (by an obvious linear homotopy) and so this is a $\mathbb{Z}^2$-simplex. hence, *the map π necessarily is surjective.*

In fact, one has the following obvious (modulo elementary homology theory):

4.3.A Proposition. *Let $\pi' : S \to \Delta$ be a continuous map sending each face $S_k = \pi^{-1}(\Delta_k)$ of S to Δ_k. Then π' is onto.*

4.3.B Basic example. Let S be the space of sequences $s = s_0, s_1, \dots, s_k, \dots$ of non-negative L_q-functions on V, such that the sum

$$\sigma = \sum_k \int s_k(v)dv$$

satisfies

$$0 < \sigma < \infty ,$$

and define $\pi : S \to \Delta$ by

$$\pi : s \longmapsto \left(\int s_0/\sigma, \int s_1/\sigma, \dots, \int s_k/\sigma \dots\right) .$$

If the implied measure μ on V is continuous, as we shall always assume below, then this is a $\mathbb{Z}_2$-simplex. Important subsimplices in S are:

(a) $S_\chi \subset S$, where every s_k equals 0 or 1, i.e., s_k is the characteristic function of the set where $s_k = 1$.

(b) $S_{co} \subset S_\chi$, where the implied subsets *cover* V.

(c) $S_{pa} \subset S_{co}$, where the subsets *partition* V, that is $\sum_k s_k = 1$.

Denote by $S(k)$ the set of sequences with $s_j = 0$ for $j > k$ and look at the induced $\mathbb{Z}_2$-simplex structure over $\Delta^k \subset \Delta$.

4.3.B$_1$ Proposition. *Let $T \subset S(k)$ be $\mathbb{Z}_2$-subsimplex (over Δ^k) and x a bounded function on V. Then there exists $s = (s_0, \ldots, s_k) \in T$, such that*

$$\int_V s_0 x = \int_V s_1 x = \cdots = \int_V s_k x .$$

Proof: We may assume T is compact which provides a constant C, such that

$$\sigma_i = C + \int_V s_i x > 0$$

for all $i = 0, \ldots, k$ and $s \in T$. Then 4.2.B applies to the map $T \to \Delta^k$ defined by $s \to (\sigma_0/\sigma, \ldots, \sigma_k/\sigma)$, for $\sigma = \sum_{i=0}^{k} \sigma_i$.

4.4 $\mathbb{Z}_2$-simplices in $P \times \Delta$. Let $\bar{f} : P^k \times \Delta \to P$ be a continuous map which admits a lift to a continuous map $S^k \times \Delta \to S$, where S^k and S are the spheres double covering P^k and P respectively. Then by the elementary homology theory the pull-back $T = \bar{f}^{-1}(P') \subset P^k \times \Delta$ of every *k-coessential* subset $P' \subset P$ (i.e., coess $P' \leq k$) is a $\mathbb{Z}_2$-simplex for the projection $T \to \Delta$. In fact the same conclusion remains valid for every k-essential subset $Q \subset P^\infty$ instead of P^k. This leads to the following unification of 4.2.A and 4.3.B$_1$.

4.4.A. *Let $T \subset S(k)$ be a $\mathbb{Z}_2$-simplex (over Δ^k) in the space of sequences of subsets $V_0, \ldots, V_k$ in V and let Q be a $(k+1)$-essential (i.e.,* ess $Q \geq k+1$*) set of continuous functions on V. Then there exist a function $x \in Q$ and a sequence $(U_0, \ldots, U_k) \in T$, such that*

(1) the zero level $x^{-1}(0) \subset V$ equidivides all $U_0, \ldots, U_k$ (in the sense of 4.2.A).

(2) For a given $p < \infty$

$$\int_{U_0} |x|^p = \int_{U_1} |x|^p = \ldots = \int_{U_k} |x|^p .$$

4.4.A$_1$ Remarks. (a) one can replace (1) by the following

$$(1') \qquad \sup_{v \in U_j} x(v) = - \inf_{v \in U_j} x(v) \quad \text{for} \quad j = 0, \ldots, k .$$

In fact one can require any "equidivision property" in-so-far as the "division" is continuous in $x \in Q$.

(b) Suppose each open subset $U_j(t)$ is continuous in $t \in T$ for the Hausdorff metric is the space of subsets and assume that

$$\mu(U_j(t)) = 0 \Longleftrightarrow \operatorname{Diam} U_j(t) = 0$$

for all $t \in T$ and $j = 0, \ldots, k$. Then the condition (2) can be replaced by

$$\sup_{v \in U_0} x(v) = \sup_{v \in U_1} x(v) = \ldots = \sup_{v \in U_k} x(v) .$$

In fact, one can use here any notion of size of x on U_j, satisfying an obvious continuity condition.

4.4.B Spectra for Δ-dim. We have seen in 2.2 and 3.4.F how the above proposition is used to bound from below the spectrum of $L_1 d/L_\infty$ on the unit interval. To obtain a similar bound for a domain $V \subset \mathbb{R}^n$ (say for $L_1 d/L_q$ and $q = \frac{n}{n-1}$ or for $L_1 \partial^n / L_\infty$) one needs a $\mathbb{Z}_2$-simplex of partitions into "sufficiently round" subsets. One can also us coverings rather than partitions if one controls the multiplicity.

To be able to use our spectral language we say that a set Q of coverings of V by $k+1$ subsets $V_0, \ldots, V_k$ has Δ-dim $Q > k$ if it contains (and hence is) a $\mathbb{Z}_2$-simplex over Δ^k. Now for every energy $E = E(s)$ we can define the Δ-dim-spectrum of E. Here are some interesting energies.

(a)
$$E^\circ(s) = \sup_{0 \le j \le k} (\operatorname{Diam} V_j)/(\operatorname{Vol} V_j)^{1/n}$$

(b)
$$E^\lambda(s) = \sup_{0 \le j \le k} \lambda_1(V_j)^{-1} ,$$

where λ_1 is the first eigenvalue of a pertinent energy on V_j, say of $L_q \partial^r / L_\infty$ on V_j. (One can generalize this by using any λ_i for $i \geq 1$.)

(c) $E^\mu(s)$ = the measure theoretic multiplicity of the covering, that is the L^∞-norm of the sum of the characteristic functions of U_j.

Questions. What are the spectra of $E^\circ + E^\mu$ and of $E^\lambda + E^\mu$? What are the spectra of E° and E^λ on the space of partitions (i.e., for $E^\mu = 1$)?

Example. For every compact smooth domain $V \subset \mathbb{R}^2$ one can easily construct a k-simplex of *partitions* for all $k = 1, 2, \ldots$ such that $E^\circ(s) \asymp k$. It is unlikely that one can make $E^\circ \asymp 1$, (i.e, bounded) but something like $E^\circ(s) \asymp k^{1/2}$ might be possible.

Remark. The energy $E^\lambda + E^\mu$ (or E^λ on partitions) is designed to bound from below the spectrum of a pertinent energy on functions x on V but it is unclear how sharp such a bound might be. In other words we want to know how close E^λ is to the dual of E from where λ_1 (or λ_i) comes.

References

[Fl]$_1$ A. Floer, A relative Morse Index for the symplectic action, Preprint (1987).

[Fl]$_2$ A. Floer, A relative Morse Index for the symplective action, to apear.

[F-S] G. Folland and E. Stein, Estimates on $\overline{\partial}$ and the Heisenberg group, Comm. Pure-Appl. Math. 27 (1974), 429-522.

[G-M] S. Gallot and D. Meyer, D'un résultat hilbertien a un principle de compariason entre spectres. Applications.

[Gr]$_1$ M. Gromov, Homotopical effects of dilatation, J.D.G. 13 (1978), 303-310.

[Gr]$_2$ M. Gromov, Filling Riemannian manifolds, J.D.G. 18 (1983), 1-147.

[Gr]$_3$ M. Gromov, Partial differential relation, Springer-Verlag, 1986.

[Gr]$_4$ M. Gromov, Paul Lévy isoperimetric inequality, Preprint (1980).

[Gr-Mi] M. Gromov and V. Milman, A topological application of isoperimetric inequality, Am. J. Math. 105 (1983), 843-854.

[G-L-P] M. Gromov, J. Lafontaine and P. Pansu, Structure metriques pour les variétès reimanniennes, Cedic/Fernard Nathan, Paris 1981.

[I-T] A. Ioffe and V. Tichomirov, Duality for convex functions and extremal problems, Uspecky 6:144 (1968), 51-117.

[Ka] D. Kazhdan, Arithmetc Varieties, In "Proc. on Lie Groups", Budapest (1971), 151-217.

[Kaŝ] B. Kaŝin, Some results on estimating width, Pure ICM-1983, 977-981, Warszawa 1984.

[M] D. Meyer, Un lemma de geometrie hilbertienne, C.R.Ac.Sci. 295 (1982), 467-469.

[M-H] V. Mazia and V. Havin, Non-linear potential thoery, Uspecky 27:6 (1972), 67-138.

[Mi-Sch] V. Milman and G. Schechtman, Asymptotic theory of finite dimensional normed spaces. Lecture Notes in Math. 1200 (1986) Springer-Verlag.

[Pa]$_1$ P. Pansu, Une inegalité isoperimetrique sur le groupe d'Heisenberg. C.R.Ac.Sci. (1982).

[Pa]$_2$ P. Pansu, Croissance do boules et des geodesiques fermèe dans la subvarietes, Erg. Theory and Dyn. Systems 3 (1983), 415-445.

[Pi] A. Pincus, n-width in approximation theory, Springer-Verlag 1985.

[Ru] W. Rudin, L^2-appriximation by partial sum of orthogonal developments, Duke Math J. 19:1 (1952), 1-4.

[Ste] C. Stechkin, On optimal approximation of given classes of functions by polynomials, Uspecky 9:1 (1954), 133-134.

[St] R. Strichartz, Sub-Riemannian geometry, J.D.G. 24 (1986), 221-263.

[V] N. Varapoulos, Chaines de Markov et inègalités isopèrimètriques, C.R.Ac.Sci. 298 (10) (1984), 233-235.

[Z] E. Zehnder, Some perspectives in Hamiltonian systems, preprint.

SOME USEFUL FACTS ABOUT BANACH SPACES

J. Lindenstrauss

Hebrew University

Dedicated to the memory of D.P. Milman (1912-1982).

0. Introduction

This paper is based on a talk given at the D.P. Milman memorial conference. One of the most well known and influential results of D.P. Milman is the Krein-Milman theorem. In view of the wide spread use of this theorem in many areas of mathematics I decided to devote my talk at the conference to describing some results in Banach space theory which can be applied without having first to master complicated definitions or heavy machinery.

The idea behind the study of abstract spaces and in particular infinite-dimensional linear spaces is that such a setting makes it possible to consider a-priori complicated objects (like functions) as single points. One can apply geometric notions and intuition to the study of sets of such "points". In many cases the most interesting sets which appear in applications turn out to be convex sets. Of course, in order for such an approach to become really useful one has to have at one's disposal theorems (hopefully non trivial) on the behaviour of abstract spaces. There are several classical results of this type which appear over and over again in applications and which are included in all graduate texts on functional analysis, like the Hahn-Banach and the Krein-Milman theorems. My purpose here is to mention several other general results of this nature with some emphasis on results related to the Krein Milman theorem. These results, while known to experts in Banach space theory are not always well known to the general mathematical public.

Some of the results I describe below are old while others are new and even very recent. Some of the results are very difficult and deep while others turn out to have a relatively simple proof. Most of the results I mention were for several years before their proofs rather well known open problems. They inspired a lot of related research work and their solution was at the time quite surprising. The word "useful" in the title of this paper is naturally not well defined and perhaps even somewhat controversial. Some of the results mentioned below have already been used quite often while others have as yet no real application. I believe that most will find nice applications in the future and will also find a place in the next generation of textbooks on general functional analysis.

Before I proceed I should make it clear that this short paper should not by any means be considered as a survey of "applicable Banach space theory". Many of the central directions in the research activity of Banach space theory today are not discussed in this paper. Actually the directions not treated here contain perhaps the deepest and most striking results in the subject. They are also currently the source of most of the applications of Banach space theory to other areas of mathematics. The so called local theory leads to many applications to classical convexity theory and other areas (in the present volume several such examples can be found). The local theory has by now an extensive machinery, and in order to use it well one really has to know quite a lot. Another direction which I will practically not mention here is the study of the structure of classical Banach spaces and other special spaces of e.g., differentiable or analytic functions. This study, which is of much interest in itself, often leads also to nice results, problems and new viewpoints in classical analysis.

In this paper I chose to present some results on the following topics.

1. Topology
2. Coordinates
3. Representation theorems
4. Nice norms
5. Convexity
6. Differentiability
7. A special object.

I shall assume below that the Banach spaces are over the real field. I do this just in order to be specific; the results hold also for spaces over the complex field. The main object of study will be infinite dimensional and separable Banach spaces.

1. Topology

The basic facts on the topological structure of Banach spaces and their convex subsets are the following two theorems.

Theorem 1.1 (Kadec, Anderson). *Every separable infinite dimensional Banach space X is homeomorphic to $R^{\aleph_0}$ (the product of countable copies of the real lines). Also every closed and* *convex subset of X with non empty interior (e.g., the unit ball $\{x; \|x\| \leq 1\}$) is homeomorphic to $R^{\aleph_0}$. The same is true of the unit sphere $\{x; \|x\| = 1\}$ of X.*

Theorem 1.2 (Keller). *Every convex compact and infinite dimensional subset K of a Banach space is homeomorphic to the Hilbert cube $I^{\aleph_0}$ (the product of countable copies of $[0,1]$).*

These two theorems are now part of an extensive and deep theory called infinite-dimensional topology. We refer to [B.P] for an exposition of this theory as well as to references to the original papers. It should be pointed out that both 1.1 and 1.2 hold in a more general setting. In 1.1 it suffices to assume that X is an infinite dimensional separable Frechét space while in 1.2 we may take as K any compact convex metrizable and infinite dimensional set in a locally convex Hausdorff space (in the case of 1.2 this extension to a more general setting is trivial).

Another basic result of topological nature is

Theorem 1.3 (Schauder, Tychonoff). *Every continuous map f from a compact convex set K in a locally convex Hausdorff space into itself has a fixed point.*

This theorem is besides the Krein-Milman theorem the best known and most widely used among the theorems mentioned in this paper. Theorem 1.3 follows from the Brouwer fixed point theorem in finite dimensions by a simple limiting procedure. If we use the structure theorem 1.2 this limiting procedure becomes completely trivial in the metrizable case but of course it is much harder to prove 1.2 than 1.3.

There is a kind of converse to 1.3 which is due to Klee [Kl]. Quite recently this converse was proved in a stronger and more quantitative way.

Theorem 1.4 (Lin-Sternfeld [L.S]). *Let K be a closed non compact convex set in a Banach space. Then there is a Lipschitz map $f : K \to K$ which fails to have even an approximate fixed point i.e.,*

$$\inf\{\|x - f(x)\|\ ,\ x \in K\} > 0\,.$$

This theorem, together with the classical argument connecting the Brouwer fixed point theorem with the non existence of a retraction from the ball onto the sphere in R^n, yields the following corollary (which was actually proved a little earlier).

Corollary (Benyamini-Sternfeld [B.S]). *For every infinite dimensional Banach space X there is a Lipschitz retraction from the unit ball of X onto the unit sphere.*

The fact that there is a Lipschitz retraction and not only a continuous one is of special interest. In the context of Theorem 1.1 the maps constructed in its proof are only continuous and one cannot in general obtain Lipschitz continuous maps. Two infinite dimensional separable Banach spaces are rarely Lipschitz equivalent or even only uniformly homeomorphic (cf [B] for a recent survey related to this question).

2. Coordinates

We say that a sequence $\{x_n\}_{n=1}^{\infty}$ in a Banach space X is a *Schauder basis* of X if every $x \in X$ has a unique representation of the form $x = \sum_{n=1}^{\infty} \lambda_n x_n$ with $\{\lambda_n\}_{n=1}^{\infty}$ scalars.

A simple application of the closed graph theorem shows that if such a representation exists then the coefficients $\{\lambda_n\}_{n=1}^{\infty}$ depend linearly and continuously on x, i.e., $\lambda_n = x_n^*(x)$ with $\{x_n^*\}_{n=1}^{\infty}$ in the dual space X^*. The functionals $\{x_n^*\}_{n=1}^{\infty}$, which clearly satisfy $x_n^*(x_m) = \delta_{n,m}$, are called the *biorthogonal functionals* to $\{x_n\}_{n=1}^{\infty}$. Unfortunately, as proved first by P. Enflo, not every space has a Schauder basis (for references and a detailed discussion of this point see [L.T1] and [L.T2]). The strongest known positive result on the existence of coordinates in the general case is

Theorem 2.1 (Ovsepian-Pelczynski [O.P], [P]). *Let X be a separable infinite dimensional Banach space and let $\varepsilon > 0$. Then there exist vectors $\{x_n\}_{n=1}^{\infty}$ in X and $\{x_n^*\}_{n=1}^{\infty}$ in X^* so that*

(i) $x_n^*(x_m) = \delta_{n,m}$

(ii) $\|x_n\| = 1$, $\|x_n^*\| \leq 1 + \varepsilon$ *for all n.*

(iii) X is the closed linear span of $\{x_n\}_{n=1}^{\infty}$.

(iv) The $\{x_n^*\}_{n=1}^{\infty}$ are total over X, i.e., $x_n^*(x) = 0$ for all n implies that $x = 0$.

The $\{x_n^*\}_{n=1}^{\infty}$ can be used to represent X as a sequence space. Conditions (i) and (iii) imply that the sequences which have only finitely many non vanishing coordinates belong to X and are dense there. Condition (ii) says that x_n and x_n^* behave well as far as their norms are concerned. Ideally one would like to have also $\|x_n^*\| = 1$ for all n. In finite dimensional spaces this can be done (if $\dim X = m < \infty$ we look of course only for $\{x_n\}_{n=1}^{m}$ and $\{x_n^*\}_{n=1}^{m}$). This is the so called Auerbach lemma (see [L.T1], p.16). In infinite dimensions it seems to be unknown whether this can always be achieved.

3. Representation theorems

There are several theorems which represent general separable Banach spaces or important special classes of spaces as subspaces of rather concrete spaces. Theorems of this kind enable us in some situations to prove theorems or do constructions valid for general spaces while working with rather concrete objects. The first two of the theorems we mention have already been applied very often; the third theorem is very recent. There are also theorems which represent general spaces as quotient spaces of concrete spaces. However, representations as quotient spaces turn out to be less useful than the subspace representation theorems.

Theorem 3.1 (Banach-Mazur [B.M]). *Every separable Banach space is isometric to a subspace of $C(0,1)$ (the space of continuous functions on $[0,1]$ with the supremum norm).*

Since $C(0,1)$ has a Schauder basis it follows in particular that every separable Banach space X can be represented as a subspace of a space with a Schauder basis.

The second theorem involves an important special class of Banach spaces, namely Banach lattices. A *Banach lattice* is a Banach space which is also a vector lattice so that $\|x\| \le \|y\|$ whenever $|x| < |y|$ where there absolute value $|x|$ of an element x in a vector lattice is defined by $|x| = x \vee (-x)$. (A norm on a vector lattice which satisfies this condition is called a *lattice norm.*) A separable Banach lattice X is *order continuous* if whenever $\{x_n\}_{n=1}^{\infty}$ is a sequence decreasing to 0 (i.e., 0 is the greatest lower bound of $\{x_n\}_{n=1}^{\infty}$) we have $\|x_n\| \downarrow 0$. The following theorem has its roots in Kakutani's work on abstract L and M spaces. In the final form it appears in [L.T2].

Theorem 3.2. *Let X be a separable order continuous Banach lattice. Then there is a probability space (Ω, Σ, μ) and a (in general not closed) subspace $\tilde{X}$ of $L_1(\Omega, \Sigma, \mu)$ and a lattice norm $\|\ \|_{\tilde{X}}$ on $\tilde{X}$ so that*

(i) There is a linear isometry which preserves also the order from X onto $(\tilde{X}, \|\ \|_{\tilde{X}})$.

(ii) $L_\infty(\Omega, \Sigma, \mu) \subset \tilde{X}$ and is dense in $\tilde{X}$; $\tilde{X}$ is dense in $L_1(\Omega, \Sigma, \mu)$ in the L_1 norm.

(iii) $f \in \tilde{X}$, $g \in L_1(\Omega, \Sigma, \mu)$, $|g| \le |f| \Rightarrow g \in \tilde{X}$

(iv) $\|f\|_1 \le \|f\|_{\tilde{X}} \le 2\|f\|_\infty$ whenever $f \in L_\infty(\Omega, \Sigma, \mu)$

(v) The dual of the isometry given in (i) maps X^* onto the lattice $\tilde{X}^*$ of all Σ measurable function g so that

$$\|g\|_{\tilde{X}^*} = \sup\Big\{ \int_\Omega fg\,d\mu \ ;\ \|f\|_{\tilde{X}} \le 1 \Big\} < \infty\,.$$

The third result we mention in this section is a representation theorem (or perhaps it is better to call it an embedding theorem) for spaces such that their duals are also separable. A Schauder basis $\{x_n\}_{n=1}^{\infty}$ of a Banach space X for which X^* is the closed linear span of the biorthogonal functions $\{x_n^*\}_{n=1}^{\infty}$ is called a *shrinking basis*. Obviously a shrinking basis can exist only in spaces X for which X^* is separable. For a general Schauder basis X^* is, as is easily seen from the definitions, only the w^* closed linear span of $\{x_n^*\}_{n=1}^{\infty}$. (It follows from this remark that in a reflexive Banach space every Schauder basis is shrinking.)

Theorem 3.3 (Zippin [Z]). *Every Banach space with a separable dual is isometric to a subspace of a space with a shrinking basis. Every separable reflexive space is isometric to a subspace of a reflexive space with a Schauder basis.*

4. Nice Norms

In this section we shall present some renorming theorems, i.e., theorems which allow us to replace the given norm in a space by another equivalent norm which has some desirable property. In all the theorems stated here the change to a "good" norm can be done by an arbitrary small perturbation. In other words given $\varepsilon > 0$ the new norm $|||\cdot|||$ can be chosen so that $\|x\| \le |||x||| \le (1+\varepsilon)\|x\|$ for every x in the space. We shall however ignore this point in the statement of the theorems and just talk of an equivalent renorming.

A simple and common requirement on a norm is that it be *strictly convex* i.e., that $\|x+y\| = \|x\| + \|y\|$, $x \ne 0$ implies that $y = \lambda x$ for some $\lambda > 0$ or alternatively, that there should be no line segments on the unit sphere.

Theorem 4.1 (Klee [K2]). *Every separable Banach space X has an equivalent norm so that with it both X and X^* are strictly convex.*

Of course the norm in X^* will be the natural dual norm $\|x^*\| = \sup\{x^*(x)\ ;\ \|x\| \le 1\}$. The fact that X^* is strictly convex implies in particular that for every $0 \ne x \in X$ there is a unique $x^* \in X^*$ such that $\|x^*\| = 1$ and $x^*(x) = \|x\|$. In other words, through every point in the unit sphere of X there is a unique supporting hyperplane to the unit ball of X. A space which has this property of uniqueness of supporting hyperplanes is called *smooth*.

Quite often one needs quantitative versions of strict convexity. The quantitative versions which is often available is *local uniform convexity*. A norm is called locally uniform convex if $\|x_n\| = \|y\| = 1$, $n = 1, 2, \ldots$ and $\lim_n \|x_n + y\| = 2$ imply $\|x_n - y\| \to 0$. If a space has such a norm then obviously norm and weak convergence of sequences coincide on the unit sphere.

Theorem 4.2 (Kadec, Klee, Asplund). *Every separable Banach space X has an equivalent locally uniformly convex norm. If X^* is separable the norm can be chosen so that also the norm in X^* becomes locally uniformly convex.*

We refer to [D] for a detailed discussion of this theorem as well as references to the original papers. It is interesting to note that Kadec proved the first part of the theorem above in order to use it for proving a special case of Theorem 1.1. The contribution of Asplund to Theorem 4.2 was by inventing an averaging method (actually an interpolation procedure) which allows one to combine two norms which have two "good" properties into a single norm which has both properties. Here it was used to get a single norm which makes both X and X^* locally uniformly convex. The assumption that X^* be separable is essential; it follows easily from the Bishop- Phelps theorem (Theorem 5.1 below) that if X is separable and X^* has a locally uniformly convex norm then X^* has also to be separable.

If X is a Banach lattice one wants naturally an equivalent norm which is also a lattice norm.

Theorem 4.3 (Davis, Ghoussoub, Lindenstrauss [D.G.L]). *Let X be a separable Banach lattice. The following are equivalent*

(i) X is order continuous

(ii) X has an equivalent locally uniformly convex lattice norm

(iii) X has an equivalent lattice norm so that with it norm and weak convergence coincide on the unit sphere.

The proof of Theorem 4.3 is based on the representation theorem 3.2.

In many applications it is desirable to have even a stronger property than local uniform convexity, namely *uniform convexity*. A Banach space is said to be uniformly convex if for every $\varepsilon > 0$ there is a $\delta(\varepsilon) > 0$ so that

$$\|x\| = \|y\| = 1 \ , \ \|x - y\| \geq \varepsilon \Longrightarrow \|x + y\| \leq 2 - \delta(\varepsilon) \ .$$

Uniformly convex norms are however harder to get than locally uniform convex norms. It was shown by D.P. Milman and Pettis (independently) that every uniformly convex Banach space is reflexive (cf. [L.T2], p.61). Moreover, even separable reflexive spaces may fail to have an equivalent uniformly convex norm. It is perhaps an unfortunate coincidence in terminology that while the question of existence of a locally uniform convex norm belongs to global Banach space theory, the question of existence of a uniformly convex norm belongs to the local theory of Banach spaces, i.e., depends on the structure of finite dimensional subspaces of Banach space. The rest of this section is a short detour into local theory (however, as mentioned in the introduction the main body of this vast area is not discussed in this paper).

Theorem 4.4 (James [J2], Enflo [E]). *Let X be a Banach space. The following assertions are equivalent*

(i) X has an equivalent uniformly convex norm

(ii) X^* has an equivalent uniformly convex norm

(ii) There exists an integer n and an $\varepsilon > 0$ so that whenever $\{x_i\}_{i=1}^n$ are unit vectors in X then for some $1 \le k \le n$, $\|x_1 + x_2 + \cdots + x_k - x_{k+1} - \cdots - x_n\| \le n - \varepsilon$.

In view of the equivalence of (i) and (ii) and the Asplund averaging method mentioned above it follows that if X is uniformly convex we can find an equivalent norm so that with it both X and X^* are uniformly convex.

Another addition of Theorem 4.4 involves the function $\delta(\varepsilon)$ appearing in the definition of uniform convexity, called the *modulus of convexity* of X (or the norm in X). Naturally one would like to have as large a $\delta(\varepsilon)$ as possible. It is known (cf [L.T2], p.68) that the "most" uniform convex space is the Hilbert space, i.e., always $\delta(\varepsilon) \le 2 - (4 - \varepsilon^2)^{1/2} = \varepsilon^2/4 + O(\varepsilon^4)$. On the other hand one has

Theorem 4.5 (Pisier [Pi]). *A uniformly convex space has an equivalent norm so that in this norm the modulus of convexity satisfies $\delta(\varepsilon) \ge c\varepsilon^p$ for some $c > 0$ and $p < \infty$.*

A proof (due to Maurey) of theorem 4.5, which is somewhat simpler than the original one, can be found in [Be].

To conclude this section let us mention in connection with Theorem 4.3 that a uniformly convex space which is also a Banach lattice has an equivalent uniformly convex lattice norm. This fact is discussed in detail in [L.T2].

5. Convexity

The Krein Milman theorem states that every compact convex set in a locally convex Hausdorff space is the closed convex hull of its extreme points. It is important here (even in the Banach space context) to allow compactness not only in the norm topology but in other topologies (e.g., the w or w^* topologies). In many cases it is possible to determine explicitly the extreme points of an abstractly defined convex set. Very often the extreme points turn out to have an especially nice or simple form. The Krein Milman theorem is thus a very useful tool for presenting many convex sets. Most of this section is devoted to the statement of refinements of the Krein Milman theorem in many directions: The notion of representing a convex set in terms of their extreme points can be made more precise than just saying "closed convex hull"; A special subset of the extreme points (which may be easier to describe than all extreme points) usually suffices and in many situations the compactness assumption can be relaxed. We start by a reformulation of the Krein Milman theorem.

Theorem 5.1 (Krein-Milman). *Let K be a compact convex set in a locally convex Hausdorff space. Then for every $x \in K$ there is a probability measure μ_x (regular Borel measure) on the closure K_0 of the set $\operatorname{ext} K$ of extreme points of K so that $x = \int_{K_0} y\,d\mu_x(y)$ (in the sense that $x^*(x) = \int_{K_0} x^*(y)d\mu_x(y)$ for every continuous linear functional x^*).*

For this statement of the Krein Milman theorem as well as for the proofs of the following two theorems and references to the original papers we refer to [Ph].

Theorem 5.2 (Choquet). *Let K be a compact convex and metrizable set in a locally convex Hausdorff space. Then for every $x \in K$ there is a probability measure μ_x on K so that $x = \int_K y\,d\mu_x(y)$ and $\mu_x\{\operatorname{ext} K\} = 1$.*

The difference between 5.2 and 5.1 is that while in 5.2 we have that μ_x is concentrated on $\operatorname{ext} K$, 5.1 ensures only that μ_x is concentrated on the closure of $\operatorname{ext} K$. The set $\operatorname{ext} K$ is in general not closed; an example where the closure of $\operatorname{ext} K$ is much larger than $\operatorname{ext} K$ is the subject of section 7 below. For metrizable K the set $\operatorname{ext} K$ is always a Borel set (actually a G_δ set) and thus the requirement $\mu_x\{\operatorname{ext} K\} = 1$ makes sense.

We shall next define the notion of a *simplex*. To a convex set K in a linear space X corresponds a convex cone $\widetilde{K} = \{(\lambda x, \lambda)\ ;\ x \in K\ ,\ \lambda \geq 0\}$ in $X \oplus R$. The cone defines a partial order in $X \oplus R$ by putting $u \leq v$ iff $v - u \in \widetilde{K}$. We shall consider here only bounded and closed convex sets K in locally convex Hausdorff spaces X. The set K is called a simplex if the order determined by $\widetilde{K}$ is a lattice order (on the linear subspace $\widetilde{K} - \widetilde{K}$ of $X \oplus R$). Alternatively, K is a simplex iff for every $x \in X$ the intersection of K with $K + x$ is of the form $y + \lambda K$ with $\lambda \geq 0$ and $y \in X$ (provided the intersection is non empty). In finite dimensional spaces simplexes are the usual objects (a segment on the line, a triangle in the plane ect.). In infinite dimensions simplexes are more varied and complex objects (see sections 7) but still the simplest class of convex sets.

Theorem 5.3 (Choquet-Meyer). *A compact convex metrizable set K is a simplex if and only if for every x in K the measure μ_x given by Theorem 5.2 is uniquely determined.*

In the setting of Theorem 5.2 we cannot pass from $\operatorname{ext} K$ to a smaller set. Indeed, if $x \in \operatorname{ext} K$ then clearly the only μ_x which can work in 5.2 is the Dirac measure at x. However, if we return to the original formulation of the Krein Milman theorem (or even to the setting of 5.1) we can replace ext K by a special subset of it in some cases. A point $x \in K$ is called an *exposed point* of the convex set K if there is a continuous linear functional x^* such that $x^*(x) > x^*(y)$ for every $y \notin x$ in K. Obviously every exposed point of K is an extreme point and easy 2-dimensional examples show that the converse is false. In infinite dimensions we can pass to a further natural subset. Let K be a convex set in a Banach space X. A point $x \in K$ is called a *strongly exposed point* of K if there is an $x^* \in X^*$ such that whenever $\{y_n\}_{n=1}^{\infty} \in K$ and $x^*(y_n) \to x^*(x)$ then $\|y_n - x\| \to 0$.

Another concept we need in order to state below a generalization of the Krein Milman theorem is that of the *Radon Nikodym property* (R.N.P. for short). Its purpose is to allow the extension of the Krein Milman theorem also to some non compact settings. There are two extensive books on the R.N.P. ([D.U] and [Bo]) and we refer to them for proofs, more details and references to the original papers concerning the next three theorems.

Theorem 5.4. *The following three assertions concerning a closed convex set C in a Banach space X are equivalent*

(i) Every function $F : [0,1] \to X$ which satisfies a Lipschitz condition and for which $(F(t) - F(s))/(t - s) \in C$ for all $t \neq s$ in $[0,1]$, is differentiable a.e.

(ii) Every bounded convex subset A of C has arbitrary small slices, i.e., for every $\varepsilon > 0$ there exists an $x^* \in X^*$ and a λ so that the set $\{x \,;\, x \in A \,;\, x^*(x) > \lambda\}$ has a positive diameter which is smaller than ε (unless A is a single point).

(iii) If (Ω, Σ, ν) is a probability space and ν a vector measure on (Ω, Σ) with values in X which is of finite variation, absolutely continuous with respect to μ and satisfies $\nu(B)/\mu(B) \in C$ for all $B \in \Sigma$ with $\mu(B) \neq 0$, then ν has a Radon Nikodym derivative with respect to μ, i.e., there is a measurable $f : \Omega \to C$ so that for all $B \in \Sigma$

$$\nu(B) = \int_B f(\omega)\, d\mu(\omega) \ .$$

In view of condition (iii) any set which satisfies the equivalent conditions of 5.4 is called an R.N.P. set (if $C = X$ the entire Banach space we say that X has R.N.P.). Important examples of R.N.P. sets (besides the quite obvious ones of norm compact convex sets) are w compact convex sets and norm closed convex subsets in conjugate Banach spaces (like ℓ_1 for example). These later sets need not be compact in any reasonable topology.

Theorem 5.5 (Phelps, Bourgain). *A closed convex set C in a Banach space X is an R.N.P. set if and only if every closed bounded convex subset K of C is the closed convex hull of its strongly exposed points.*

The important and non trivial part of 5.5 is of course the only if assertion.

The theorem of Choquet (5.2 above) can also be extended to non compact convex sets provided R.N.P. holds.

Theorem 5.6 (Edgar). *Let K be a bounded closed convex set which is an R.N.P. set in a separable Banach space. Then for every $x \in K$ there is probability measure μ_x on K so that $x = \int_K y\, d\mu_x(y)$ and $\mu_x\{\text{ext}\, K\} = 1$.*

The set $\text{ext}\, K$ can be shown to be universally measurable so the statement above makes sense. As in the compact case the uniqueness of μ_x for every x characterizes simplexes (this is due to Bourgin-Edgar and Saint Raymond, cf. [Bo] for details).

We conclude this section with two theorems which are at least implicitly involved in the proofs of the theorems above and which are useful in many other contexts.

Theorem 5.7 (Bishop-Phelps [B.Ph]). *Let C be a closed bounded convex set in a Banach space. Then the set of $x^* \in X^*$ which attain their maximum on C are norm dense in X^*.*

Theorem 5.8 (James [J.1]). *Let C be a closed convex set in a Banach space. Then every $x^* \in X^*$ attains its maximum on C if (trivially) and only if C is weakly compact.*

These theorems are the starting points for applications of the notions discussed in this section to optimization problems. For a recent study of this direction we refer to [G.M].

6. Differentiability

There are various natural notions of differentiability for functions defined on infinite dimensional Banach space. The most useful notions are Frechèt differentiability and Gateaux differentiability. A function F defined on an open set in a Banach space X with values in a Banach space Y is said to be *Frechèt differentiable* at x_0 if there is a bounded linear operator

$T : X \to Y$ so that $F(x_0 + x) = F(x_0) + Tx + o(\|x\|)$ as $x \to 0$. The notion of Gateaux differentiability is weaker, F as above is *Gateaux differentiable* at x_0 if there is a T as above so that for every $x \in X$, $\lim_{\lambda \to 0} (F(x_0 + \lambda x) - F(x_0))/\lambda = Tx$.

The following theorem follows quite directly from the definitions (cf [D] for details).

Theorem 6.1.a. *The norm (i.e., the function $x \to \|x\|$ from X to R) in a Banach space X is Gateaux differentiable at every $x \neq 0$ if and only if X is smooth.*

b. *If X^* is locally uniformly convex then the norm is Frechèt differentiable at every $x \neq 0$.*

It follows from Theorems 6.1 as well as 4.1 and 4.2 that a separable space X has an equivalent norm which is Gateaux differentiable at every $x \neq 0$ and if X^* is separable we get a norm which is even Frechèt differentiable for $x \neq 0$. We note also that it follows easily from Theorem 5.7 that for a separable space X the norm cannot be Frechèt differentiable away from 0 unless X^* is separable.

The main result on differentiability is the following very recent result.

Theorem 6.2 (Preiss [Pr]). *Assume that the norm in a Banach space X is Frechèt (resp. Gateaux) differentiable at every $x \neq 0$. Then any real valued function defined on an open set of X and satisfying a Lipschitz condition is Frechèt (resp. Gateaux) differentiable at a dense set. Thus in particular every real valued Lipschitz function on a separable space X (resp. a space with separable dual) is Gateaux (resp. Frechèt) differentiable at a dense set.*

Thus, in particular, every real valued Lipschitz function on a separable space X (resp. a space with separable dual) is Gateaux (resp. Frechèt) differentiable at a dense set.

The situation for functions with values in a Banach space is somewhat different. Lipschitz functions between Banach spaces are often not Frechèt differentiable at any point. This is true even for natural maps in Hilbert space. For example, if g is a continuous map from R into itself, the map $f(t) \to g(f(t))$ on $L_2(0,1)$ say, can be Frechèt differentiable at a single point only if g is affine ([V]). (Note that if g is a Lipschitz map from R to R then $f \to g \circ f$ is a Lipschitz map from $L_2(0,1)$ into itself.) For Gateaux differentiability we have however.

Theorem 6.3 (Aronszajn, Christensen, Mankiewicz). *Let X be a separable Banach space and let Y have the R.N.P. Then any Lipschitz function from open set of X into Y is Gateaux differentiable at a dense set.*

For a discussion of this theorem (as well as a proof and references to the original papers) we refer to [B].

7. A special object

Every decent mathematical theory starts with examples which direct the growth of the theory. In Banach space theory such examples were Hilbert space, L_p spaces, $C(K)$ spaces and others. The developments of many theories produce however also examples which are especially nice and were unexpected when the theory started. Banach space theory has also examples of this kind. In this section we present one such example which arose while working in Banach space theory.

We mentioned the concept of compact metrizable simplexes in section 5. While for every n there is up to affine equivalence a unique n-dimensional simplex the situation is different in infinite dimensions. The compact simplexes K whose extreme points form a closed set are easy to analyse. These are just the probability measures on a compact Hausdorff (compact metric if the simplex is metrizable) space M with the usual w^* topology of measures. The probability measures on M_1 are affinely equivalent to the probability measures on M_2 if and only M_1 is homeomorphic to M_2. The "number" of compact metrizable simplexes with a closed set of extreme point is thus the same as the number of compact metric spaces (up to homeomorphism). Not all compact metrizable simplexes have a closed set of extreme points. Poulsen [Po] showed by a simple and elegant construction that there is even a metrizable simplex with a dense set of extreeme points. His construction is highly non canonical; nevertheless it turns out that such a simplex is unique and has very nice properties. Before we state the theorem we recall that a *face* F of a convex set K is a closed convex subset so that $x, y \in K$, $(x+y)/2 \in F$ implies that $x, y \in F$ (thus a point is a face iff it is an extreme point). It is trivial to verify that a face of a simplex is again a simplex.

Theorem 7.1 (Lindenstrauss, Olsen and Sternfeld [L.O.S]). *There is up to affine equivalence a unique metrizable compact simplex P whose extreme points form a dense subset. This simplex has the following properties:*

(i) It is highly homogeneous. If F_1 and F_2 are two proper faces of P, any affine homeomorphism from F_1 onto F_2 can be extended to an affine automorphism of P.

(ii) The set $\operatorname{ext} P$ is homeomorphic to $R^{\aleph_0}$. More precisely, there is a homeomorphism from $[0,1]^{\aleph_0}$ onto P which carries $(0,1)^{\aleph_0}$ onto $\operatorname{ext} P$.

(iii) It is universal. Any metrizable compact simplex is affinely homeomorphic to some face of P.

Let us point out that property (i) implies in particular that if $\{x_i\}_{i=1}^n$ and $\{y_i\}_{i=1}^n$ are two n-tuples of distinct extreme points of P then there is an affine automorphism τ of P so that $\tau(x_i) = y_i$, $1 \leq i \leq n$.

Such a nice and unique object has to have a role also in other areas of mathematics. And it does. For instance it turns out that the state space of the models of statistical mechanics introduced by Ruelle is the Poulsen simplex (cf e.g., [B.R], p. 398).

To every simplex K corresponds a canonical Banach space: $A(K)$ = the space of continuous affine real valued functions on K with the supremum norm. The study of the Poulsen simplex P is dominated by the Banach space properties of the space $A(P)$.

References

[B] Y. Benyamini, The uniform classification of Banach spaces, Longhorn notes, University of Texas 1984-85.

[Be] B. Beauzamy, Introduction to Banach spaces and their goemetry, North Holland Mathematical Studies 68. sectond edition, 1985.

[Bo] R.D. Bourgin, Geometric aspects of convex sets with the Radon Nikodym Property, Springer Lecture Notes #993, 1983.

[B.M] S. Banach and S. Mazur, Zur Theorie der linearen Dimension, Studia Math. 4 (1933), 110-112.

[B.P] C. Bessaga and A. Pelczynski, Selected topics in infinite dimensional topology, Monografie Matematyczne 58, Warsaw 1975.

[B.Ph] E. Bishop and R.R. Phelps, The support functionals of a convex set, Proc. Symp. Pure Math. AMS 7 (1963), 27-36.

[B.R] O. Bratteli and D.W. Robinson, Operator algebras and quantum statistical mechanics I, Springer Verlag 1979.

[B.S] Y. Benyamini and Y. Sternfeld, Spheres in infinite dimensional normed spaces are Lipschitz contractible, Proc. AMS 88 (1983), 439-445.

[D] J. Diestel, Geometry of Banach spaces, selected topics, Springer Lecture Notes #485, 1975.

[D.G.L] W. Davis, N. Ghoussoub and J. Lindenstrauss, A lattice renorming theorem and applications to vector valued processes, Trans. AMS 263 (1981), 531-540.

[D.U] J. Diestel and J.J. Uhl, Vector measures, Math. Surveys 15 AMS, 1977.

[E] P. Enflo, Banach spaces which can be given an equivalent uniformly convex norm, Israel J. Math. 13 (1973), 281-288.

[G.M] N. Ghoussoub and B. Maurey, H_δ embeddings in Hilbert space and optimization on G_δ sets, Memoirs of the AMS, #349, 1986.

[J1] R.C. James, Weakly compact sets, Trans. AMS 113 (1964), 129-140.

[J2] R.C. James, Some self dual properties of normed linear spaces, Ann. Math. Studies 69 (1972), 159-175.

[K1] V. Klee, Some topological properties of convex sets, Trans. AMS 78 (1955), 30-45.

[K2] V. Klee, Some new results on smoothness and rotundity in normed linear spaces, Math. Ann 139 (1959), 51-63.

[L.O.S] J. Lindenstrauss, G. Olsen and Y. Sternfeld, The Poulsen simplex, Ann. Inst. Fourier 28 (1978), 91-114.

[L.S] P.K. Lin and Y. Sternfeld, Convex sets with the Lipschitz fixed point property are compact, Proc. AMS 93 (1985), 633-639.

[L.T1] J. Lindenstrauss and L. Tzafriri, Classical Banach spaces, Vol 1, Sequence spaces, Springer Verlag 1977.

[L.T2] J. Lindenstrauss and L. Tzafriri, Classical Banach spaces, Vol 2, Function spaces, Springer Verlag 1979.

[O.P] R.I. Ovsepian and A.Pelczynski, The existence in every separable Banach space of a fundamental total and bounded biorthogonal system. Studia Math. 54 (1975), 149-159.

[P] A. Pelczynski, Any separable Banach space admits for every $\varepsilon > 0$ fundamental and total biorthogonal sequences bounded by $1+\varepsilon$. Studia Math. 55 (1976), 295-304.

[Ph] R.R. Phelps, Lectures on Choquet's theorem, Van Nostrand Math. Studies 1966.

[Pi] G. Pisier, Martingales with values in uniformly convex spaces, Israel J. Math. 20 (1975), 326-350.

[Po] E.T. Poulsen, A simplex with dense extreme boundary, Ann. Inst. Fourier 11 (1961), 83-87.

[Pr] D. Preiss, Differentiability of Lipshcitz functions on Banach spaces (to appear).

[V] M.M Vainberg, Some questions of differential calculus in linear spaces, Uspehi Mat. Nauk 7 (50) (1952), 55-102.

[Z] M. Zippin, Banach spaces with separable duals, (to appear in Trans. AMS).

HOMOGENEOUS BANACH SPACES

William B. Johnson[1]

Texas A & M University
USA

An infinite dimensional Banach space is said to be *homogeneous* provided it is isomorphic to each of its infinite dimensional subspaces. Still unsolved is Banach's problem whether every homogeneous Banach space is isomorphic to L_2. Indeed, the only significant progress on this problem seems to be work on the approximation property by Enflo, Davie, Figiel and Szankowski, which, when combined with the Maurey-Pisier-Krivine theorem, yields that a homogeneous Banach space has for every $\varepsilon > 0$ cotype $2+\varepsilon$ and type $2-\varepsilon$ (see Theorem 1.g.6 in [LT]).

Recently Pisier [P] introduced an interesting class of Banach spaces called *weak Hilbert spaces* which, among other properties, have for every $\varepsilon > 0$ cotype $2+\varepsilon$ and type $2-\varepsilon$. It turns out that the concept of weak Hilbert space can be used to give some information on Banach's problem:

Theorem. *Suppose that X is a Banach space such that*

1. *X is homogeneous*
2. *X^* is homogeneous*
3. *X has the gl-property.*

Then X is isomorphic to a Hilbert space.

Recall that a Banach space X is said to have the *gl-property* provided every absolutely summing operator from X into L_2 factors through L_1. Gordon and Lewis [GL] proved that a Banach space which has an unconditional basis also has the *gl*-property, so the theorem connects Banach's problem to another famous conjecture; namely, whether every infinite dimensional Banach space has a subspace which has an unconditional basis. However, it seems likely to me that even if this latter conjecture is false, every infinite dimensional Banach space has an infinite dimensional subspace which has the *gl*-property.

I have no idea whether the homogeneous property is self-dual.

---

[1] Supported in part by NSF DMS-8500764

According to Pisier [P], a Banach space X is a weak Hilbert space if X has non-trivial type and both X and X^* have weak cotype 2 in the sense of Milman-Pisier [MP]; i.e., for some constant C, every finite dimensional subspace E of the space contains a smaller subspace F which is C-isomorphic to the Euclidean space of dimension $n = \dim F$ and $n > (1/C) \dim E$.

Lemma 1. *If every infinite dimensional subspace of the Banach space X has the gl-property, then X has weak cotype 2.*

Proof: By Corollary 3.7 of [FJ], it suffices to prove that there is a constant C so that $gl(E) < C$ for every finite dimensional subspace E of X, where $gl(C)$ is the supremum of $\gamma_1(T)$, the L_1-factorization norm of T, as T ranges over all operators from X into L_2 with $\pi_1(T) < 1$, and where $\pi_1(T)$ is the absolutely summing norm of T. But if no constant C "works" for X, then also no constant "works" for any finite codimensional subspace of X, so a standard construction yields a finite dimensional decomposition $(E_n)_{n=1}^{\infty}$ for a subspace Y of X so that $gl(E_n) \to \infty$ as $n \to \infty$. Since the E_n's are uniformly complemented in Y, $gl(Y) = \infty$, contradicting the assumption that Y has the gl-property.

Lemma 1 implies that a Banach space which satisfies the assumptions of the theorem is a weak Hilbert space. To complete the proof of the theorem, we need to check that Hilbert space is the only homogeneous weak Hilbert space.

Following Pisier [P], we say that a Banach space X is *asymptotically Hilbertian* provided there is a constant C so that, for every n, X contains a finite codimensional subspace all of whose n-dimensional subspaces are C-isomorphic to ℓ_2^n. Pisier gives in [P] a nice proof that every weak Hilbert space is asymptotically Hilbertian. (Pisier credits the proof to me, but in fact the argument he gives is a considerable improvement over the one I outlined to him. Also, Pisier [P] properly comments that J. Bourgain made an important observation that eliminated a restrictive extra hypothesis, but he modestly fails to mention that Pisier himself provided me with valuable assistance in proving the result.)

In view of this last-mentioned result, we can complete the proof of the theorem by proving:

Lemma 2. *If X is an asymptotically Hilbertian Banach space which embeds isomorphically into each of its infinite dimensional subspaces, then X is isomorphic to L_2.*

Proof: Recall [LP] that X is isomorphic to a Hilbert space if there is a constant C such that every finite dimensional subspace of X is C-isomorphic to the Euclidean space of its dimension. Thus if X is not isomorphic to L_2, there are finite dimensional subspaces $(E_n)_{n=1}^{\infty}$ of X so that for every n, the isomorphism constant from E_n to $\ell_2^{d(n)}$ (where $d(n) = \dim E_n$) is larger that

n^2. Since X is asymtotically Hilbertian, there is a constant C and a basic sequence $(x_n)_{n=1}^{\infty}$ in X with basis constant at most 2 so that for every m, all $d(m)$-dimensional subspaces of $\text{span}(x_n)_{n=m}^{\infty}$ are C-isomorphic to $\ell_2^{d(m)}$. Since $\text{span}(x_n)_{n=1}^{m}$ is $2m$-isomorphic to ℓ_2^m under the natural basis-to-basis mapping, every $d(m)$-dimensional subspace of $\text{span}(x_n)_{n=1}^{\infty}$ is at most $9\max(C,m)$-isomorphic to $\ell_2^{d(m)}$, which means that X cannot isomorphically embed into $\text{span}(x_n)_{n=1}^{\infty}$.

References

[F] T. Figiel and W.B. Johnson, Large subspaces of ℓ_n^{∞} and estimates of the Gordon-Lewis constant, Israel J. Math. 37 (1980), 92-112.

[GL] Y. Gordon and D.R. Lewis, Absolutely summing operators and local unconditional structures, Acta Math. 133 (1974), 27-48.

[LP] J. Lindenstrauss and A. Pełczynski, Absolutely summing operators in L_p spaces and their applications, Studia Math. 29 (1986) 275-326.

[JT] J. Lindenstrauss and L. Tzafriri, Classical Banach Spaces II, Function spaces, Ergebnisse 97, Springer-Verlag (1979).

[MP] V. Milman and G. Pisier, Banach spaces with a weak cotype 2 property, Israel J. Math. 54 (1986), 139-158.

[P] G. Pisier, Weak Hilbert spaces.

[S] A. Szankowski, Subspaces without approximation property, Israel J. Math. 30 (1978), 123-129.

AN APPROACH TO POINTWISE ERGODIC THEOREMS

J. Bourgain
IHES
France

1. Introduction

This exposé is closely related to recent work of the author (see [B1,2,3]) on the extension of Birkhoff's ergodic theorem to certain subsets of the integers. It was proved in [B2] that given a dynamical system $(\Omega, \mathcal{B}, \mu, T)$ and a polynomial $p(n)$ of the variable $n \in \mathbb{Z}$ taking integer values, then the ergodic means

$$A_N f = \frac{1}{N} \sum_{0 \leq n < N} T^{p(n)} f \tag{1}$$

Converge almost surely ($N \to \infty$) assuming f a function of $L^2(\Omega, \mu)$. A (partial) L^p-theory ($p < 2$) has been developed in [B3]. Of course, these facts may be reformulated in the [B-L1] terminology of "universally good sets of integers".

For the sake of simplicity, only L^2-theory will be considered here, except in the context of the usual ergodic theorem. L^p-analogues obtained by adapting the argument of [B3] are not worked out here and only the conclusion will be mentioned.

The ideas appearing in [B2] may be summarized as follows

(a) Reduction of the general problem to statements about the shift S on $\mathbb{Z}$ (which are of a "finite" quantitative nature)

(b) Proof of certain maximal inequalities, relative to the shift, by Fourier-analysis methods

(c) Use of the Hardy-Littlewood circle method and in particular estimates on the relevant exponential sums.

As I oberved in [B2], this approach should rather be considered as a general method than the solution to some isolated problems.

The purpose of this exposé is two-fold. First, we will discuss (a) in the context of the standard Birkhoff theorem. Our proof of the pointwise and maximal ergodic theorem is related to [K-W], although different. In particular, in order to illustrate ideas, it will be shown how to avoid the invariance of the limit. When dealing with subsets of $\mathbb{Z}$, this invariance is indeed not available. Thus the pointwise ergodic theorem is not an immediate consequence of the maximal ergodic theorem (except in special cases, cf. [B2]). Secondly, we work one more example out, namely the case of the set $\mathcal{P}$ of prime numbers. Thus

Theorem 1. *Denoting $\mathcal{P}_N = \{p \mid p = \text{prime} \leq N\}$ and $|\mathcal{P}_N|$ its cardinality, the ergodic means*

$$A_N f = |\mathcal{P}_N|^{-1} \sum_{p \in \mathcal{P}_N} T^p f \tag{2}$$

converge almost surely for $f \in L^2(\Omega, \mu)$.

Although arithmetically less elementary than in the case of the set of squares $\{n^2 \mid n = 1, 2, \ldots\}$ for instance, the argument for the primes slightly simplifies because of better estimates of the corresponding exponential sums at rational points. The information on prime exponential sums used here may be found in [D].

2. Aspects of Birkhoff's Ergodic Theorem

Let $(\Omega, \mathcal{B}, \mu, T)$ be a dynamical system. In this section we consider the usual ergodic averages $A_N f = \frac{1}{N} \sum_{1 \leq n \leq N} T^n f$. We discuss their convergence properties, partly keeping in mind possible extensions to subsets of $\mathbb{Z}$.

(2.I) Mean convergence.

The sequence of complex polyomials $P_N(z) = \frac{1}{N} \sum_{n=1}^{N} z^n$ pointwise converges on the unit circle (to 0 except for $z = 1$). Consequently, by general spectral theory of unitary operators, $A_N f$ converges in $L^2(\mu)$ whenever $f \in L^2(\mu)$. The main point here is the existence of a spectral measure. The Herglotz-Bochner theorem assures indeed the existence of a positive Radon measure ν on the circle $\mathbb{T}$, such that

$$\langle T^n f, f \rangle = \hat{\nu}(n) = \int_0^1 e^{-2\pi i n\theta} \nu(d\theta) \tag{3}$$

implying that the map $L^2(\pi, \nu) \to L^2(\Omega, \mu)$ mapping the nth character $e^{2\pi i n\theta}$ on $T^n f$ is an isometry. Thus the convergence of $A_N f$ in $L^2(\Omega, \mu)$ is equivalent to the convergence of $P_N(z)$ in $L^2(\pi, \nu)$.

This is clearly an L^2-theory. In general, given a subset Λ of the positive integers, the pointwise convergence on the unit circle of the sequence of polynomials

$$P_N(z) = \frac{1}{|\Lambda \cap [1, N]|} \sum_{\substack{1 \leq n \leq N \\ n \in \Lambda}} z^n \tag{4}$$

is equivalent to a mean ergodic theorem for the set Λ. In the case of "arithmetic sets", this test is particularly useful since the convergence of $P_N(z)$ given by (4) is closely related to the phenomena of uniform distribution. For instance, if Λ is the set of squares $\{n^2 \mid n = 1, 2, \ldots\}$, we have

$$P_N(e^{2\pi i\alpha}) \longrightarrow 0 \quad \text{if} \quad \alpha \quad \text{is irrational}$$

and

$$P_N(e^{2\pi i\alpha}) \quad \to S(q,a) = \frac{1}{q}\sum_{r=0}^{q-1} e^{2\pi i \frac{a}{q} r^2} \quad \text{for} \quad \alpha = \frac{a}{q}$$

(the Gauss-sums).

It is not surprising that the (stronger) almost-sure convergence properties result from a finer analysis of these exponential sums. The sequence $\Lambda \subset \mathbb{Z}_+$ is "ergodic" provided $P_N(z) \to 0$ for $z \in \mathbb{T} \setminus \{1\}$. This property implies mean convergence of $A_N f$ to $\int_\Omega f d\mu$, assuming T ergodic (this is the case for $\Lambda = \mathbb{Z}$, but not if $\Lambda = \{n^2 \mid n = 1, 2, \ldots\}$ for instance).

Constructions due to A. Bellow and B. Weiss [W] give an example of an ergodic sequence Λ (in fact a union of disjoint intervals) which does not satisfy the pointwise ergodic theorem, even for bounded measurable functions.

(2.II) Maximal ergodic theorems.

Let again

$$A_N f = \frac{1}{N}\sum_{n=1}^{N} T^n f$$

and define the "maximal function"

$$f^* = \sup_{N=1,2,\ldots} |A_N f|$$

There are the L^p-inequalities $(1 < p \leq \infty)$

$$\|f^*\|_{L^p(\Omega,\mu)} \leq C(p)\|f\|_{L^p(\Omega,\mu)} \tag{5}$$

and the weak-type inequality

$$\|f^*\|_{L^{1,\infty}(\Omega,\mu)} \leq C\|f\|_{L^1(\Omega,\mu)} \tag{6}$$

where $\|g\|_{L^{1,\infty}} = \sup_{\lambda>0} \lambda\mu[|g| > \lambda]$ and C, $C(p)$ are absolute.

Let us give a simple proof of (5), (6) by deriving them from the shift model $(\mathbb{Z}, S)$. In the case of the shift, the weak-type property (6) easily follows from geometric covering properties of integer-intervals, similarly as for the Hardy-Littlewood maximal function on the real line. Once (6) is obtained, the L^p-inequalities follow from the Marcinkiewicz interpolation theorem. Consider now the case of a general dynamical system (Ω, μ, T). Of course it suffices to prove inequalities (5), (6) (with fixed constants) for a "restricted" maximal function

$$\overline{f} = \sup_{1 \le N \le \overline{N}} A_N f \qquad (f \ge 0) \tag{7}$$

where $\overline{N}$ is an arbitrarily chosen positive integer. Take an integer $J \gg \overline{N}$ and for fixed $x \in \Omega$, consider the orbit

$$x, Tx, T^2x, \dots, T^J x \, .$$

From the function f, define the function φ on $\mathbb{Z}$ as follows

$$\left.\begin{aligned} \varphi(j) &= f(T^j x) \quad \text{if} \quad 0 \le j \le J \\ &= 0 \qquad \text{otherwise} \end{aligned}\right\} \tag{8}$$

Thus $A_N\varphi(j) = A_N f(T^j x)$ provided $0 \le j < J - N$ and hence, with the definition (7),

$$\overline{\varphi}(j) = \overline{f}(T^j x) \quad \text{for} \quad 0 \le j < J - \overline{N} \tag{9}$$

The inequality $\|\overline{\varphi}\|_{\ell^p(\mathbb{Z})} \le \|\varphi^*\|_{\ell^p(\mathbb{Z})} \le C(p)\|\varphi\|_{\ell^p(\mathbb{Z})}$ then immediately implies by (8), (9)

$$\sum_{0 \le j < J - \overline{N}} |\overline{f}(T^j x)|^p \le C(p)^p \sum_{0 \le j \le J} |f(T^j x)|^p \tag{10}$$

Integrating (10) in $x \in \Omega$ with respect to the measure μ yields

$$\sum_{0 \le j < J - \overline{N}} \|T^j \overline{f}\|^p_{L^p(\Omega,\mu)} \le C(p)^p \sum_{0 \le j \le J} \|T^j f\|^p_p$$

is since T is measure preserving

$$\|\overline{f}\|_p \le C(p) \frac{J}{J - \overline{N}} \|f\|_p$$

hence

$$\|f^*\|_p \le C(p)\|f\|_p \, .$$

One can deal similarly with the weak-type inequality (6). Assume $f \in L^1(\Omega, \mu)$, $\lambda > 0$ and let $\Omega_\lambda = [\overline{f} > \lambda]$, χ its indicator function. Given $x \in \Omega$, let φ be defined as above and let $|I|$ stand for the cardinality of a (finite) subset I of $\mathbb{Z}$. The shift inequality gives thus

$$\|\overline{\varphi}\|_{\ell^{1,\infty}(\mathbb{Z})} \le C\|\varphi\|_{\ell^1(\mathbb{Z})}$$

and, by (9)

$$\lambda|\{0 \le j < J - \overline{N} \mid \overline{f}(T^j x) > \lambda\}| \le C \sum_{0 \le j \le J} f(T^j x)$$

hence

$$\sum_{0 \le j < J - \overline{N}} \chi(T^j x) \le \frac{C}{\lambda} \sum_{0 \le j \le J} f(T^j x) . \tag{11}$$

Again integrating

$$\lambda \mu(\Omega_\lambda) \le C \frac{J}{J - \overline{N}} \|f\|_1$$

from where (6) easily follows.

Presently, the covering argument leading to weak-type inequalities seems not available when dealing with particular *subsets* of $\mathbb{Z}$, such as the squares or the primes. In these cases, we were unable so far to develop an L^1-theory. The L^2 and L^p-inequalities (in certain ranges of p) were obtained by making essential use of Fourier-transform methods. This is a similar approach as in differentiation problems in Euclidean analysis involving lower dimensional manifolds.

(2.III) Almost-sure Convergence.

By the maximal inequality and a standard truncation argument, the almost sure convergence of $A_N f$ for f in $L^1(\Omega, \mu)$ reduces to functions in $L^1 \cap L^\infty$. Denote F the L^2-limit of $(A_N f)$ and, for given $\varepsilon > 0$, let N_ε satisfy

$$\|F - A_{N_\varepsilon} f\|_2 < \varepsilon .$$

By the invariance of the limit (since the ergodic means relate to the full set of positive integers) and the maximal inequality

$$\| \sup_N | F - A_N(A_{N_\varepsilon} f)\|_2 < C\varepsilon . \tag{12}$$

Since

$$|A_N(A_{N_\varepsilon} f) - A_N f| \le 2 \frac{N_\varepsilon}{N} \|f\|_\infty$$

it follows from (12)

$$\| \overline{\lim} |F - A_N f| \|_2 < C\varepsilon \text{ , hence } \overline{\lim_N} |F - A_N f| = 0 \text{ almost sure .}$$

This discussion completes the proof of Birkhoff's theorem. It is clear that the argument given in (2.III) does not apply when dealing with the more general averages

$$A_N f = \frac{1}{|\Lambda \cap [1, N]|} \sum_{\substack{1 \le n \le N \\ n \in \Lambda}} T^n f \tag{13}$$

corresponding to a *subset* Λ of $\mathbb{Z}_+$.

If the eigenfunctions of T generate a dense subspace of L^2, the almost sure convergence of $A_N f$ for f of class L^p, $p \leq 2$, is implied by the pointwise convergence of the sequence $p_N(z)$, $|z| = 1$ given by (4) and the maximal inequality

$$\|f^*\|_p \leq C\|f\|_p \quad ; \qquad f^* = \sup |A_N f| \quad \text{given by (13)}$$

This is the case for instance for the model $(\Omega, T) = (\mathbb{T}, R_a)$, $R_a x = x + a$.

In the remainder of this section, an alternative method is explained for the purpose of proving the theorem stated in the introduction (this is also the argument appearing in [B2], section 7 when dealing with sets $\Lambda = \{p(n) \mid n = 1, \dots\}$).

Take f in $L^1(\Omega, \mu)$, $|f| \leq 1$. For $\varepsilon > 0$, consider the subset

$$Z_\varepsilon = \{[(1-\varepsilon)^n] \mid n = 1, 2, \dots\} \tag{14}$$

of $\mathbb{Z}_+$. Clearly, for each $N \in \mathbb{Z}_+$, there is $N' \in Z_\varepsilon$ such that

$$|A_N f - A_{N'} f| \leq 2\varepsilon \ .$$

Thus to prove the almost sure convergence of $(A_N f)$, it suffices to show that there is no $\varepsilon > 0$ and no sequence of positive integers N_j, $N_{j+1} > 2N_j$, such that

$$\|\mathcal{M}_j f\|_2 > \varepsilon \quad \text{where} \quad \mathcal{M}_j f = \sup_{\substack{N_j \leq N \leq N_{j+1} \\ N \in Z_\varepsilon}} |A_N f - A_{N_j} f| \tag{15}$$

In fact, a more quantitative statement is shown

$$\sum_{1 \leq j \leq J} \|\mathcal{M}_j f\|_2 \leq o(J)\|f\|_2 \tag{16}$$

for J large (depending on ε appearing in the definition of $\mathcal{M}_j$). Since (16) only involves finitely many iterates of T, the general case reduces again to the shift (Z, S). The argument is similar as in proving the maximal inequality, see (2.II).

To prove (16) for the shift, again Fourier analysis is used. The main estimates appearing in the proof of the maximal inequality (for instance in the case of Theorem 1) reappear in proving (16).

3. Use of Fourier Analysis

The dual group of $\mathbb{Z}$ are the real numbers mod 1, i.e., $\mathbb{R}/\mathbb{Z} = \mathbb{T}$. Aften $\mathbb{T}$ will be seen as $[0,1]$, with identification of the endpoints.

The Fourier transform of a function f in $\ell^2(\mathbb{Z})$ is given by

$$\hat{f}(\alpha) = \sum_{j\in\mathbb{Z}} f(j)e^{-2\pi ij\alpha} \qquad (L^2 - \text{convergence})$$

and conversely

$$f(j) = \int_0^1 \hat{f}(\alpha)e^{2\pi ij\alpha}d\alpha$$

Parseval's identity states that

$$\|f\|_2^2 \equiv \sum_{j\in\mathbb{Z}} |f(j)|^2 = \int_0^1 |\hat{f}(\alpha)|^2 d\alpha \, .$$

Let $f * g$ stand for the convolution of f, g in $\ell^2(\mathbb{Z})$, whose Fourier-transform $(f * g)\hat{}(\alpha) = \hat{f}(\alpha)\hat{g}(\alpha)$.

We prove some lemmas for later use (these lemmas also appear in [B2']).

Lemma 1. *Let $0 \le \varphi \le 1$ be a (smooth) bumpfunction on $\mathbb{R}$ vanishing outside a τ neighborhood of 0. Assume $k \in L^1(\mathbb{R})$ satisfying the maximal inequality $\|\sup_{t>0} |f * k_t|\|_{L^2(\mathbb{R})} \le c(k)\|f\|_{L^2(\mathbb{R})}$ where as usual $k_t(x) = \frac{1}{t}k\left(\frac{x}{t}\right)$* (17)

Let R be a set of rational points $\theta \in [0,1]$ which can be given a common denominator Q satisfying $Q\tau < \frac{1}{100}$. Given $f \in \ell^2(\mathbb{Z})$, denote for $t > 0$

$$A_t f(x) = \sum_{\theta\in R} \int \hat{k}(t(\alpha - \theta))\hat{f}(\alpha)e^{2\pi i\alpha x}\varphi(\alpha - \theta)d\alpha \tag{18}$$

Then the maximal operator

$$Mf = \sup_{t>0} |A_t f|$$

satisfies the inequality

$$\|Mf\|_{\ell^2(\mathbb{Z})} \le 4c(k)\|f\|_{\ell^2(\mathbb{Z})} \tag{19}$$

Proof: The argument is straightforward. Rewrite $A_t f$ as

$$A_t f(x) = \int_{-\infty}^{\infty} \hat{k}(t\beta)\varphi(\beta)\{\sum_{\theta\in R} \hat{f}(\theta + \beta)e^{2\pi i(\theta+\beta)x}\}d\beta$$

and put $x = Qy + z$ where $y \in \mathbb{Z}$, $z \in \{0, 1, \ldots, Q-1\}$. Thus by hypothesis on R and Q

$$A_t f(x) = \int \widehat{k}(t\beta)\varphi(\beta)F(z,\beta)e^{2\pi i\beta Qy}d\beta \tag{20}$$

where

$$F(z,\beta) = \sum_{\theta \in R} \hat{f}(\theta + \beta)e^{2\pi i(\theta+\beta)z} \tag{21}$$

We have

$$\|Mf\|_2^2 = \sum_{z=0}^{Q-1} \left\| \sup_{t>0} \left| \int \widehat{k}(t\beta)\varphi(\beta)F(z,\beta)e^{2\pi i\beta Qy}d\beta \right| \right\|^2_{\ell^2(\mathbb{Z},dy)} \tag{22}$$

and evaluate for fixed z the corresponding term in (22). The purpose of what follows is to replace $\ell^2(\mathbb{Z})$ and $L^2(\mathbb{R})$-norms. Denote ς the best constant in the a priori inequality

$$\left\| \sup_{t>0} \left| \int \widehat{k}(t\beta)\varphi(\beta)F(\beta)e^{2\pi iQ\beta y}d\beta \right| \right\|_{\ell^2(\mathbb{Z})} \leq \varsigma \left\{ \int |F(\beta)|^2\varphi(\beta)^2 d\beta \right\}^{1/2} \tag{23}$$

For $0 \leq u \leq 1$, estimate the left member of (3.8) as

$$\begin{aligned} &\left\| \sup_{t>0} \left| \int \widehat{k}(t\beta)\varphi(\beta)F(\beta)e^{2\pi iQ\beta(y+u)}d\beta \right| \right\|_{\ell^2(\mathbb{Z})} \\ &+ \left\| \sup_{t>0} \left| \int \widehat{k}(t\beta)\varphi(\beta)F(\beta)(e^{2\pi iQ\beta u} - 1)e^{2\pi iQ\beta y}d\beta \right| \right\|_{\ell^2(\mathbb{Z})} \end{aligned} \tag{24}$$

Use (23) to evaluate the second term in (24) as

$$\varsigma\left(\int |F(\beta)|^2 |e^{2\pi iQ\beta u} - 1|^2 \varphi(\beta)^2 d\beta \right)^{1/2} \leq \varsigma(2\pi Q\tau)\left(\int |F(\beta)|^2 \varphi(\beta)^2 d\beta \right)^{1/2} \tag{25}$$

Integrating the first term of (24) in $u \in [0, 1]$, one obtains by (17) and change of variable $y' = Qy$ (from Parseval's identity).

$$\left\| \sup_{t>0} \left| \int \widehat{k}(t\beta)\varphi(\beta)F(\beta)e^{2\pi i\beta Qy}d\beta \right| \right\|_{L^2(\mathbb{R})} \leq c(k)Q^{-1/2}\left(\int |F(\beta)|^2\varphi(\beta)^2 d\beta \right)^{1/2}. \tag{26}$$

Hence, from (24), (25) and (26), it follows that

$$\begin{aligned} &\varsigma \leq (2\pi Q\tau)\varsigma + Q^{-1/2}c(k) \\ &\Rightarrow \varsigma \leq 2Q^{-1/2}c(k) \end{aligned} \tag{27}$$

since $Q\tau < \frac{1}{100}$.

Applying (23) with $F(\beta) = F(z, \beta)$ gives following estimate on (22)

$$\begin{aligned}\|Mf\|_2^2 &\le 4Q^{-1}c(k)^2 \int \varphi(\beta)^2 \sum_{z=0}^{Q-1} \Big|\sum_{\theta\in R} \hat{f}(\theta+\beta)e^{2\pi i(\theta+\beta)z}\Big|^2 d\beta \\ &= 4c(k)^2 \int \varphi(\beta)^2 \sum_{\theta\in R} |\hat{f}(\theta+\beta)|^2 d\beta \\ &\le 4c(k)^2 \Big(\int_{-1}^{2} |\hat{f}(\alpha)|^2 d\alpha\Big)^{1/2}\end{aligned}$$

using the fact that the translates φ_θ $(\theta \in R)$ are disjointly supported, since $Q\tau < \frac{1}{100}$. Hence, again by Parseval

$$\|Mf\|_2 \le 4c(k)\|f\|_2$$

proving Lemma 1.

In proving almost sure convergence and in particular inequality (16), the following lemma will be useful (this estimate will apply to individual major arcs in the circle method approach).

Lemma 2. *Denote K the indicator function of $[0,1]$, $0 \le \varphi \le 1$ a function on $\mathbb{R}$ supported by $\left[-\frac{1}{100}, \frac{1}{100}\right]$ and t_k an increasing sequence of positive numbers satisfying $t_k < \frac{1}{2}t_{k+1}$. Denote for $f \in \ell^2(\mathbb{Z})$ and $t > 0$*

$$A_t f(x) = \int_{-\infty}^{\infty} \hat{f}(\alpha)\varphi(\alpha)\widehat{K}(t\alpha)e^{2\pi i\alpha x} dx \tag{28}$$

Then

$$\Big(\sum_k \Big\| \sup_{t_k \le t \le t_{k+1}} |A_t f - A_{t_k} f| \Big\|_2^2\Big)^{1/2} < C\|f\|_2 \tag{29}$$

where $\| \quad \|_2$ stands for the $\ell^2(\mathbb{Z})$-norm, $\widehat{K}$ the Fourier-transform of K as a function on $\mathbb{R}$.

Lemma 3. *Assume $K \in L^1(\mathbb{R})$, and the derivative of $\widehat{K}$ satisfying*

$$|\lambda||(\widehat{K})'(\lambda)| < C, \qquad \lambda \in \mathbb{R} \tag{30}$$

Defining $A_t f$ by (28), one has

$$\Big\| \sup_{t>0} |A_t f| \Big\| \le C\gamma(K)\Big(\int |\hat{f}(\alpha)|^2 \varphi(\alpha)^2 d\alpha\Big)^{1/2} \tag{31}$$

where $\gamma(K) = \sum_{j\in\mathbb{Z}} \Big(\sup_{2^j \le |\lambda| \le 2^{j+1}} |\widehat{K}(\lambda)|\Big)^{1/2}$ and $\| \quad \|_2$ again refers to the $\ell^2(\mathbb{Z})$-norm.

Proof of Lemma 3. If one replaces $\| \quad \|_{\ell^2(\mathbb{R})}$ by $\| \quad \|_{L^2(\mathbb{Z})}$, Lemma 3 may be found in [B4], for instance.

To pass from the reals to the integers, denote again ς a constant satisfying the a priori inequality

$$\|\sup_{t>0}|A_t f|\|_2 \leq \varsigma\Big(\int |\hat{f}(\alpha)|^2\varphi(\alpha)^2 d\alpha\Big)^{1/2} \tag{32}$$

For $0 \leq u \leq 1$, write $A_t f(x) = A_t f(x+u) + A_t f_u(x)$, letting $\widehat{f_u}(\alpha) = \hat{f}(\alpha)[1-e^{2\pi i\alpha u}]$. Integration in u gives

$$\|\sup_{t>0}|A_t f|\|_2 \leq \|\sup_{t>0}|A_t f|\|_{L^2(\mathbb{R})} + \sup_{u\in[0,1]}\|\sup_{t>0}|A_t f_u|\|_2 \tag{33}$$

The first term in the right member of (33) is bounded by the right member of (31). By hypothesis on φ, for $0 \leq u \leq 1$

$$\|\sup_{t>o}|A_t f_u|\|_2 \leq \varsigma\Big(\int |\hat{f}(\alpha)|^2|1-e^{2\pi i\alpha u}|^2\varphi(\alpha)^2 d\alpha\Big)^{1/2} \leq \frac{2\pi}{100}\varsigma\Big(\int |\hat{f}(\alpha)|^2\varphi(\alpha)^2 d\alpha\Big)^{1/2}$$

Consequently, $\varsigma \leq C\gamma(K) + \frac{1}{2}\varsigma$, proving the lemma.

Proof of Lemma 2. Define for $\ell = 0, 1, 2, \ldots$

$$g_\ell(\alpha) = \hat{f}(\alpha) \quad \text{if} \quad t_{\ell+1}^{-1} \leq |\alpha| > t_\ell^{-1}$$
$$= 0 \quad \text{otherwise}$$

For fixed k, estimate by the triangle inequality

$$\|\sup_{t_k\leq t\leq t_{k+1}}|A_t f - A_{t_k} f|\|_2 \leq \sum_\ell \|\sup_{t_k\leq t\leq t_{k+1}}\Big|\int g_\ell(\alpha)\varphi(\alpha)\big[\widehat{K}(t\alpha) - \widehat{K}(t_k\alpha)\big]e^{2\pi i\alpha x}d\alpha\Big|\|_2$$

and evaluate the ℓ-term distinguishing the cases $\ell = k$, $\ell < k$, $\ell > k$. Notice first that for $K = \chi_{[0,1]}$, (30) holds and

$$|\widehat{K}(\lambda)| \leq C|\lambda|^{-1}, \quad \text{for } |\lambda| \geq 1 \tag{35}$$

$$|1 - \widehat{K}(\lambda)| \leq C|\lambda| \tag{35'}$$

Case $\ell = k$. Estimate $\|\sup_{t>0}|\int g_k(\alpha)\varphi(\alpha)\widehat{K}(t\alpha)e^{2\pi i\alpha x}d\alpha|\|_2$ by Lemma 3 to get

$$C\gamma(K)\Big(\int |g_k(\alpha)|^2\varphi(\alpha)(^2 d\alpha\Big)^{1/2} \tag{36}$$

where $\gamma(K)$ is clearly bounded by (35), (36).

Case $\ell < k$. Consider a smooth function $0 \le \psi \le 1$ on $\mathbb{R}$ satsifying $\psi(\lambda) = 1$ for $|\lambda| \le \frac{1}{2}$ and $\psi(\lambda) = 0$ for $|\lambda| \ge 1$. Estimate the ℓ-term as

$$2\Big\| \sup_{t \ge t_k} \Big| \int g_\ell(\alpha)\varphi(\alpha)\widehat{K}(t\alpha)e^{2\pi i\alpha x}d\alpha\Big|\Big\|_2 \tag{37}$$

By definition of g_ℓ, $\widehat{K}$ may clearly be replaced by $\widehat{K}_1$ defined by

$$\widehat{K}_1(\alpha) = \widehat{K}(\alpha)\left[1 - \psi\left(\frac{t_{\ell+1}}{t_k}\alpha\right)\right] \tag{38}$$

and (37) is thus bounded by

$$c\gamma(K_1)\Big(\int |g_\ell(\alpha)|^2\varphi(\alpha)^2 d\alpha\Big)^{1/2} \tag{39}$$

Also, by (35)

$$\gamma(K_1) \le C\left[\frac{t_{\ell+1}}{t_k}\right]^{1/2} \tag{40}$$

Case $\ell > k$. Take the function ψ as above. Writing

$$\widehat{K}(t\alpha) - \widehat{K}(t_k\alpha) = \big[\widehat{K}(t\alpha) - 1\big] - \big[\widehat{K}(t_k\alpha) - 1\big]$$

estimate ℓ-term in (34) as

$$2\Big\| \sup_{t \le t_{k+1}} \Big| \int g_\ell(\alpha)\varphi(\alpha)\big[1 - \widehat{K}(t\alpha)\big]e^{2\pi i\alpha x}d\alpha\Big|\Big\|_2 \tag{41}$$

and use again definition g_ℓ to replace $1 - \widehat{K}$ by $\widehat{K}_1$ given by

$$\widehat{K}_1(\alpha) = \big[1 - \widehat{K}(\alpha)\big]\psi\left(\frac{t_\ell}{2t_{k+1}}\alpha\right) \tag{42}$$

Estimate (41) as

$$C\gamma(K_1)\Big(\int |g_\ell(\alpha)|^2\varphi(\alpha)^2 d\alpha\Big)^{1/2} \tag{43}$$

where now from (35′)

$$\gamma(K_1) \le C\left[\frac{t_{k+1}}{t_\ell}\right]^{1/2} \tag{44}$$

Collecting estimates (34), (36), (39), (40), (43), (44), one finds

$$\begin{aligned}\Big\| \sup_{t_k \le t \le t_{k+1}} |A_t f - A_{t_k} f|\Big\|_2 &\le C\Big(\int |g_k(\alpha)|^2\varphi(\alpha)^2 d\alpha\Big)^{1/2} \\ &+ C\sum_{\ell<k}\left(\frac{t_{\ell+1}}{t_k}\right)^{1/2}\Big(\int |g_\ell(\alpha)|^2\varphi(\alpha)^2 d\alpha\Big)^{1/2} \\ &+ C\sum_{\ell>k}\left(\frac{t_{k+1}}{t_\ell}\right)^{1/2}\Big(\int |g_\ell(\alpha)|^2\varphi(\alpha)^2 d\alpha\Big)^{1/2}\end{aligned}$$

where, by hypothesis on the sequence t_k, $\frac{t_\ell}{t_k} < 2^{-(k-\ell)}$ for $k \geq \ell$. Hence, applying Cauchy-Schwarz

$$\Big\| \sup_{t_k \leq t \leq t_{k+1}} |A_t f - A_{t_k} f| \Big\|_2^2 \leq C \int |g_k(\alpha)|^2 \varphi(\alpha)^2 d\alpha$$
$$+ C \sum_{\ell < k} 2^{-\frac{k-\ell}{2}} \int |g_\ell(\alpha)|^2 \varphi(\alpha)^2 d\alpha$$
$$+ C \sum_{\ell > k} 2^{-\frac{\ell-k}{2}} \int |g_\ell(\alpha)|^2 \varphi(\alpha)^2 d\alpha$$

and summation over k gives therefore the estimate

$$C \sum_k \int |g_k(\alpha)|^2 \varphi(\alpha)^2 d\alpha \leq C \int |\hat{f}(\alpha)|^2 \varphi(\alpha)^2 d\alpha \leq C \|f\|_2^2 .$$

This completes the proof of Lemma 2.

4. Prime Number Exponential Sums

This section is based on some results which the reader may find in [D]. Recall the following (standard) notations

$\mu =$ Moebius function

$\Lambda(k) = \log p$ if k is a power of the prime p

$= 0$ otherwise

$\phi(q) =$ number of Dirichlet characters to the modulus q

If $q = 2^\alpha p_1^{\alpha_1} p_2^{\alpha_2} \ldots$, then $\phi(q) = \phi(2^\alpha)\phi(p_1^{\alpha_1})\phi(p_2^{\alpha_2}) \ldots$, where

$\phi(2^\alpha) = 2^{\alpha-1}$

$\phi(p^\alpha) = p^{\alpha-1}(p-1)$, p a prime > 2 .

Proposition 4. *Denote*

$$S_N(\alpha) = \sum_{n \leq N} \Lambda(n) e^{2\pi i n \alpha} \tag{45}$$

$$|S_N(\alpha)| \leq C(Nq^{-1/2} + N^{4/5} + N^{1/2} q^{1/2})(\log N)^4 \tag{46}$$

provided

$$\left|\alpha - \frac{a}{q}\right| \leq \frac{1}{q^2} \quad \text{where} \quad (a, q) = 1$$

If

$$\left|\alpha - \frac{a}{q}\right| \leq (\log N)^{20} N^{-1} \quad \text{where} \quad 1 \leq a \leq q\,, \quad (a,q) = 1 \quad \text{and} \quad q \leq (\log N)^{20}\,, \tag{47}$$

then

$$S_N(\alpha) = \frac{\mu(q)}{\phi(q)} T\left(\alpha - \frac{a}{q}\right) + 0(N \exp(-c\sqrt{\log N})) \tag{48}$$

where

$$T(\beta) = \sum_{1 \leq k \leq N} e^{2\pi i k \beta}$$

Observe that

$$\phi(q) \geq c_\tau q^{1-\tau} \quad \text{for all} \quad \tau > 0 \tag{49}$$

Coming back to Theorem 1 and considering the shift model $(\mathbb{Z}, S)$, to which the problems may be reduced by the discussion in section 2, rewrite

$$A_N f = |P_N|^{-1} \sum_{p \in P_N} S^p f = f * K_N$$

letting

$$K_N = |P_N|^{-1} \sum_{p \in P_N} \delta_{\{p\}}$$

Here P_N stands for the set of primes $\leq N$ and $\delta_{\{p\}}$ for the Dirac measure at the point p. Clearly

$$\left\| \frac{\log N}{N} \sum_{p \in P_N} \delta_{\{p\}} - \frac{1}{N} \sum_{k \leq N} \Lambda(k) \delta_{\{k\}} \right\|_{\ell^1(\mathbb{Z})} \leq$$

$$\frac{\log N}{N} \left|\{k \leq N \mid k \text{ is a proper power of a prime number}\}\right| + \frac{1}{N} \sum_{p \in P_N} |\log N - \log k| < \frac{C}{\log N}.$$

Hence, the Fourier transform of K_N satisfies

$$\left|\widehat{K}_N(\alpha) - \frac{1}{N} S_N(-\alpha)\right| \leq C(\log N)^{-1} \tag{50}$$

(This rather crude estimate suffices.)

Denote η a smooth function of $\mathbb{R}$, $0 \leq \eta \leq 1$, $\eta(\lambda) = 1$ if $|\lambda| \leq \frac{1}{4}$ and $\eta(\lambda) = 0$ for $|\lambda| > \frac{1}{2}$. For $s \geq 0$, denote $\eta_s(\lambda) = \eta(10^s \lambda)$. Let also

$$v_N(\beta) = \int_0^1 e^{-2\pi i N \beta x} dx = \widehat{\chi}_{[0,1]}(N\beta)$$

satisfying

$$|v_n(\beta)| \le C[1 + N|\beta|]^{-1} \tag{51}$$

Define

$$\widehat{L}_N(\alpha) = v_N(\alpha)\eta(\alpha) + \sum_{s=1}^{\infty} \sum_{2^{s-1} \le q < 2^s} \frac{\mu(q)}{\phi(q)} \sum_{\substack{1 \le a \le q \\ (a,q)=1}} v_N\left(\alpha - \frac{a}{q}\right) \eta_s\left(\alpha - \frac{a}{q}\right) \tag{52}$$

Here α is a real number modulo 1.

Note that for fixed s, the function $\eta_s\left(\alpha - \frac{a}{q}\right)$, $2^{s-1} \le q < 2^s$, $1 \le a \le q$, $(a,q) = 1$, are disjointly supported.

Lemma 5. *There is a uniform estimate*

$$|\widehat{K}_N(\alpha) - \widehat{L}_N(\alpha)| \le C(\log N)^{-1} \tag{53}$$

Proof: Fix α and apply Dirichlet's principle to find $1 \le a \le q$, $(a,q) = 1$ such that

$$\left|\alpha - \frac{a}{q}\right| \le \frac{1}{qQ} \tag{54}$$

where

$$q \le Q \equiv N(\log N)^{-20} \tag{55}$$

Let $\beta = \alpha - \frac{a}{q}$.

Case 1. $q \le (\log N)^{20}$ (i.e., α belongs to a "major" arc). By (54), (55) and hypothesis, (47) holds. Hence, by (50)

$$\widehat{K}_N(\alpha) = \frac{\mu(q)}{\phi(q)}\left[\frac{1}{N}\sum_{1 \le k \le N} e^{-2\pi i k\beta}\right] + O\left(\frac{1}{\log N}\right) = \frac{\mu(q)}{\phi(q)} v_N(\beta) + O\left(\frac{1}{\log N}\right) \tag{56}$$

Now examine $\widehat{L}_N(\alpha)$. For $\frac{a'}{q'} \ne \frac{a}{q}$, it follows from (54)

$$\left|\alpha - \frac{a'}{q'}\right| \ge \left|\frac{a}{q} - \frac{a'}{q'}\right| - Q^{-1} \ge \frac{1}{qq'} - Q^{-1}$$

Hence $\left|\alpha - \frac{a'}{q'}\right| > N^{-1/2}$ unless $q' > N^{1/3}$. In the first alternative, by (51)

$$\left|v_N\left(\alpha - \frac{a'}{q'}\right)\right| \le CN^{-1/2}$$

while, in the second alternative, by (49), $\phi(q') > N^{-1/4}$. Consequently

$$\widehat{L}_N(\alpha) = \frac{\mu(q)}{\phi(q)} v_N(\beta)\eta_s(\beta) + O(N^{-1/4}) \tag{57}$$

when $2^{s-1} \le q < 2^s$ and hence $\eta_s(\beta) = 1$ since $|\beta| < N^{-1/2} < \frac{1}{4}10^{-s}$ by hypothesis on q. Thus (52) follows as a consquence of (56), (57).

Case 2. $q > (\log N)^{20}$. By (54), (55), (46), $|S_N(\alpha)| \leq CN(\log N)^{-6}$. Hence, by (49)

$$|\widehat{K}_N(\alpha)| \leq C(\log N)^{-1} . \tag{58}$$

By hypothesis on q, $\phi(q) < (\log N)^{-1}$. For $(a', q') = 1$, $\frac{a}{q} \neq \frac{a'}{q'}$, by (54)

$$\left|\alpha - \frac{a'}{q'}\right| \geq \left|\frac{a}{q} - \frac{a'}{q'}\right| - \frac{1}{qQ} \geq \frac{1}{qq'} - \frac{1}{qQ} .$$

Assume $q' \leq (\log N)^4$. Then $\left|\alpha - \frac{a'}{q'}\right| > \frac{1}{2qq'}$ and by (51), (55)

$$\left|v_N\left(\alpha - \frac{a'}{q'}\right)\right| \leq C\left(\frac{N}{qq'}\right)^{-1} \leq (\log N)^{-1} \tag{59}$$

Thus $|\widehat{L}_N(\alpha)| < (\log N)^{-1} + (\log N)^{-1}\sum_s^1 2^{-s/2} + \sum_s^2 2^{-s/2}$, where $\sum^1$ refers to values of s with $2^s \leq (\log N)^4$ and $\sum^2$ to the remaining values. Hence

$$|\widehat{L}_N(\alpha)| \leq C(\log N)^{-1} \tag{60}$$

which together with (58) again implies (53). This completes the proof of Lemma 5.

5. Proof of Theorem 1: The Maximal Inequality

Considering the shift model, we prove the inequality

$$\|\sup_N |f * K_N|\|_2 \leq C\|f\|_2 \tag{61}$$

where

$$K_N = \frac{1}{|P_N|}\sum_{p\in P_N} \delta_{\{p\}} .$$

Of course, in proving (61), we may assume $f \geq 0$ and restrict to diadic values of N, i.e., $N \in Z = \{2^n \mid n = 1, 2, \ldots\}$. Let L_N be given by (52). Lemma 5 will allow us to substitute K_N and L_N. Write

$$\sup_{N\in Z} |f * K_N| \leq \sup_{N\in Z} |f * L_N| + \Big(\sum_{n\in Z} |f * (K_N - L_N)|^2\Big)^{1/2}$$

and, since $\|f * (K_N - L_N)\|_2 \leq \|\widehat{K}_N - \widehat{L}_N\|_\infty \|f\|_2$,

$$\|\sup_{N\in Z} |f * K_N|\|_2 \leq \|\sup_{N\in Z} |f * L_N|\|_2 + \Big(\sum_{N\in Z} \|\widehat{K}_N - \widehat{L}_N\|_\infty^2\Big)^{1/2}\|f\|_2 \tag{62}$$

By (53), $\|\widehat{K}_N - \widehat{L}_N\|_\infty \leq C(\log N)^{-1}$. Hence, the second term of (62) is bounded by $C\|f\|_2$. Estimate the first term using the definition (52) of $\widehat{L}_N$ and the triangle inequality. Thus

$$(f * L_N)(x) = \int_0^1 \hat{f}(\alpha)\widehat{L}_N(\alpha)e^{2\pi i\alpha x}d\alpha\,,$$

$$\sup_N |(f * L_N)(x)| \leq \sup_N \left| \int \hat{f}(\alpha)v_N(\alpha)\eta(\alpha)e^{2\pi i\alpha x}d\alpha\right| + \tag{63}$$

$$\sum_{s=1}^{\infty} \sum_{2^{s-1}\leq q\leq 2^s} \frac{1}{\phi(q)} \sup_N \left| \sum_{\substack{1\leq a\leq q\\(a,q)=1}} \int \hat{f}(\alpha)v_N\left(\alpha - \frac{a}{q}\right)\eta_s\left(\alpha - \frac{a}{q}\right)e^{2\pi i\alpha x}d\alpha\right| \tag{64}$$

For $s = 1, \ldots$ and $2^{s-1} \leq q < 2^s$, write

$$\sup_N \left| \sum_{\substack{1\leq a\leq q\\(a,q)=1}} \int \hat{f}(\alpha)v_N\left(\alpha - \frac{a}{q}\right)\eta_S\left(\alpha - \frac{a}{q}\right)e^{2\pi i\alpha x}d\alpha\right| \leq$$

$$\sup_{t>0} \left| \sum_{\substack{1\leq a\leq q\\(a,q)=1}} \int \hat{f}_q(\alpha)\hat{k}\left(t\left(\alpha - \frac{a}{q}\right)\right)\eta_s\left(\alpha - \frac{a}{q}\right)e^{2\pi i\alpha x}d\alpha\right| \tag{65}$$

where $k(\lambda) = \chi_{[0,1]}(\lambda)$ and $\hat{f}_q(\alpha) = \sum_{\substack{1\leq a\leq q\\(a,q)=1}} \eta_{s-1}\left(\alpha - \frac{a}{q}\right)\hat{f}(\alpha)$.

Apply Lemma 1 to estimate (65), letting $R = \left\{\frac{a}{q} \mid 1 \leq a \leq q\,,\ (a,q) = 1\right\} = R_q$, with common denominator q. The function η_s is supported by a $\tau = \frac{1}{2}10^{-s}$ neighborhood of 0, thus $\tau q < \frac{1}{100}$. Thus (65) is bounded by $C\|f_q\|_2$. By summation, it now follows from (63) and (64) that

$$\left\|\sup_N |f * L_N|\right\|_2 \leq C\|f\|_2 + C\sum_{s=1}^{\infty} \sum_{2^{s-1}\leq q\leq 2^s} \frac{1}{\phi(q)} \left\{ \sum_{\substack{1\leq a\leq q\\(a,q)=1}} \int |\hat{f}(\alpha)|^2\eta_{s-1}\left(\alpha - \frac{a}{q}\right)^2 d\alpha\right\}^{1/2}$$

which by (49) and the Cauchy-Schwarz inequality is bounded by

$$C\|f\|_2 + C\sum_{s-1}^{\infty} 2^{-s/4} \left\{ \sum_{2^{s-1}\leq q\leq 2^s} \sum_{\substack{1\leq a\leq q\\(a,q)=1}} \int |\hat{f}(\alpha)|^2\eta_{s-1}\left(\alpha - \frac{a}{q}\right)^2 d\alpha\right\}^{1/2} \leq C\|f\|_2$$

Recalling (62), we proved that

$$\left\|\sup_{N\in Z} |f * K_N|\right\|_2 \leq C\|f\|_2$$

and thus the maximal ergodic theorem for the primes in the model $(\mathbb{Z}, S)$, hence in general, by the discussion in the second section.

7. Proof of Theorem 1: Almost Sure Convergence

Going back to the reduction of the problem in the last part of section 2, we have to prove (16)

$$\sum_{1\leq j\leq J} \|\mathcal{M}_j f\|_2 \leq o(J)\|f\|_2 \tag{66}$$

where $f \in \ell^2(\mathbb{Z})$, $|f| \leq 1$ and $\mathcal{M}_j f = \sup_{\substack{N_j\leq N\leq N_{j+1}\\ N\in Z_\varepsilon}} |(f * K_N) - (f * K_{N_j})|$.

Here $Z_\varepsilon = \{[(1+\varepsilon)^n]\,;\, n = 1,2,\dots\}$ and N_j an increasing sequence satsifying $N_{j+1} > 2N_j$. Fix a positive integer J. First replace the K_N by L_N, writing

$$\mathcal{M}_j f \leq \sup_{\substack{N_j\leq N\leq N_{j+1}\\ N\in Z_\varepsilon}} |(f * L_N) - (f * L_{N_j})| + 2\left\{\sum_{\substack{N_j\leq N\leq N_{j+1}\\ N\in Z_\varepsilon}} |(f * K_N) - (f * L_N)|^2\right\}^{1/2}$$

and

$$\sum_{j=1}^{J} \|\mathcal{M}_j f\|_2^2 \leq \sum_{1\leq j\leq J} \Big\| \sup_{N_j\leq N\leq N_{j+1}} |(f * L_N) - (f * L_{N_j})|\Big\|_2^2 + \sum_{N\in Z_\varepsilon} |(f * K_N) - (f * L_N)|^2 \tag{67}$$

Again by Parseval's identity and (53), the ℓ^2-norm of (67) is at most

$$\left(\sum_{N\in Z_\varepsilon} \|\widehat{K}_N - \widehat{L}_N\|_\infty^2\right) \|f\|_2^2 \leq \left(\sum_{n=1}^{\infty} (n\varepsilon)^{-2}\right) \|f\|_2^2 \leq c\varepsilon^{-2}\|f\|_2^2\,.$$

Thus it remains to show that (66) is bounded by $o(J)\|f\|_2^2$ for J large. Fix an integer $\bar{s}$ (which will depend on J) and put

$$\widehat{M}_N(\alpha) = v_N(\alpha)\eta(\alpha) + \sum_{s=1}^{\bar{s}} \sum_{2^{s-1}\leq q<2^s} \frac{\mu(q)}{\phi(q)} \sum_{\substack{1\leq a\leq q\\ (a,q)=1}} s_N\left(\alpha - \frac{a}{q}\right)\eta_s\left(\alpha - \frac{a}{q}\right) \tag{68}$$

Coming back to the estimates in the previous section, we see that

$$\Big\|\sup_N |f * (L_N - M_N)|\Big\|_2 \leq C\left(\sum_{s>\bar{s}} 2^{-s/4}\right)\|f\|_2 \leq C2^{-\bar{s}/4}\|f\|_2 \tag{69}$$

Hence (66) is at most

$$\sum_{1\leq j\leq J} \Big\| \sup_{N_j\leq N\leq N_{j+1}} |(f * M_N) - (f * M_{N_j})|\Big\|_2^2 + CJ2^{-s/2}\|f\|_2^2 \tag{70}$$

The first term of (70) is estimated by writing $\widehat{M}_N$ as sum of at most $4^{\bar{s}}$ terms, given by (68), and applying triangle inequality. This gives the bound $4^{\bar{s}}$. I, letting I be an upper bound for the expressions

$$\sum_{1\leq j\leq J} \Big\| \sup_{N_j\leq N\leq N_{j+1}} |A_N f - A_{N_j} f| \Big\|_2^2 \tag{71}$$

where

$$A_N f(x) = \int \hat{f}(\alpha) v_N(\alpha - \theta)\eta_s(\alpha-\theta) e^{2\pi i \alpha x} d\alpha \tag{72}$$

Here $v_N(\lambda) = \widehat{K}(N\lambda)$, $K = \chi_{[0,1]}$. Making a change of variable $\alpha' = \alpha - \theta$, Lemma 2 applies and (30) gives the bound $C\|f\|_2^2$. Hence (70) is at most $C(4^{\bar{s}} + J2^{-\bar{s}/2})\|f\|_2^2 = o(J)\|f\|_2^2$, for appropriate choice of $\bar{s}$. This completes the proof.

8. Further Comments

(1) The particular sets $\Lambda \subset \mathbb{Z}_+$ considered in [B2] and $\Lambda = \{\text{primes}\}$ have a common feature about the behavior of the corresponding exponential sums

$$\widehat{K}_N(\alpha) = \frac{1}{|\Lambda \cap [1,N]|} \sum_{\substack{1\leq n\leq N \\ n\in\Lambda}} e^{2\pi i n\alpha} \tag{73}$$

Roughly speaking, $\widehat{K}_N(\alpha)$ is "small" except for α in a major arc, thus α close to a rational point with small denominator. Moreover, for α close to such a point $\frac{a}{q}$, $(a,q) = 1$, $\widehat{K}_N(\alpha)$ may be identified with an expression of the form

$$\widehat{K}_N(\alpha) \approx \Phi(q,a)\widehat{k}\left(N\left(\alpha - \frac{a}{q}\right)\right) \tag{74}$$

where k is a function on $\mathbb{R}$ satisfying the Hardy-Littlewood maximal inequality

$$\Big\| \sup_{t>0} |f * k_t| \Big\|_{L^{1,\infty}(\mathbb{R})} \leq C\|f\|_{L^1(\mathbb{R})} \tag{75}$$

The function k relates to the density of Λ while the numbers $\Phi(q,a)$ are related to irregular distribution of Λ in the residue classes. Moreover, there is an estimate

$$|\Phi(q,a)| \leq \lambda(q) \quad \text{for} \quad (a,q) = 1$$

with $\lambda(q)$ being multiplicative, i.e.,

$$q = 2^{\alpha} p_1^{\alpha_1} p_2^{\alpha_2} \ldots \Longrightarrow \lambda(q) = \lambda(2^{\alpha})\lambda(p_1^{\alpha_1})\lambda(p_2^{\alpha_2}) \tag{76}$$

Let $\delta, \nu > 0$ satisfy

$$\lambda(q) \leq C q^{-\delta} \tag{77}$$

and

$$\lambda(p) \leq C p^{-\nu} \quad \text{if} \quad p \quad \text{is a prime} \tag{78}$$

In the case of $\Lambda = \{\text{primes}\}$, $k = \chi_{[0,1]}$ and $\Phi(q, a) = \frac{\mu(q)}{\phi(q)}$. Hence we may take for δ, ν any number < 1.

For $\Lambda = \{\text{squares}\}$, $k \sim x^{-1/2}\chi_{[0,1]}$ and $\Phi(q, a)$ is given by the Gauss sum

$$S(q, a) = \frac{1}{q}\sum_{r=0}^{q-1} e^{2\pi i r^2 \frac{a}{q}}$$

For δ, ν, we may take the value $\frac{1}{2}$.

Already in the case of the squares, there are some changes needed in the previous exposition for $\Lambda = \{\text{primes}\}$, when proving the corresponding pointwise ergodic theorem. This is due to a worse estimate (77). The main modification is an application of Lemma 1 taking for R larger sets

$$R = \left\{ \frac{a}{q} \mid 1 \leq a < q \ , \ (a, q) = 1 \ , \ q \subset A \right\}$$

rather than just

$$R = R_q = \left\{ \frac{a}{q} \mid 1 \leq a \leq q \ , \ (a, q) = 1 \right\}$$

for a single value of q. (see [B1,2]).

In the context of a general polynomial $p(n)$ of degree t, Hua's estimate gives $\delta = \frac{1}{t}$ and the estimate of A. Weil $\nu = \frac{1}{2}$. The fact of having this "better" estimate for prime modulus is essential in the argument given in [B2], as soon as $t \geq 4$.

Presently, I am able to prove an L^2-theorem provided (77) holds with some $\delta > 0$ and (78) with $\nu > \frac{1}{4}$.

(2) L^p-Estimates, $p < 2$.

Such estimates are obtained by interpolation arguments, using the L^2-estimates and a more careful analysis of the Fourier multipliers used in this process (see [B3], where the case $\Lambda = \{n^t \mid n = 1, 2, \ldots\}$ was considered). A straightforward adaptation of [B3] to the more abstract setting described above gives the following conclusion:

If (77) holds with some $\delta > 0$ and (78) with $\nu > \frac{1}{4}$, assuming (76), then Λ satisfies the pointwise ergodic theorem for functions of class L^p, provided

$$p > \frac{1}{2} + \left(\frac{1}{4} + \frac{1}{2\nu}\right)^{1/2} \tag{79}$$

Presently, this restriction seems mainly technical. In particular, Theorem 1 of this paper extends to L^p-functions, for $p > \frac{1+\sqrt{3}}{2}$.

(3) The Wiener-Wintner theorem is the following "strong form" of Birkhoff's theorem. Given a dynamical system $(\Omega, \mathcal{B}, \mu, T)$, f say a bounded μ-measurable function on Ω, the ergodic means ($|z| = 1$)

$$\frac{1}{N} \sum_{1 \le n \le N} z^n T^n f \tag{80}$$

converge pointwise except on a μ negligible set which is independent of the point z on the unit circle (at least may be taken as such). The reader may find a proof of this result in [B-L,2]. Presently, I do not know of an analogue result for "arithmetic sets" Λ, as considered in [B2] or this paper. The question was brought to my attention by B. Weiss.

References

[B1] J. Bourgain, On pointwise ergodic theorems for arithmetic sets, CRA Sc. Paris, t305, Ser 1, 397-402, 1987.

[B2] J. Bourgain, On the maximal ergodic theorems for certain subsets of the integers, Israel J. Math. 61 (1988).

[B3] J. Bourgain, On the pointwise ergodic theorems on L^p for arithmetic sets, Israel J. Math. 61 (1988).

[B4] J. Bourgain, On high dimensional maximal functions associated to convex sets. American J. Math. 108, 1986, 1467-1476.

[B-L,1] A. Bellow, V. Losert, On sequences of desity zero in Ergodic Theory, Contemporary Math. 26, 1984, 49-60.

[B-L,2] A. Bellow, V. Losert, The weighted pointwise ergodic theorem and the individual ergodic theorems along subsequences. TAMS 288, 1985, 307-355.

[D] H. Davenport, Multiplicative number theory, Springer-Verlag 1980.

[KW] Y. Katznelson, B. Weiss, A simple proof of some ergodic theorems, Israel J. Math., 42, N4, 1982.

[W] B. Weiss, private communications.

SOME REMARKS ON THE GEOMETRY OF CONVEX SETS

Keith Ball

Department of Pure Mathematics and Mathematical Statistics
Cambridge University
Cambridge, England

Abstract

We prove a strengthening of Santalo's inequality for the unit balls of normed spaces with 1-unconditional bases and observe that all central sections of the unit cube in $\mathbf{R}^n$ (for $n \geq 10$) have smaller volume than those of the Euclidean ball of volume 1.

0. Introduction

In this note we discuss two results concerning symmetric convex sets in $\mathbf{R}^n$. The first, and more important of the two, provides a strengthening of Santalo's inequality, for a special class of convex sets; namely, the unit balls of spaces with 1-unconditional bases. In our second remark, we observe that a previously published result of the present author yields a natural solution to a problem of Busemann and Petty [6].

Santalo's inequality, [13], states that if C is a symmetric convex set in $\mathbf{R}^n$ and

$$C^0 = \{y \in \mathbf{R}^n : |\langle x, y\rangle| \leq 1 \ , \ \text{for all } x \in C\}$$

is its polar, then

$$|C| \cdot |C^0| \leq v_n^2$$

where v_n is the volume of the Euclidean ball of radius 1 in $\mathbf{R}^n$. Hensley showed that in questions concerning the volumes of sections of convex sets, it is critical to obtain bounds on expressions such as

$$\int_C |x|^2 d\mu(x)$$

for symmetric convex sets C in $\mathbf{R}^n$, where $|x|$ is the usual Euclidean norm of the vector x and μ is Lebesgue measure in $\mathbf{R}^n$. Our first result provides an upper bound on an expression of this type involving C and C^0, for unit balls of spaces with 1-unconditional bases.

Before embarking upon our exposition, we should clarify, geometrically, the restriction imposed on the convex sets with which we deal. A basis $e_1, \ldots, e_n$ of an n-dimensional normed space is 1-unconditional if the norm $\|\sum \varepsilon_i \lambda_i e_i\|$ is independent of a choice of signs $\varepsilon_1, \ldots, \varepsilon_n$ for each scalar sequence (λ_i). Thus, a convex set $C \subset \mathbf{R}^n$ is the unit ball of such a space (represented on $\mathbf{R}^n$) if there is an affine transformation $\widetilde{C}$, of C, which is symmetric with respect to each of the coordinate hyperplanes.

1. Extensions of the Brunn-Minkowski Inequality and Santalo's Inequality

The purpose of this section is to prove the following.

Theorem 1. *Let $C \subset \mathbf{R}^n$ be a symmetric convex set which is the unit ball of a normed space $(\mathbf{R}^n, \|\cdot\|)$ with a 1-unconditional basis. Then*

$$\int_C \int_{C_0} \langle x, y\rangle^2 d\mu(y) d\mu(x) \leq \frac{n v_n^2}{(n+2)^2} \cdot$$

There is equality if C is an ellipsoid.

Remark. That the assertion of Theorem 1 is stronger than that of Santalo's inequality is a consequence of the fact that for any symmetric convex set C with polar C^0,

$$\frac{n|C|^{1+\frac{2}{n}}|C^0|^{1+\frac{2}{n}}}{(n+2)^2 v_n} \leq \int_C \int_{C^0} \langle x, y\rangle^2 d\mu(y) d\mu(x) .$$

A proof of this fact appears in [3]. It is a consequence of the simple Lemma 6 of [2].

The proof of Theorem 1 uses a modification of an inequality of Prékopa and Leindler, [10] and [9]. Their inequalities for integrals of functions are generalisations of the classical Brunn-Minkowski inequality for volumes of sets. A short proof of Leindler's inequality was obtained by Rinott, [11]; another short proof, and a variety of extensions of Leindler's inequality appear in Brascamp and Lieb, [4]. We shall deduce our first inequality from Leindler's, which we state as a lemma.

Lemma 2. *Suppose $f, g, m : \mathbf{R} \to [0, \infty)$ are measurable, $\lambda \in (0,1)$ and for all $x, y \in \mathbf{R}$*

$$m(\lambda x + (1-\lambda)y) \geq f(r)^\lambda g(s)^{1-\lambda} .$$

Then

$$\int_{\mathbf{R}} m(x)dx \geq \left[\int_{\mathbf{R}} f(x)dx\right]^\lambda \left[\int_{\mathbf{R}} g(x)dx\right]^{1-\lambda} . \qquad \square$$

The inequality we require, differs from Leindler's in that we replace the arithmetic mean $\lambda x + (1-\lambda)y$ by the geometric mean $x^\lambda y^{1-\lambda}$.

Lemma 3. *Suppose $f, g, m : [0,\infty) \to [0,\infty)$ are measurable, $\lambda \in (0,1)$ and for all $x, y \in [0,\infty)$*

$$m(x^\lambda y^{1-\lambda}) \geq f(x)^\lambda g(y)^{1-\lambda} .$$

Then

$$\int_0^\infty m(x)dx \geq \left[\int_0^\infty f(x)dx\right]^\lambda \left[\int_0^\infty g(x)dx\right]^{1-\lambda} .$$

Proof: Define $F, G, M : \mathbf{R} \to [0,\infty)$ by

$$\begin{aligned} F(x) &= f(e^x)e^x , \\ G(x) &= g(e^x)e^x , \\ M(x) &= m(e^x)e^x . \end{aligned}$$

Then for any x and y

$$\begin{aligned} M(\lambda x + (1-\lambda)y) &= m[(e^x)^\lambda (e^y)^{1-\lambda}](e^x)^\lambda (e^y)^{1-\lambda} \\ &\geq f(e^x)^\lambda g(e^y)^{1-\lambda}(e^x)^\lambda (e^y)^{1-\lambda} \\ &= F(x)^\lambda G(y)^{1-\lambda} . \end{aligned}$$

Therefore

$$\begin{aligned} \int_0^\infty m(x) = \int_{\mathbf{R}} M(u)du &\geq \left[\int_{\mathbf{R}} F(u)du\right]^\lambda \left[\int_{\mathbf{R}} G(u)du\right]^{1-\lambda} \\ &= \left[\int_0^\infty f(x)dx\right]^\lambda \left[\int_0^\infty g(x)dx\right]^{1-\lambda} . \end{aligned}$$

□

We are now in a position to prove Theorem 1. We use Santalo's inequality (for unit balls of spaces with 1-unconditional bases) in the proof, but it will be clear that a slight adaptation of the argument would provide a proof of this too. The general case of Santalo's inequality can be deduced from that for 1-unconditional sets by using Steiner symmetrisations as in the proof given by Saint-Raymond, [12].

Proof of Theorem 1. The case of equality is easily checked, since the expression to be dominated is invariant under linear transformations of C. If $n = 1$, the only symmetric convex sets are "ellipsoids", so assume $n > 1$.

Without loss of generality assume that the standard basis $e_1, \ldots, e_n$ of $\mathbf{R}^n$ is a normalised, 1-unconditional basis for the norm whose unit ball is C and hence also for the dual norm corresponding to C^0. Thus C and C^0 are symmetric with respect to all coordinate hyperplanes, and the vectors $e_1, \ldots, e_n$ are on the boundaries of both C and C^0.

For $H = \langle e_1 \rangle^\perp = \text{span}(e_2, \ldots, e_n)$, define $f, f^* : [0, \infty) \to [0, \infty)$ by

$$f(r) = |(H + re_1) \cap C|$$
$$f^*(r) = |(H + re_1) \cap C^0| .$$

Observe that if $r > 1$, $f(r) = f^*(r) = 0$. (1)

Suppose $r, s \in [0, 1]$ and $f(r), f^*(s) > 0$. Then the sets

$$D = H \cap (C - re_1) , \qquad E = H \cap (C^0 - se_1)$$

are the unit balls of two $(n-1)$-dimensional normed spaces (represented on H) with 1-unconditional bases; in particular D and E are symmetric convex sets in H. Now suppose $u \in D$, $v \in E$. Then since $\pm u \in D$, $\pm v \in E$,

$$re_1 \pm u \in C , \qquad se_1 \pm v \in C^0 .$$

Hence

$$rs + |\langle u, v \rangle| = \langle re_1 + \text{sgn}\langle u, v \rangle u \, , \, se_1 + v \rangle \leq 1 .$$

So $|\langle u, v \rangle| \leq 1 - rs$. Therefore $E \subset (1 - rs)D^0$, where D^0 is the polar of D in H, and hence by Santalo's inequality in $n - 1$ dimensions,

$$|D||E| \leq (1 - rs)^{n-1} {v_{n-1}}^2 .$$

Thus, we have that for all $r, s \in [0, 1]$,

$$f(r) f^*(s) \leq (1 - rs)^{n-1} {v_{n-1}}^2 . \tag{2}$$

Let $m : [0, \infty) \to [0, \infty)$ be defined by

$$m(t) = \begin{cases} t^2(1 - t^2)^{\frac{n-1}{2}} \cdot v_{n-1} & 0 \leq t \leq 1 \\ 0 & t > 1 . \end{cases}$$

Then by (1) and (2) we have that for $r, s \in [0, \infty)$,

$$f(r)^{1/2} r \cdot f^*(s)^{1/2} s \leq m(\sqrt{rs}) .$$

Hence, by Lemma 3 with $\lambda = \frac{1}{2}$,

$$\int_0^\infty m(t) dt \geq \left[\int_0^\infty f(r) r^2 dr \right]^{1/2} \left[\int_0^\infty f^*(s) s^2 ds \right]^{1/2} .$$

But this is exactly the statement that

$$\frac{1}{2}\left[\int_C \int_{C^0} x_1^2 y_1^2 \, d\mu(y) d\mu(x)\right]^{1/2} \leq \frac{v_n}{2(n+2)}$$

where x_i denotes the ith coordinate of $x \in \mathbf{R}^n$. The unconditionality assumption on C ensures that if $i \neq j$,

$$\int_C x_i x_j \, d\mu(x) = 0$$

and so

$$\int_C \int_{C^0} \langle x, y\rangle^2 d\mu(y) d\mu(x) = \sum_{i=1}^{n} \int_C \int_{C^0} x_i^2 y_i^2 \, d\mu(y) d\mu(x) \leq \frac{n v_n^2}{(n+2)^2} \, .$$

□

Remark. While Theorem 1 is stated only for the unit balls of spaces with 1-unconditional basis, it seems likely that the assertion is true for all symmetric convex sets. In contrast to the situation of Santalo's inequality, a simple symmetrisation argument seems insufficient to provide a deduction of the general case from the special.

2. Convex sets with small sections

A problem of Busemann and Petty, [6], asks: suppose C and D are symmetric convex sets in $\mathbf{R}^n$ for which

$$|H \cap C| \geq |H \cap D|$$

for every 1-codimensional subspace H of $\mathbf{R}^n$; is it then true that

$$|C| \geq |D| \, .$$

Busemann, [5], showed that the answer is yes, if D is an Euclidean ball (or if the dimension $n = 2$). Larman and Rogers, [8], showed that in general, the assertion is false if $n \geq 12$, even if C is an Euclidean ball. They constructed a random convex set D, all of whose sections have small volume.

In this section we observe that the answer is negative in dimension $n \geq 10$, in consequence of the principal result of [1]. The latter states that if $Q_n = \left[-\frac{1}{2}, \frac{1}{2}\right]^n \subset \mathbf{R}^n$ is the central unit cube in n dimensions, then for every 1-codimensional subspace H of $\mathbf{R}^n$,

$$|H \cap Q_n| \leq \sqrt{2} \, .$$

Theorem 4. *If n is sufficiently large, Q_n is the central unit cube in $\mathbf{R}^n$ and C is a central Euclidean ball of appropriate radius, then*

$$|H \cap C| \geq |H \cap Q_n|$$

for every 1-codimensional subspace H of $\mathbf{R}^n$, but

$$|C| < 1 = |Q_n| .$$

In fact this is the case if $n \geq 10$.

Proof: Since $|H \cap Q_n| \leq \sqrt{2}$ for every 1-codimensional subspace H, it suffices to show that, for the appropriate n, the Euclidean ball of volume 1 in $\mathbf{R}^n$ has central sections whose volume is greater than $\sqrt{2}$.

Let the volume of a 1-codimensional section of the Euclidean ball of volume 1 be

$$a_n = \frac{v_{n-1}}{v_n^{\frac{n-1}{n}}} .$$

Then

$$a_n = \frac{\Gamma\left[\frac{n}{2}+1\right]^{\frac{n-1}{n}}}{\Gamma\left[\frac{n-1}{2}+1\right]} .$$

By Stirling's formula, $a_n \to \sqrt{e} \quad (n \to \infty)$ proving the first assertion.

Now $a_{10} = \frac{32.(120)^{\frac{9}{10}}}{945.\sqrt{\pi}} = 1.420$ correct to 3 decimal places. So $a_{10} > \sqrt{2}$.

That $a_n > \sqrt{2}$ if $n \geq 10$ follows from the fact that the sequence (a_n) is increasing. We give a brief sketch of one possible proof of this fact in the appendix. □

The problem of Busemann and Petty remains open for dimension n between 3 and 9. It seems likely that there are examples of convex sets with small sections in these dimensions as well.

It is a widespread conjecture that up to an absolute constant the problem has an affirmative answer: i.e., that there is a constant $\delta > 0$, independent of n, such that if C is a symmetric convex set of volume 1 in $\mathbf{R}^n$, then there is a 1-codimensional subspace H of $\mathbf{R}^n$ for which $|C \cap H| \geq \delta$. This conjecture is related to many questions concerning convex sets; some examples appear in [3].

Appendix

We give a sketch of a proof that the sequence (a_n), defined above, is increasing.

Lemma 5. *Let a_n be the volume of a central 1-codimensional section of the Euclidean ball of volume 1 in $\mathbf{R}^n$. Then (a_n) is an increasing sequence.*

Proof: By the convexity of the function $t \mapsto t^{-1}$ on $(0, \infty)$, we have

$$\log\left(1 + \frac{1}{x}\right) \geq \frac{2x+1}{2x(x+1)}, \qquad x > 0. \tag{1}$$

In the other direction

$$\log\left(1 + \frac{1}{x-1}\right) \geq \frac{1}{x} + \frac{1}{2x^2} > \frac{2x+1}{2x^2+1}, \qquad x > 1. \tag{2}$$

Differentiating and using (1), we find that the function

$$x \mapsto \left(\frac{x+1}{x}\right)^{2x+1}$$

is decreasing on $(0, \infty)$, while by (2), the function

$$x \mapsto \left(\frac{x}{x-1}\right)^{2x-1-\frac{1}{x}}$$

is increasing on $(1, \infty)$.

These imply that if $n \geq 3$

$$\left(\frac{n-1}{n-2}\right)^{\frac{n-3}{2n-5}} \left(\frac{n+1}{n}\right)^{\frac{n}{2n-1}} < \frac{n}{n-1}. \tag{3}$$

Now for $n \geq 0$ let $I_n = \int_{-\pi/2}^{\pi/2} \cos^n \theta d\theta$, and for $n \geq 1$ define

$$u_n = \left(\frac{I_{n-1}}{I_n}\right)^{n-1}, \qquad x_n = \left(\frac{u_{n+1}}{u_n}\right)^{\frac{1}{2n-1}} = \frac{I_n}{I_{n-1}}\left(\frac{n+1}{n}\right)^{\frac{1}{2n-1}}.$$

Then inequality (3) says exactly that $x_n < x_{n-2}$ for $n \geq 3$: since $x_n \to 1$ $(n \to \infty)$, this implies that $x_n \geq 1$ for all $n \geq 1$. So the sequence (u_n) is increasing. (4)

Now $a_n = \frac{v_{n-1}}{v_n^{\frac{n-1}{n}}} = \frac{v_n^{\frac{1}{n}}}{I_n}$ and in consequence we have the relation $a_n = u_n^{\frac{1}{n}} \cdot a_{n-1}^{\frac{n-1}{n}}$ for $n \geq 2$. (5)

Observe that $a_1 = 1 < u_2$: assuming inductively that $a_{n-1} < u_n$, we see from relation (5) that $a_n < u_n$ and then from (4), that $a_n < u_{n+1}$. Hence, $a_{n-1} < u_n$ for all $n \geq 2$. But then using (5) again we obtain

$$a_n > a_{n-1}^{\frac{1}{n}} \cdot a_{n-1}^{\frac{n-1}{n}} = a_{n-1}, \qquad n \geq 2$$

as desired. □

References

1. K.M. Ball, Cube slicing in R^n, Proc. Amer. Math. Soc. 97 #3 (1986), 465-473.
2. K.M. Ball, Logarithmically concave functions and sections of convex sets, Studia Math, to appear.
3. K.M. Ball, Isometric problems and sections of convex sets, Dissertation, University of Cambridge, England (1986).
4. Herm Jan Brascamp and Elliott H. Lieb, On extensions of the Brunn-Minkowski and Prékopa-Leindler theorems, including inequalities for log. concave functions, and with an application to the diffusion equation, J.F.A. 4 (1976), 366-389.
5. H. Busemann, Volumes in terms of concurrent cross-sections, Pacific J. Math. 3 (1953), 1-12.
6. H. Busemann and C.M. Petty, Problems on convex bodies, Math. Scand. 4 (1956), 88-94.
7. D. Hensley, Slicing convex bodies – bounds for slice area in terms of the bodies' covariances, Proc. Amer. Math. Soc. 79 #4 (1980), 619-625.
8. D.G. Larman and C.A. Rogers, The existence of a centrally symmetric convex body with central sections that are unexpectedly small, Mathematika 22 (1975), 164-175.
9. L. Leindler, On a certain converse of Hölder's inequality. II, Acta Sci. Math. 33 (1972), 217-223.
10. A. Prékopa, Logarithmic concave measures with application to stochastic programming, Acta Sci. Math. 32 (1971), 301-316.
11. Y. Rinott, On convexity of measures, Thesis, Weizmann Inst., Rehovot, Israel, 1973.
12. J. Saint-Raymond, Sur le volume des corps convexes symétriques, Séminaire d'initiation à l'analyse, 20e Année, 1980/1, Exp. #11, eds. G. Choquet, M. Rogalski, J. Saint-Raymond (Publ. Math. Univ. Pierre et Marie Curie 46), Univ. Paris VI, Paris, 1981.
13. L.A. Santalo, Un invariante afin para los cuerpos convexos del espacio de n dimensiones, Portugal math. 8, Fasc. 4 (1949).

ON FINITE DIMENSIONAL HOMOGENEOUS BANACH SPACES

J. Bourgain[1]

IHES, France

1. Introduction

The purpose of this note is to prove the following fact, answering a question of V. Milman.

Proposition. *There is a constant $c_1 > 0$ such that if all $[c_1 n]$-dimensional subspaces of a normed space $X, \| \ \|$ are isomorphic up to a constant K, i.e., the Banach-Mazur distance*

$$d(E,F) < K$$

whenever E, F are subspaces of X, $\dim E = [c_1 n] = \dim F$, then the Hilbertian distance

$$d_X = d(X, \ell_n^2) \leq \varphi(K) \ .$$

The method described below yields an explicit value for the constant c_1. This value may be considerably improved by reorganizing and refining the argument. For different reasons there seem to be difficulties in letting $c_1 \to 1$. Applying Theorem 6.2 of [FLM], the previous proposition has the following corollary.

Corollary. *Let $k \leq c_1 n$ and $d(E,F) \leq K$ whenever E, F are k-dimensional subspaces of X. Then*

$$d(E, \ell_k^2) \leq \varphi(K)$$

if E is a k-dimensional subspace of X and

$$d_X \leq \Psi(K)^{\frac{\log n}{\log k}} \ .$$

This result is in a sense an isomorphic version of Gromov's work related to the Banach problem on n-dimensional convex symmetric bodies which k-dimensional sections are affinely equivalent [Gr]. It follows from Gromov's theorem that for a given k the function $\varphi(K) \to 1$ if $K \to 1$. The methods used are of a different nature here. In fact we need a considerable amount of concepts, techniques and results from finite dimensional Banach space theory. the main reference for this is the book of Milman and Schechtman [M-S]. Through the paper, we use the letter c for numerical constants.

---

[1] IHES, 35 Route de Chartes, 91440 Bures-sur-Yvette

2. Enlarging Hilbert Subspaces

In this section, we will use an argument due to Milman and Pisier [M-P] in order to prove the following.

Lemma 1. *Assume $0 \leq \beta \leq \alpha \leq \frac{1}{3}$ and every $[\alpha n]$-dimensional subspace of the n-dimensional normed space X contains a further $[\beta n]$-dimensional K-Hilbertian subspace. Then there is a subspace X_0 of X, $\dim X_0 > \frac{n}{3}$ satisfying*

(1) $$d_{X_0} \leq c(K)\beta^{-1}(\log d_X)\ .$$

Sketch of Proof. The existence of the "ℓ-ellipsoid" and the estimate on the K-convexity constant $K(X) \leq c \log d_X$ permit us to assume

(2) $$MM_* \leq c \log d_X$$

where

$$M = \int_{S_{n-1}} \|x\| \sigma(dx) \quad ; \quad M_* = \int_{S_{n-1}} \|x\|_* \sigma(dx)\ .$$

By a result of [M] (see also [B-M]), there is a subspace X_1 of X, $\dim X_1 > \frac{9}{10} n$ satisfying

(3) $$\|x\| \geq c M_*^{-1} |x| \qquad \text{for} \qquad x \in X_1\ .$$

If X_2 is a subspace of X_1, $\dim X_2 < \frac{n}{3}$, there exists by hypothesis a subspace Y of $X_1 \cap X_2^{\perp}$, $\dim Y = [\beta n]$ and $d_Y \leq K$. Since $M_Y \leq c\beta^{-1/2} M$, where

$$M_Y = \int_{S_{n-1} \cap Y} \|x\| \sigma_{m-1}(dx) \quad ; \quad m = [\beta n]$$

the fact that Y is Hilbertian permits us to find a subspace Z of Y, $\dim Z \sim \dim Y$ where now

(4) $$\|x\| \leq c(K) \beta^{-1/2} M |x| \qquad \text{for} \qquad x \in Z\ .$$

This enables us to perform an iterated construction of orthogonal subspaces

$$Z_1 \perp Z_2 \perp \cdots \perp Z_t \quad ; \quad t \sim \left[\frac{1}{\beta}\right]$$

of X_1, each of which satisfies (4).

Defining

$$X_0 = Z_1 \oplus \cdots \oplus Z_t$$

we get for $x = x_1 + \cdots + x_t \in X_0$, by (3), (4)

$$cM_*^{-1}|x| \le \|x\| \le \|x_1\| + \cdots + \|x_t\| \le c\beta^{-1/2}M(|x_1| + \cdots + |x_t|) \le c(K)\beta^{-1/2}t^{1/2}M|x| .$$

Hence by (2)

$$d_{X_0} < c\beta^{-1/2}t^{1/2}MM_* \le c(K)\beta^{-1}\log d_X .$$

Lemma 2. *Let $\alpha \le \frac{1}{3}$ and $X, \|\ \|$ an n-dimensional normed space satisfying $d(E,F) \le K$ whenever E, F are $[\alpha n]$-dimensional subspaces of X.*

Let M stand for the mean of $\|\ \|$ with respect to an ellipsoid contained in the unit ball of X. Thus assume the euclidean norm $|\ \ |$ satisfies $\|x\| \le |x|$ and let

$$M = \int_{S_{n-1}} \|x\|\sigma(dx) .$$

Then

$$d_X \le c(\alpha, K)M^{-5} . \tag{5}$$

Proof: By Theorem 2.6 of [FLM], there is a Hilbertian subspace of dimension $m = [\beta n]$ where $\beta = cM^2$. It is clear from our assumption on the subspaces of X that the hypothesis of Lemma 1 is satisfied. Thus X has a subspace X_0, $\dim X_0 > \frac{n}{3}$, which euclidean distance is bounded by

$$d_{X_0} \le c(K)M^{-2}(\log d_X) . \tag{6}$$

Using our hypothesis again, it follows that any $[\alpha n]$-dimensional subspace X_0 of X satisfies (6). Hence, by Kwapien's estimate of euclidean distance by type 2, co-type 2 constants $T_2(X)$ and $C_2(X)$ and the fact that these constants may essentially be evaluated on n-vectors, one gets

$$d_X \le T_2(X)C_2(X) \le 4T_2^{(n)}C_2^{(n)}(X) \le c(\alpha)c(K)^2M^{-4}(\log d_X)^2$$

(where $c(\alpha)$ may be evaluated accurately again from the results of [FLM]). This implies (5).

3. A Conceptualization of Gluskin's Construction

Gluskin [G1] proved by random considerations that ℓ_n^∞ has a pair of subspaces E, F satisfying

$$\dim E = m = \dim F \quad , \quad d(E,F) > cm \qquad (m \sim n) .$$

In this section, we use the same idea to perform a similar construction in a more abstract context.

We first introduce an appropriate euclidean structure. In general the distance ellipsoid (to euclidean space) and maximal volume ellipsoid contained in the unit ball B_X of X may be different. It is shown, however, in [B-M2] that by melting these two ellipsoids together and an appropriate affine transformation, one may realize the following properties:

(7) $$\|x\| \leq |x| \leq 2d_X\|x\| \quad \text{and} \quad \pi_2(I : X \longrightarrow \mathcal{R}^n \,,\, |\ \ |) \leq 2\sqrt{n}\,.$$

Since $\|I : \mathcal{R}^n \,,\, |\ \ | \to X\| \leq 1$, $\pi_2(I^{-1}) \leq 2\sqrt{n}$, the argument in [D-M-T] gives an orthonormal system $\{x_i\}_{1\leq i\leq \frac{n}{2}}$ satisfying $\|x_i\| \geq \frac{1}{4}$. Consider norming functionals $\{x_i^*\}_{1\leq i\leq \frac{n}{2}}$, i.e., $\|x_i^*\|_* \leq 1$ and $\langle x_i, x_i^*\rangle = \frac{1}{4}$. It is then possible to find an orthonormal basis $\{e_i\}_{i=1}^n$ where for $1 \leq i \leq \frac{n}{100}$

(8) $$x_i^* = a_{ii}e_i + \sum_{j<i} a_{ij}e_j \,, \qquad a_{ii} > \frac{1}{100}\,.$$

In order to carry out a probabilistic approach, it is easier to work with the following Gaussian substitute of the orthogonal group $O(n)$. Fix $m \sim n$, $m < \frac{n}{2}$ and let P be the orthogonal projection on $[e_1, \ldots, e_m]$. Let

$$\Omega = \left\{G = \frac{1}{\sqrt{n}}(g_{ij})_{1\leq i,j\leq n}\right\}$$

be the space of Gaussian random matrices equipped with natural probability measure μ and define

$$\Omega_0 = \left\{G \in \Omega \mid \|G\|_{2\to 2} \leq C \,,\ PG(B_2^{(n)}) \supset cB_2^{(m)}\right\}.$$

For appropriate constants $0 < c < C < \infty$, one can ensure that $\mu(\Omega_0) > \frac{1}{2}$. The restriction $\|G\|_{2\to 2} \leq C$ is ensured by Chevet's inequality (cf. [D-M-T]) and the second condition obtained by straightforward verification ($m < \frac{n}{2}$).

Our purpose is to produce by statistical considerations $G_1, G_2 \in \Omega_0$ such that

(9) $$d(PG_1(B_X^0), PG_2(B_X^0)) \geq \lambda$$

where λ will be specified later.

Here $d(K_1, K_2)$ stands for the Banach-Mazur distance of the normed space $\mathcal{R}^m$, $\|\ \ \|_1$ and $\mathcal{R}^m$, $\|\ \ \|_2$ with respective unit balls K_1, K_2 (and thus is a linear invariant notion).

From (9), one thus gets m-dimensional quotient spaces of X^* which distance is at least λ.

For $G \in \Omega_0$, evaluate the volume of $K = PG(B_X^\circ)$ by Urysohn's inequality (see [M]). There is following volume-ratio estimate

$$Vr(K) \equiv \left\{ \frac{\mathrm{Vol}_m(K)}{\mathrm{Vol}_m B_2^{(m)}} \right\}^{1/m} \leq \int_{S^{(m-1)}} \|G^* x\| \sigma_{m-1}(dx) =$$

$$= \frac{c}{\sqrt{m}} \int \Big\| \sum_{1 \leq i \leq m} g_i(\omega) G^* e_i \Big\| d\omega \leq \frac{c}{\sqrt{m}} \|G\|_{2\to 2} \int \Big\| \sum_{1 \leq i \leq n} g_i(\omega) e_i \Big\| d\omega \,.$$

Hence

(10) $$Vr(K) \leq c\sqrt{\frac{n}{m}} M \,.$$

For K fixed and $A \in GL(\mathcal{R}^m)$, evaluate using (8)

$$\mu[G \in \Omega_0 \mid PG(B_X^\circ) \subset A(K)] \leq \mu\Big[G \in \Omega \mid \frac{1}{100} PGe_i + \sum_{j<i} a'_{ij} PGe_j \in A(K)\,, \quad i \leq \frac{n}{100}\Big]$$

where $PGe_i = \frac{1}{\sqrt{n}}(g_{i,1}, \dots, g_{i,m})$ are independent vectors.

Hence, by (10), we get the following measure estimate

(11) $$\left[\frac{\mathrm{Vol}\,(c\sqrt{\frac{n}{m}} A(K))}{\mathrm{Vol}\, B_2^{(m)}} \right]^{n/100} \leq \left[c\frac{n}{m} (\det A)^{1/m} M \right]^{mn/100} .$$

For $G \in \Omega_0$, we clearly have

$$\frac{c}{d_X} B_2^{(m)} \subset PG(B_X^\circ) \subset c B_2^{(m)} \,.$$

Therefore, in distance considerations about the $PG(B_X^\circ)$, we may restrict ourselves to linear transformations $A \in GL(\mathcal{R}^m)$ belonging to some cd_X^{-1}-net $\mathcal{E}$ in the $(1 + d_X)^3$-unit ball of $\ell_m^2 \check{\otimes} \ell_m^2$. There is such a net $\mathcal{E}$ satisfying

(12) $$\mathrm{card}(\mathcal{E}) \leq (c d_X)^{4m^2} \,.$$

A well-known probabilistic reasoning based on estimates (11), (12) then tells that there are $G_1, G_2 \in \Omega_0$ satisfying (9), provided the condition

$$\left[C \left(\frac{n}{m}\right)^2 M^2 \lambda \right]^{nm/100} \cdot (C d_X)^{8m^2} \leq \frac{1}{10}$$

or, for $m = \tau n$,

$$\lambda \leq c\tau^2 (1 + d_X)^{-800\tau} M^{-2}$$

holds.

This gives the following conclusion:

Lemma 3. *For $0 \leq \tau < 1$, X has two subspaces E, F*

$$\dim E = m = \dim F\ , \qquad m = [\tau n]$$

satisfying

$$d(E,F) \geq c\frac{\tau^2}{M^2}(1+d_X)^{-C\tau}\ .$$

4. Proof of the Proposition

Assume all $[\alpha n]$-dimensional subspaces of the n-dimensional normed space X isomorphic up to a fixed constant K. (This statement is linearly invariant.) We assume $\alpha \leq \frac{1}{3}$. By Corollary 2

$$d_X \leq c(\alpha, K)M^{-5}\ . \tag{14}$$

Taking $\tau = \alpha$ in Lemma 3, it follows from the homogeneity hypothesis that

$$K \geq c\frac{\alpha^2}{M^2}(1+d_X)^{-C\alpha}\ .$$

Clearly (14),(15) imply boundedness of d_X (as a function of K), provided the number α is smaller than some numerical constant c_1.

References

[Gr] M. Gromov, On a geometric conjecture of Banach, Izv. Akad. Nauk SSSR, Ser. Mat. 31 (1967), 1105-1114.

[B-M] J. Bourgain and V. Milman, New volume ratio properties for convex symmetric bodies in $\mathbb{R}^n$. Invent. Math. 88 (1987), 319-340.

[B-M2] J. Bourgain and V. Milman, Distances between normed spaces, their subspaces and quotient spaces, Integral Equ. and Oper. Th. 9 (1986), 31-46.

[D-M-T] W. Davis, V. Milman and N. Tomczak-Jaegermann, The distance between certain n-dimensional Banach spaces, Israel J. Math 39, 1-2 (1981), 1-15.

[F-L-M] T. Figiel, J. Lindenstrauss and V.D. Milman, The dimension of almost spherical sections of convex bodies, Acta Math. 39 (1977), 53-94.

[Gl] Gluskin, The diameter of the Minkowski compactum is roughly equal to n, Functional Anal. Appl. 15 (1981), 72-73.

[M-S] V.D. Milman and G. Schechtman, Asymptotic theory of finite dimensional normed spaces, Springer LNM, 1200.

[M-P] V.D. Milman and G. Pisier, Banach spaces with a weak cotype 2 property, Israel J. Math. 54, 2 (1986), 139-158.

[M] V.D. Milman, Geometrical inequalities and mixed volumes in local theory and Banach spaces, Asterique 131 (1985), 373-400.

VECTOR-VALUED HAUSDORFF-YOUNG INEQUALITIES AND APPLICATIONS

J. Bourgain

(IHES)

1. Introduction and Results

My purpose in this paper is to develop further an earlier work [B1] on Hausdorff-Young type inequalities in B-convex Banach spaces. The main result in [B1] is the following fact.

Proposition 1. *If X is a B-convex Banach space, then there exist some $p > 1$ and $q < \infty$ and constants $M < \infty$, $\delta > 0$, such that*

$$\left\{\int \Big\| \sum_{\gamma\in\Gamma} x_\gamma \gamma(t)\Big\|^2 dt\right\}^{1/2} \leq M\Big(\sum \|x_\gamma\|^p\Big)^{1/p} \tag{1}$$

and

$$\left\{\int \Big\| \sum_{\gamma\in\Gamma} x_\gamma \gamma(t)\Big\|^2 dt\right\}^{1/2} \geq \delta\Big(\sum \|x_\gamma\|^q\Big)^{1/q} \tag{2}$$

whenever $(x_\gamma)_{\gamma\in\Gamma}$ is a finitely supported sequence of elements of X and Γ the dual of a compact abelian group G.

The reader may like to consult [B1] for the proof of Proposition 1 and background material. Based on Proposition 1, the following will be shown here.

Proposition 2. *For $1 \leq p_1 < q'$, $p' < q_1 \leq \infty$ with p, q satisfying (1), (2)*

$$\Big\|\sum_{\gamma\in\Gamma} x_\gamma \gamma\Big\|_{L^{p_1'}(G)} < M_1\Big(\sum \|x_\gamma\|^{p_1}\Big)^{1/p_1} \tag{3}$$

and

$$\Big\|\sum_{\gamma\in\Gamma} x_\gamma \gamma\Big\|_{L^{q_1'}(G)} \geq \delta_1\Big(\sum \|x_\gamma\|^{q_1}\Big)^{1/q_1} \tag{4}$$

where G is either a torus or the Cantor group.

Proposition 3. *If X is a B-convex Banach space, there is $1 < p < 2$ such that for X-valued functions on the real $\mathbb{R}$*

$$\|\hat{f}\|_{L_X^{p'}(\mathbb{R})} \leq C\|f\|_{L_X^p(\mathbb{R})} \tag{5}$$

where

$$\hat{f}(\gamma) = \int_{-\infty}^{\infty} f(t) e^{-2\pi i \lambda t} dt .$$

Note that from the standard duality reasoning, it suffices to derive (3) from (2). I do not have a general argument for an abstract compact abelian group, but will treat the case of the Cantor group and the circle (or, more generally, the connected case) separately. I believe it is possible to handle the general case invoking the structure theorem for compact abelian groups and combining the methods for $D_\infty = \{1, -1\}^{\mathbb{N}}$ and $\Pi = \mathbb{R}/\mathbb{Z}$.

Inequality (5) is self-dual and will be derived form the circle case. For application to interpolation theory, see [Pe] (cf. also [Mi]). My interest in this subject has been revived after some recent discussions with A. Pelczynski related to multipliers acting on the Hardy spaces $H^1_X(\Pi)$ or $H^1_X(\mathbb{R})$, where X is a B-convex space. Applications of Proposition 2 and 3 in this context are based on the atomic decomposition theorem (which is also valid in the general vector-valued setting, see [B2]) and will appear in the last section of this paper.

Remarks.

(a) We will use inequality (2) in the following formulation. Let G be a finite group, $|G| = N$ and Γ its dual. Let $G = \{t_1, \dots, t_N\}$ and $\Gamma = \{\gamma_1, \dots, \gamma_N\}$ be enumerations of their elements and let $(a_{ij})_{1 \le i,j < N}$ be the matrix defined as $a_{ij} = \gamma_i(t_j)$. Then

$$\left(\sum_{i=1}^{N} \Big\| \sum_{j=1}^{N} a_{ij} \xi_j \Big\|^q \right)^{1/q} \le C N^{1/2} \left(\sum_{j=1}^{N} \|\xi_j\|^2 \right)^{1/2} \tag{6}$$

where $\xi_j \in X$.

Here and in what follows, C will stand for a constant.

To derive (6) from (2), just consider the function f on G given by $f(t_j) = \xi_j$.

(b) In order to prove (3), interpolation is used as follows. First, it is shown that for q satisfying (6)

$$\Big\| \sum_{\gamma \in \Lambda} x_\gamma \gamma \Big\|_q \le C |\Lambda|^{1/q'} \tag{7}$$

whenever Λ is a finite subset of Γ and $\|x_\gamma\| \le 1$. Thus the map

$$\ell^{q',1}_X(\Gamma) \longrightarrow L^q_X(G) : \{x_\gamma\} \longrightarrow \sum x_\gamma \gamma \tag{8}$$

is bounded, implying (3) for $p_1 < q'$, by the Marcinkiewicz interpolation theorem.

(c) Hausdorff-Young inequalities for D_∞ are of particular interest in geometry of Banach spaces (cf. [P]). I include a comment here, not of importance for what follows, but clarifying one point about the dependence of the exponent p in (1). This remark shows that the problem raised in [P] (remarque 8) has a negative answer. It is not possible to express p as a function $p = p(p_X, q_X) > 1$ involving only the type and cotype exponents of the space X. The estimate obtained in [B1] expresses p as a function of $\overline{p}$, a type exponent for the dual space X^*, and the corresponding constant $T_{\overline{p}}(X^*)$. The next example shows that the dependence of p also on type and cotype constants is unavoidable.

Fix n and consider the finite Cantor group $D_n = \{1, -1\}^n$ which dual Γ are the Walsh functions $w_S(\varepsilon) = \prod_{k \in S} \varepsilon_k$, $|S| \leq n$. For $2 \leq r < \infty$, consider the interpolation space

$$X_r = [L^r(G)\ ,\ \mathcal{F}(\ell^{r'}(\Gamma))]_{1/2} \tag{9}$$

$r' = \frac{r}{r-1}$ and where $\mathcal{F}(\ell^{r'}(\Gamma))$ refers to the $\ell^{r'}(\Gamma)$-norm on the Fourier coefficients, i.e.,

$$\|f\|_{\mathcal{F}(\ell^{r'}(\Gamma))} = \Big(\sum_{|S| \leq n} |\hat{f}(S)|^{r'}\Big)^{1/r'} .$$

It results from the classical Hausdorff-Young inequality that $\|f\|_r \leq \|f\|_{X_r}$. The Fourier-transform clearly yields and isometry of X_r and its dual X_r^*. Since $T_2(L^r(G)) \leq \sqrt{r}$ and $T_{r'}(\ell^{r'}(\Gamma)) = 1$, the interpolation property for type yields

$$T_{\overline{p}}(X_r) \leq r^{1/4} \quad \text{where} \quad \overline{p} = \frac{4}{3 - \frac{2}{r}} \tag{10}$$

X_r is an invariant function space on G and hence, taking $x_S = w_S \in X_r$,

$$\|x_S\| = 1$$

$$\Big\| \sum_{|S| \leq n} w_S(\varepsilon) x_S \Big\| = \Big\| \sum x_S \Big\| \geq \Big\| \sum w_s \Big\|_{L^r(G)} = 2^{n/r'} .$$

Hence X_r may only satisfy the inequality (1) for $p \leq r'$.

It is worthwhile to note that the estimate obtained in [B1] is essentially the correct one (as shown by the previous example) for $r \to \infty$. Indeed, it was shown in [B1] that one may take any

$$p < \frac{[CT_{\overline{p}}(X^*)]^{\overline{p}'}}{[CT_{\overline{p}}(X^*)]^{\overline{p}'} - 1} \tag{11}$$

where C is numerical.

Finally, the example also shows that there is no "good " substitute for the estimate (see [P], Lemma 3)

$$\Big\| \sum_{|S| \leq n} w_S x_S \Big\|_{L^2_X} \leq C T_p(X) C_q(X) 2^{n(\frac{1}{p} - \frac{1}{q})} \Big(\sum \|x_S\|^2\Big)^{1/2}$$

when $\frac{1}{p} - \frac{1}{q} \geq \frac{1}{2}$ (unless in special cases, for instance when X is a Banach lattice).

2. The Case of the Cantor Group

In this section, I prove the inequality (7), i.e.,

$$\|\Sigma_\Lambda \xi_\gamma \gamma\|_q \leq C|\Lambda|^{1/q'} \tag{7}$$

for Λ a finite subset of the Walsh functions on $G = D_\infty$ and $\|\xi_\gamma\| \leq 1$. Let k be a postitive integer to be defined later. I will use the following identity

Lemma 4. $$\sum_{S\subset\{1,\dots,k\}} w_s(\varepsilon)\left[\prod_{j\notin S}\frac{1+\gamma(x_j)}{2}\prod_{j\in S}\frac{1-\gamma(x_j)}{2}\right] = \gamma\Big(\sum_{\varepsilon_j=-1} x_j\Big) \tag{12}$$
whenever $\gamma \in G^*$, $x_1, \dots, x_k \in G$ *and* $\varepsilon \in D_k = \{1, -1\}^k$.

Proof: Define for $T \subset \{1, \dots, k\}$ the element ε_T of D_k by $\varepsilon_T^j = 1$ if $j \notin T$ and $\varepsilon_T^j = -1$ if $j \in T$. The left member of (12) then clearly equals

$$2^{-k}\sum_{\substack{|S|\leq k\\ |T|\leq k}} w_s(\varepsilon)(-1)^{|S\cap T|}\prod_{j\in T}\gamma(x_j) = 2^{-k}\sum_{\substack{|S|\leq k\\ |T|\leq k}} w_s(\varepsilon)w_s(\varepsilon_T)\prod_{j\in T'}\gamma(x_j) = \prod_{j\in T'}\gamma(x_j)$$

where T' is such that $\varepsilon_{T'} = \varepsilon$. This proves the Lemma.

Next, write, by Lemma 4

$$\|\Sigma_\Lambda \xi_\gamma \gamma\|_q = 2^{-k/q}\left(\sum_{\varepsilon\in D_k}\Big\|\sum_{\gamma\in\Lambda}\gamma\Big(\sum_{\varepsilon_j=-1} x_j\Big)\xi_\gamma\gamma\Big\|_q^q\right)^{1/q} =$$

$$2^{-k/q}\left\{\int_G\left[\sum_{\varepsilon\in D_k}\left\|\sum_{|S|\leq k} w_s(\varepsilon)\left(\sum_{\gamma\in\Lambda}\gamma(y)\xi_\gamma\left(\prod_{j\notin S}\frac{1+\gamma(x_j)}{2}\prod_{j\in S}\frac{1-\gamma(x_j)}{2}\right)\right)\right\|^q dy\right]\right\}^{1/q} \tag{13}$$

Fix $y \in G$ and apply (6) with the character-matrix $\big(w_s(\varepsilon)\big)_{\varepsilon\in D_k, |S|\leq k}$ obtained from the group

$$G_0 = \{w_s \mid |S| \leq k\}\ ,\ \Gamma_0 = G_0^* = D_k\ ,\ N = 2^k\ .$$

Since $\|\xi_\gamma\| \leq 1$, we have

$$\left\|\Sigma_\Lambda\gamma(y)\xi_\gamma\left(\prod_{j\notin S}\frac{1+\gamma(x_j)}{2}\prod_{j\in S}\frac{1-\gamma(x_j)}{2}\right)\right\| \leq \Sigma_\Lambda\prod_{j\notin S}\frac{1+\gamma(x_j)}{2}\prod_{j\in S}\frac{1-\gamma(x_j)}{2} \tag{14}$$

Substituting this additional estimate in the right member of (6) gives thus

$$\|\Sigma_\Lambda \xi_\gamma \gamma\|_q \leq C 2^{k(-\frac{1}{q}+\frac{1}{2})} \left[\sum_{|S|\leq k} \left(\Sigma_\Lambda \prod_{j\notin S} \frac{1+\gamma(x_j)}{2} \prod_{j\in S} \frac{1-\gamma(x_j)}{2} \right)^2 \right]^{1/2} \tag{15}$$

where $x_1, \ldots, x_k$ are arbitrary elements of G. Integration and the fact that

$$\begin{aligned} \|\Sigma_\Lambda \prod_{j\notin S} (1+\gamma(x_j)) \prod_{j\in S} (1+\gamma(x_k))\|_{L^2(G^k)} &\leq |\Lambda| + \| \sum_{1\leq |T|\leq k} \sum_{\gamma\in\Lambda} (\prod_{j\in T} \gamma(x_j))\|_{L^2(G^k)} \\ &\leq |\Lambda| + \left((2^k-1)|\Lambda|\right)^{1/2} \end{aligned} \tag{16}$$

(assuming Λ does not contain the trivial character) gives

$$\|\Sigma_\Lambda \xi_\gamma \gamma\|_q \leq C 2^{-k\frac{1}{q}} \left[|\Lambda| + 2^{k/2}|\Lambda|^{1/2}\right] .$$

Choosing $2^k \sim |\Lambda|$ gives now (7).

This completes the proof of Proposition 2 when G is the Cantor group. The argument is easily adaptable to groups $G = \Pi(\mathbb{Z}/k\mathbb{Z})$ but the estimates will depend on k. In order to get uniform estimates, the method presented in the next section has to be combined with the previous one. I do not intend to carry the details out in this paper.

3. The Case of the Torus

Let in this section G be a compact connected abelian group and Λ be a finite subset of Γ. Our purpose is to derive an equality (7) from (6). In this case the character matrix (a_{jk}) will be given by $a_{jk} = \exp 2\pi i \frac{jk}{N}$ where $N = |\Lambda|$.

Denote $R(N)$ the smallest constant satisfying

$$\|\sum_{\gamma\in\Lambda} \xi_\gamma \gamma\|_q \leq |R(N) N^{1/q'} \tag{17}$$

whenever $\Lambda \subset \Gamma$, $|\Lambda| = N$ and $\{\xi_\gamma \mid \gamma \in \Lambda\}$ a sequence of vectors in the unit ball of the given space X. Let $R^*(N) = \max_{n\leq N} R(n)$.

Assume $\Lambda_0 \subset \Lambda$, $x \in G$ and $\{j(\gamma) \mid \gamma \in \Lambda_0\}$ different integers in $[o, N[$ satisfying

$$|\gamma(x) - e^{2\pi i j(\gamma) N^{-1}}| \leq \frac{2\pi}{N} \quad \text{for} \quad \gamma \in \Lambda_0 \tag{18}$$

Thus also, for $k = 1, \ldots, K$; $K < N$

$$|\gamma(kx) - e^{2\pi i k j(\gamma) N^{-1}}| < \frac{K}{N} 2\pi \quad , \quad \gamma \in \Lambda_0 \tag{19}$$

Estimate

$$\|\sum_{\gamma \in \Lambda} \xi_\gamma \gamma\|_q \leq K^{-1/q} \Big(\sum_{0 \leq k < K} \|\sum_{\gamma \in \Lambda_0} \gamma(kx) \xi_\gamma \gamma\|_q^q \Big)^{1/q} + \|\sum_{\gamma \in \Lambda \setminus \Lambda_0} \xi_\gamma \gamma\|_q$$

$$\leq K^{-1/q} \Big(\sum_{o \leq k \leq K} \|\sum_{\gamma \in \Lambda_0} e^{2\pi i k j(\gamma) N^{-1}} \xi_\gamma \gamma\|_q^q \Big)^{1/q} + 2\pi R(N) \frac{K}{N} N^{1/q'} + R^*(N)(N - |\Lambda_0|)^{1/q'} \tag{20}$$

by definition of the numbers $R(N)$ and by (19).

The sequence $j(\gamma)$, $\gamma \in \Lambda_0$ is an enumeration of a subset E of $\{0, 1, \ldots, N-1\}$. For fixed $y \in G$, define for $j \in \{0, 1, \ldots, N-1\}$

$$\eta_j = \gamma(Y) \xi_\gamma \quad \text{if} \quad j(\gamma) = j \quad \text{for some} \quad \gamma \in \Lambda_0$$

$$= 0 \quad \text{if} \quad j \notin E$$

With these notations, by (6)

$$\Big(\sum_{o \leq k \leq K} \|\sum_{\gamma \in \Lambda_0} e^{2\pi i k j(\gamma) N^{-1}} \xi_\gamma \gamma(y) |^q \Big)^{1/q} =$$

$$\Big(\sum_{o \leq k \leq K} \|\sum_{0 < j < N} e^{2\pi i j k N^{-1}} \eta_j|^q \Big)^{1/q} \leq C N^{1/2} (\Sigma \|\eta_j\|^2)^{1/2} < CN$$

Substitution in (20) then gives

$$\|\Sigma_\Lambda \xi_\gamma\|_q \leq C K^{-1/q} N + 2\pi R(N) \frac{K}{N} N^{1/q'} + R^*(N)(N - |\Lambda_0|)^{1/q'} \tag{21}$$

Our next purpose is to extract a proportional subset Λ_0 of Λ satisfying (18).

Lemma 5. *There is an absolute constant c such that given $\Lambda \subset \Gamma$, $|\Lambda| = N$, there is a subset E of $\{0, 1, \ldots, N-1\}$, $|E| > cN$ and elements $\{\gamma_j \mid j \in E\}$ of Λ, such that for some $x \in G$*

$$|\gamma_j(x) - e^{2\pi i j N^{-1}}| < \frac{2\pi}{N} \qquad (j \in E) \tag{22}$$

Taking $\Lambda_0 = \{\gamma_j \mid j \in E\}$, (21) them implies that

$$\|\Sigma_\Lambda \xi_\gamma \gamma\|_q \leq C K^{-1/q} N + 2\pi R(N) \frac{K}{N} N^{1/q'} + (1 - c) R^*(N) N^{1/q'}$$

hence

$$R(N) \leq C K^{-1/q} N^{1/q} + 2\pi R(N) \frac{K}{N} + (1 - c) R^*(N)$$

For appropriate choice of K, it follows that $R(N) \leq C' + (1 - c') R^*(N)$. This *inequality clearly implies boundedness of* $R^*(N)$, hence $R(N)$. Hence (7) holds and thus Proposition 2 is proved then G is a torus.

Proof of Lemma 5. Denote $\varsigma_j = e^{2\pi i \frac{j}{N}}$ $(o \leq j < N)$ with N^{th} roots of unity. Consider the functions $\varphi_j(z)$, $|z| = 1$ on the unit circle

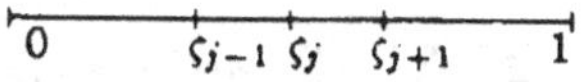

$\varphi_j(\varsigma_j) = 1$

and given by

$$\varphi_j(x) = \sum_{k=-\infty}^{\infty} \widehat{\varphi}(k)\varsigma_j^{-k} z^k \qquad (|z| = 1)$$

where

$$\widehat{\varphi}(o) = \frac{1}{N} \quad \text{and} \quad \widehat{\varphi}(k) = \frac{N}{\pi^2 k^2} \sin^2 \frac{\pi k}{N}, \qquad k \neq o \tag{23}$$

Defining for $x \in G$

$$\lambda_j(x) = \# \left\{ \gamma \in \Lambda \mid |\gamma(x) - \varsigma_j| < \frac{2\pi}{N} \right\}$$

clearly

$$\lambda_j(x) \geq \sum_{\gamma \in \Lambda} \varphi_j(\gamma(x)) = \sum_{k=-\infty}^{\infty} \widehat{\varphi}(k)\varsigma_j^{-k} \left[\sum_{\gamma \in \Lambda} \gamma(kx) \right] \tag{24}$$

Assume again Λ does not contain the trivial character. Thus

$$\int_G \lambda_j(x) ds \geq \widehat{\varphi}(o)|\Lambda| = 1 \quad \text{and} \quad \int_G \left[\sum_{o \leq j < N} \lambda_j(x) \right] ds \geq N \tag{25}$$

Next we show that

$$\int_G \left[\sum_{o \leq j < N} \lambda_j(x)^2 \right] dx \leq CN \tag{26}$$

Since $\lambda_j(x) \leq \sum_{|j'-j| \leq 1} \sum_{\gamma \in \lambda} \varphi_{j'}(\gamma(x))$, it follows from (24)

$$\sum_{0 \leq j < N} \lambda_j(x)^2 \leq 9 \sum_{o \leq j < N} \Big(\sum_{\gamma \in \Lambda} \varphi_j(\gamma(x)) \Big)^2$$

$$= 9 \sum_{0 \leq j < N} \left\{ \sum_{\ell=-\infty}^{\infty} \frac{1}{(|\ell| + 1)N} \sum_{0 \leq r < N} \varsigma_j^{-r} \alpha_{\ell,r}(x) \right\}^2 \tag{27}$$

letting

$$\alpha_{\ell,r}(x) = (|\ell| + 1)N\widehat{\varphi}(\ell N + r) \left[\sum_{\gamma \in \Lambda} \gamma((\ell N + r)x) \right].$$

Estimate (27) by Cauchy-Schwarz and orthogonality to get

$$CN^{-2}\sum_{\ell=-\infty}^{\infty}\sum_{o\le j<N}\Big|\sum_{o\le r<N}\varsigma_j^{-r}\alpha_{\ell,r}(x)\Big|^2 = CN^{-1}\sum_{\ell=-\infty}^{\infty}\sum_{o\le r<N}|\alpha_{\ell,r}(x)|^2$$

which, by (23), is bounded by

$$CN^{-1}\sum_{|r|<N}\Big|\sum_{\gamma\in\Lambda}\gamma(rx)\Big|^2 = CN^{-1}\sum_{\ell\neq o}\frac{1}{\ell^2}\sum_{o\le r<N}\Big|\sum_{\gamma\in\Lambda}\gamma((\ell N+r)x)\Big|^2 . \tag{28}$$

Integration in x gives then

$$CN^{-1}|\Lambda|^2 + C|\Lambda| + C\left(\sum_{\ell\neq o}\frac{1}{\ell^2}\right)|\Lambda| < CN$$

proving (26).

As a consequence of (25), (26), there is $x\in G$ such that

$$\sum_{0\le j<N}\lambda_j(x) \ge \frac{N}{2} + c\sum_{0\le j<N}\lambda_j(x)^2 . \tag{29}$$

Define $E = \{o\le j<N \mid \lambda_j(x)\neq o\}$. Since, by definition of $\lambda_j(x)$, $\sum\lambda_j(x)\le 2N$, (29) implies

$$\frac{N}{2}\le |E|^{1/2}\Big(\sum_{j\in E}\lambda_j(x)^2\Big)^{1/2}\le C\,|E|^{1/2}\Big[\sum\lambda_j(x)\Big]^{1/2}\le CN^{1/2}|E|^{1/2}$$

hence $|E|>cN$, proving Lemma 5.

4. Proof of Proposition 3

Let $1<p\le 2$ satisfy

$$\Big(\sum_{n\in\mathbb{Z}}\|\widehat{\varphi}(n)\|^{p'}\Big)^{1/p'}\le C\|\varphi\|_p \tag{30}$$

for $\varphi\in L^p_X(\Pi)$.

Identify $\Pi=\mathbb{R}/\mathbb{Z}$ and $[-\frac12,\frac12[$. Let f be supported by $]-\frac12,\frac12[$ and consider for $u\in[0,1[$ the function φ on Π given by

$$\varphi(x)=f(x)e^{-2\pi iux} .$$

Thus $\widehat{\varphi}(n)=\hat f(n+u)$ and, by (30),

$$\Big(\sum_{n\in\mathbb{Z}}\|\hat f(n+u)\|^{p'}\Big)^{1/p'}\le C\|f\|_{L^p_X(\mathbb{R})} . \tag{31}$$

Integrating (31) in $u \in [0,1]$, the inequality

$$\|\hat{f}\|_{L_X^{p'}(\mathbb{R})} \leq C\|f\|_{L_X^p(\mathbb{R})}$$

follows. To prove (5) in general, use a rescaling argument. Let thus $f \in L_X^p(\mathbb{R})$ be compactly supported and choose $\varepsilon > 0$ sufficiently small such that f_ε is supported by $]-\frac{1}{2},\frac{1}{2}[$. As usual, $f_\varepsilon(x) = \frac{1}{\varepsilon} f\left(\frac{x}{\varepsilon}\right)$ and hence $\hat{f}_\varepsilon(\lambda) = \hat{f}(\varepsilon\lambda)$. By change of variable, it follows

$$\|\hat{f}\|_{p'} = \varepsilon^{1/p'}\|\hat{f}_\varepsilon\|_{p'} \leq C\varepsilon^{1/p'}\|f_\varepsilon\|_p = C\|f\|_p \, .$$

5. An Application to Vectorvalued Hardy Spaces

The content of this section is the origin of this work and answers a question which appeared in a recent conversation with A. Pelczynski(*). We consider the space $H_X^1(\Pi)$ of functions f in $L_X^1(\Pi)$ for which $\hat{f}(n) = o$ for $n < o$. As in the scalar case, this space has the atomic decomposition property, independently of the underlying Banach space X. Thus

Proposition 6. *If $f \in H_X^1(\Pi)$, there is a sequence $\{a_\ell\}$ in $L_X^\infty(\Pi)$ and positive scalars $\{\lambda_\ell\}$ satisfying $f = \sum \lambda_\ell a_\ell$ and*

(i) Each a_ℓ is of mean 0, supported by an interval I in Π and $\|a_\ell\|_\infty \leq |I|^{-1}$

(ii) $\sum \lambda_\ell \leq C\|f\|_1$

where C is an absolute constant.

This fact was proved in [B2] using a martingale approach (see Theorem 6). The same result was obtained independently by J. Garcia-Cuerva (unpublished manuscript) by different methods.

We now restrict ourselves to B-convex spaces and let $q < \infty$ be an exponent satisfying

$$\|\hat{f}\|_{\ell_X^q(\mathbb{Z})} \leq C\|f\|_{L_X^{q'}(\Pi)} \tag{32}$$

C. Fefferman discovered a complete characterization of ℓ^1-multipliers on $H^1(\Pi)$. A sequence of scalars $(\lambda_n)_{n\geq 0}$ is an $H^1(\Pi) - \ell^1$ multiplier provided

$$\sup_m \sum_{k\geq 1} \Big(\sum_{km\leq n<(k+1)m} |\lambda_n| \Big)^2 = K^2 < \infty \, . \tag{33}$$

(*). July 1987, University of Illinois, cf. also [B-P]

See [S-W] for a proof of this result.

The sufficiency of (33) is an easy consequence of the atomic decomposition property. Adaptation of the scalar argument and substituting Parseval's identity by (32) yields

Proposition 7. *Let q satisfy (32) and $(\lambda_n)_{n\geq 0}$ a sequence of positive numbers. Then*

$$\sum_{n=0}^{\infty} \lambda_n \|\hat{f}(n)\| \leq C K_{q'} \|f\|_1 \quad \text{for} \quad f \in H^1_X(\Pi) \tag{34}$$

where K_p stands for

$$\sup_m \left\{ \sum_{k\geq 1} \left(\sum_{km\leq n<(k+1)m} \lambda_n \right)^p \right\}^{1/p} \tag{35}$$

Corollary 8. *If X is a B-convex Banach space, then $H^1_X(\Pi)$ satisfies Hardy's inequality, i.e., there is a constant C such that*

$$\sum_{n\geq 1} \frac{\|\hat{f}(n)\|}{n} \leq C\|f\|_1 \quad \text{for} \quad f \in H^1_X(\Pi)\ .$$

Proof of Proposition 7. By Proposition 6, it suffices to verify (34) when f is an atom with base-interval I centered at the origin. Denote m an integer such that $m \sim |I|^{-1}$ and estimate

$$\sum_{n\leq m} \lambda_n \|\hat{f}(n)\| + K_{q'} \left\{ \sum_{k\geq 1} \left(\max_{kn\leq n\leq (k+1)m} \|\hat{f}(n)\| \right)^q \right\}^{1/q} \geq \sum_{n=0}^{\infty} \lambda_n \|\hat{f}(n)\| \tag{36}$$

Clearly $\sum_{r\leq n\leq 2r} \lambda_n \leq K_{q'}$ for all r. Hence $\sum_{n\leq m} n\lambda_n \leq 2K_{q'}m$. (37)

Also, for $n \leq m$, since f is of mean o

$$\|\hat{f}(n)\| = \left\| \int_I f(t)[e^{-2\pi int} - 1]dt \right\| \leq C(I) \cdot \|f\|_\infty n|I| \leq c\frac{n}{m} \tag{38}$$

From (37), (38), the first term in (36) is bounded by $CK_{q'}$.

To estimate the second term in (36), choose for each k an integer

$$km \leq n_k < (k+1)m$$

and evaluate $\left(\sum_k \|\hat{f}(n_k)\|^q\right)^{1/q}$. Clearly

$$m^{1/q}\|\hat{f}(n_k)\| \leq \Big(\sum_{km\leq n<(k+1)m} \|\hat{f}(n)\|^q \Big)^{1/q} + \Big(\sum_{km\leq n<(k+1)m} \|\hat{f}(n) - f(n_k)\|^q \Big)^{1/q} \tag{39}$$

and by Hölder's inequality, for $km \leq n < (k+1)m$

$$\|\hat{f}(n) - \hat{f}(n_k)\| \leq m^{1/q'} \Big(\sum_{km \leq \ell < (k+1)m} \|\hat{f}(\ell+1) - \hat{f}(\ell)\|^q \Big)^{1/q} \tag{40}$$

Substitution of (40) in (39) and summation over k gives

$$\Big(\sum_k \|\hat{f}(n_k)\|^q \Big)^{1/q} \leq m^{-1/q} \Big(\sum_{n \geq 0} \|\hat{f}(n)\|^q \Big)^{1/q} + m^{1/q'} \Big(\sum_{\ell \geq 0} \|\hat{f}(\ell+1) - \hat{f}(\ell)\|^q \Big)^{1/q} \tag{41}$$

By (32), $\left(\sum \|\hat{f}(n)\|^q \right)^{1/q} \leq C\|f\|_{q'} = C|I|^{-1+1/q'} = Cm^{1/q}$ and

$$\Big(\sum \|\hat{f}(\ell+1) - \hat{f}(\ell)\|^q \Big)^{1/q} = \Big(\sum \|\hat{g}(\ell)\|^q \Big)^{1/q} \leq C\|g\|_{q'} \leq Cm^{-1/q'} ,$$

letting $g(t) = (e^{2\pi i t} - 1)f(t)$. Hence (41) is uniformly bounded, which completes the proof of Proposition 7.

References

[B1] J. Bourgain, A Hausdorff-Young inequality for B-convex Banach spaces, Pac. J. Math., 100, No. 2, 1982, 94-102.

[B2] J. Bourgain, Vector-valued singular integrals and the H_1-BMO duality, in "Probability Theory and Harmonic Analysis", ed. Chao/Woyczynski, Marcel Dekker, 1985, p.1-19.

[B-P] O. Blasco, A. Pelczynski, Theorems of Hardy and Paley for vectorvalued analytic functions and related classes of Banach spaces, preprint 1987, to appear in TAMS

[P] G. Pisier, Séminaire d'Analyse Fonctionnelle, 1980-81, Exp. 7, Ecole Polytechnique.

[S-W] S. Szarek, T. Wolniewicz, A proof of Fefferman's theorem on multipliers, preprint PAN, 1980.

[Mi] M. Milman, Complex interpolation and geometry of Banach spaces, Ann. Math. Pure Appl. IV, ser 136 (1984), 317-328.

[Pe] J. Petree, Sur la transformation de Fourier des fonctions á valeurs vectorielles, Rend. Sem. Mat. Univ. Padova 42 (1969), 15-26.

PROJECTION BODIES

J. Bourgain
IHES, France and
University of Illinois, Urbana

J. Lindenstrauss
Hebrew University, Jerusalem

1. A Survey

The purpose of this note is to discuss the properties of a map which assigns to each finite dimensional Banach space a finite dimensional subspace of L_1 (or dually, a zonoid). This map, which has a nice geometric meaning, goes back to Minkowski. It was the subject of quite a few papers in convexity theory, as well as differential geometry. In Banach space theory this map has not as yet received appropriate attention in spite of the fact that it generates many interesting problems and may very well have nice applications. The first section of this note is expository. In the second, we prove that the inverse of the Minkowski map is Hölder continuous in R^n, solving a problem stated in a few places in the literature (cf. e.g., the survey paper [S.W]). This result was announced in [B.L]. After we finished the work on this paper we received a recent preprint of Stefano Campi [Ca], in which the same problem was solved independently for $n = 3$. We would like to express our thanks to R. Schneider for some conversations concerning the subject of this note and for supplying us with several references.

Let K be a convex body in R^n (compact convex set with interior). The projection body $Z(K)$ of K is defined to be the body whose support functional $h_{Z(K)}(u)$ is given for every u with (Euclidean norm) $|||u||| = 1$ by

(1.1) $h_{Z(K)}(u)$ = The $(n-1)$ dimensional volume of the orthogonal projection of K on the hyperplane orthogonal to u.

If K is a polytope having $(n-1)$-dimensional faces $\{F_i\}_{i=1}^m$ with normals (i.e., unit vectors orthogonal to F_i directed outwards) $\{u_i\}_{i=1}^m$ then clearly

$$h_{Z(K)}(u) = \tfrac{1}{2}\sum_{i=1}^{m} \mathrm{Vol}_{n-1}(F_i)|\langle u_i, u\rangle| . \tag{1.2}$$

For a general convex body K we define a measure σ_K on the Euclidean unit sphere S^{n-1} which corresponds to the usual surface measure on K via the Gauss map. Explicitly, for a Borel set

$B \subset S^{n-1}$

(1.3) $$\sigma_K(B) = \mu\{x \in \partial K \ , \text{ the outer normal to } K \text{ at } x \text{ belongs to } B\}$$

where μ is the $(n-1)$-dimensional surface measure on K. Note that σ_K is well defined since the set of x for which the outer normal is not uniquely defined is of μ measure 0. For a polytope K with faces $\{F_i\}_{i=1}^m$ and normals $\{u_i\}_{i=1}^m$, σ_K is purely atomic with mass $\mathrm{Vol}_{n-1} F_i$ at u_i, $1 \le i \le m$. In terms of σ_K we can express $h_{Z(K)}(u)$ as follows

(1.4) $$h_{Z(K)}(u) = \tfrac{1}{2} \int_{S^{n-1}} |\langle u, v\rangle| d\sigma_K(v) \ .$$

It is evident from (1.4) that the norm induced by the polar to $Z(K)$ (i.e., $h_{Z(K)}(u)$) is that of a subspace of an L_1 space. Thus $Z(K)$ itself is, by definition, a zonoid (cf. [Bol], [S.W] for the basic material on zonoids and zonotopes).

If K is symmetric with respect to 0, then σ_K is clearly a symmetric measure, i.e., $\sigma_K(B) = \sigma_K(-B)$ for all $B \subset S^{n-1}$. For symmetric measures the transform (1.4) is invertible, i.e., $h_{Z(K)}(u)$ determines the measure σ_K. This statement, which (at least in the present context) is due to A.D. Aleksandrov is clearly equivalent to the following fact: The functions $u \to |\langle u, v\rangle|$, $v \in S^{n-1}$ span a dense subspace of $C_{\mathrm{sym}}(S^{n-1})$ = the space of real valued symmetric continuous functions on S^{n-1}. A particularly simple proof of this fact is the following (taken from [Ch]). Let X be the closed linear span of $|\langle u, v\rangle|$, $v \in S^{n-1}$ in $C_{\mathrm{sym}}(S^{n-1})$. For every linear transformation $T : R^n \to R^n$ we have

$$|||Tu||| = \lambda_n \int_{S^{n-1}} |\langle Tu, v\rangle| d\sigma_n(v) = \lambda_n \int_{S^{n-1}} |\langle u, T^* v\rangle| d\sigma_n(v) \in X$$

where σ_n is the normalized rotation invariant measure on S^{n-1}. In other words for every symmetric positive definite $n \times n$ matrix $A = (a_{i,j})$ we have that $F_A(u) = \big(\sum_{i,j} a_{i,j} u_i u_j\big)^{1/2} \in X$. By differentiating $F_A(u)$ with respect to $a_{i,j}$ at $A = I$ (the identity) we get that $u_i u_j \in X$ for all $1 \le i, j \le n$. Repeated differentiations of $F_A(u)$ gives that every monomial in $\{u_i\}_{i=1}^n$ of even degree belongs to X. By the Stone Weierstrass theorem we deduce that $X = C_{\mathrm{sym}}(S^{n-1})$.

A symmetric measure σ_K on S^{n-1} defines via (1.4) a zonoid $Z(K)$ which is a convex body in R^n only if σ_K is not concentrated on a proper subspace (otherwise the right-hand side of (1.4) vanishes for some $u \in S^{n-1}$). We shall call a symmetric measure on S^{n-1} which is not concentrated on a proper subspace an admissible measure. Consider the map

(1.5) $$K \longrightarrow \sigma_K \dashrightarrow Z(K) \ .$$

We just noted that the map (1.4) is a bijection from the class of all admissible measures on S^{n-1} onto all zonoids in R^n centered at the origin. A fact which goes back to Minkowski is that the map $K \to \sigma_K$ is a bijection from the class of all symmetric convex bodies in R^n centered at the origin onto all admissible measures on S^{n-1}.

The assertion of Minkowski actually applies also in the non-symmetric case. It is an elementary fact that for every convex body K in R^n the measure σ_K on S^{n-1} defined in (1.3) satisfies

(1.6) $$\int_{S^{n-1}} u d\sigma_K(u) = 0$$

(see also below) and its support has the origin in the interior of its convex hull. Minkowski proved that these are the only restrictions on σ_K, i.e., every positive measure μ on S^{n-1} for which $\int u d\mu(u) = 0$ and for which the support has 0 in the interior of its convex hull has the form σ_K for some K. The main step in the proof of this fact is the following. Given $\{u_i\}_{i=1}^m$ on S^{n-1} with $0 \in \operatorname{int} \operatorname{conv}\{u_i\}_{i=1}^m$ and positive numbers λ_i with $\sum_{i=1}^m \lambda_i u_i = 0$ there is a polytope P in R^n having faces $\{F_i\}_{i=1}^m$ with $\operatorname{vol}_{n-1} F_i = \lambda_i$ and outer normals u_i $1 \leq i \leq m$. The polytope P is obtained by considering the family $\mathcal{F}$ of polytopes of the form

$$\bigcap_{i=1}^m \{v \,;\, \langle v, u_i\rangle \leq t_i\} \qquad t_i > 0\,,\ 1 \leq i \leq m$$

which have n-dimensional volume equal to 1. Let P_0 be a polytope in $\mathcal{F}$ for which $\sum_{i=1}^m t_i\lambda_i$ attains its maximum on $\mathcal{F}$. A simple variational argument (cf., e.g., [Bo.F, p.120]) shows that $P = \rho P_0$ has the desired property for a suitable $\rho > 0$.

If μ is a general measure on S^{n-1} with a right support for which $\int u d\mu(u) = 0$ we approximate it (in the w^* topology of measures) by a sequence $\{\mu^j\}_{j=1}^\infty$ of discrete measures satisfying the two necessary conditions. By the previous step to each μ^j corresponds a polytope P^j. An application of Blaschke's selection theorem shows that there is a subsequence of $\{P^j\}_{j=1}^\infty$ which converges in the Hausdorff metric to a convex body K for which we have $\mu = \sigma_K$. (Actually, in view of the uniqueness assertion below, the passage to a subsequence is not needed.)

The map $K \to \sigma_K$ is also injective in the sense that σ_K determines K up to translation. The classical way to see this is by using the theory of mixed volumes developed by Minkowski and the Brunn Minkowski inequality. Recall (cf. [Bo.F] or [E] for this as well as (1.8) and

(1.9)) that if H and K are compact convex sets in R^n and $\lambda > 0$ then

$$\text{Vol}_n(H + \lambda K) = \sum_{k=0}^{n} \binom{n}{k} V_k(H, K)\lambda^k \tag{1.7}$$

where $V_k(H, K)$ (denoted also by $V(\overbrace{H, H, \ldots, H}^{n-k}, \overbrace{K, K, \ldots, K}^{k})$) are non-negative numbers called the mixed volumes of H and K. Clearly $V_0(H, K) = \text{Vol}_n(H)$ and $V_n(H, K) = \text{Vol}_n(K)$ and also $V_k(H, K) = V_{n-k}(K, H)$ $0 \leq k \leq n$. The quantity $V_1(H, K)$ can be expressed in terms of the support function of K and the measure σ_H:

$$V_1(H, K) = n^{-1} \int_{S^{n-1}} h_K(u) d\sigma_H(u) \tag{1.8}$$

Note that if K reduces to a single point x_0 then clearly it follows from (1.7) that $V_1(H, K) = 0$ and thus by (1.8) $\int_{S^{n-1}} \langle x_0, u\rangle d\sigma_H(u) = 0$. Since this is true for all x_0 we get (1.6) above.

The Brunn Minkowski theorem asserts that the function

$$f(\lambda) = \text{Vol}_n(\lambda H + (1 - \lambda)K)^{1/n}$$

is a concave function of λ on $[0, 1]$ and for H and K with non empty interior this function is strictly concave unless H is homothetic to K (i.e., $H = \rho K + x$ for some $\rho > 0$ and $x \in R^n$). It follows that $f'(0) \geq f(1) - f(0)$ with equality only if H is homothetic to K. By using the definition of f and (1.7) the relation $f'(0) \geq f(1) - f(0)$ translates to

$$V_1(H, K) \geq (\text{Vol}_n H)^{(n-1)/n} (\text{Vol}_n K)^{1/n} \tag{1.9}$$

with equality (for convex bodies) only if H is homothetic to K. Assume now that H and K are convex bodies with $\sigma_K = \sigma_H$. We deduce from (1.8) that

$$V_1(H, K) = n^{-1} \int_{S^{n-1}} h_K(u) d\sigma_H(u) = n^{-1} \int_{S^{n-1}} h_K(u) d\sigma_K(u) =$$
$$= V_1(K, K) = \text{Vol}_n(K)\ .$$

It follows from (1.9) that $\text{Vol}_n(K) \geq \text{Vol}_n(H)$ and by symmetry $\text{Vol}_n(K) = \text{Vol}_n(H)$. Also we have equality in (1.9) and thus H is a translate of K.

By combining all the facts mentioned above we have the following.

Theorem 1 (Minkowski). *The map $K \to Z(K)$ which assigns to every convex body its projection body is an injective map from the class of symmetric convex bodies in R^n centered at the origin onto all zonoids in R^n which have the origin as center and interior point.*

As noted by Petty [P], there is an addendum to Theorem 1 which makes the Minkowski map of special interest from the Banach space theory point of view. The projection bodies of affinely equivalent convex bodies are themselves affinely equivalent. More specifically we have

Proposition 2 (Petty). *Let K be a convex body in R^n and let $T \in \mathrm{GL}(R^n)$ with $\det T = 1$ then*

$$Z(T(K)) = (T^{-1})^*(Z(K)) . \tag{1.10}$$

Proof: It is enough to consider K a polytope, the general case will follow by continuity. Let $\{F_i\}_{i=1}^m$ be the faces of K and let $\{u_i\}_{i=1}^m$ be the normals to $\{F_i\}_{i=1}^m$ directed outward. The faces of TK are $\{TF_i\}_{i=1}^m$ and let $\{v_i\}_{i=1}^m$ be the corresponding normals. Note that since $\langle y, u\rangle = 0 \Longleftrightarrow \langle Ty, (T^{-1})^* u\rangle = 0$ we have that

$$v_i = (T^{-1})^* u_i / |||(T^{-1})^* u_i||| , \qquad 1 \le i \le m . \tag{1.11}$$

Since $\det T = 1$ we have for every i

$$\begin{aligned} \mathrm{Vol}_{n-1}(F_i) &= \mathrm{Vol}_n(F_i \times [0, u_i]) = \mathrm{Vol}_n(T(F_i \times [0, u_i])) \\ &= \mathrm{Vol}_{n-1}(TF_i)|\langle Tu_i, v_i\rangle| = \mathrm{Vol}_{n-1}(TF_i)/|||(T^{-1})^* u_i||| . \end{aligned} \tag{1.12}$$

Hence, by (1.2), (1.11) and (1.12)

$$\begin{aligned} h_{Z(T(K))}(u) &= \tfrac{1}{2}\sum_{i=1}^m \mathrm{Vol}_{n-1}(TF_i)|\langle u, v_i\rangle| = \\ &= \tfrac{1}{2}\sum_{i=1}^m \mathrm{Vol}_{n-1}(F_i)|\langle T^{-1}u, u_i\rangle| = h_{Z(K)}(T^{-1}u) . \end{aligned}$$ □

Before we proceed it is worthwhile to consider some examples of the Minkowski map.

(i) If K is the unit cube in R^n then $Z(K)$ is $2^{n-1}K$.

(ii) If K is a ball in R^n then $Z(K)$ is again a ball (with a different radius). Hence, by Proposition 2, if K is an ellipsoid the same is true for $Z(K)$.

(iii) If K is a symmetric polytope having $2m$ $(n-1)$-dimensional faces, then $Z(K)$ is a zonotope having m summands (i.e., $Z(K)$ is the Minkowski sum of m segments).

(iv) The two dimensional situation is special and much simpler than the higher dimensional case. In R^2 every symmetric convex body is a zonoid. In this case $Z(K)$ is obtained from K (in the symmetric case) by rotating by an angle of $\pi/2$ and then expanding by a factor of 2.

(v) Let Z be the zonotope $\sum_{i=1}^{m}[-u_i, u_i]$ in R^n. There is an $(n-1)$-dimensional face of Z orthogonal to some vector u if and only if the hyperplane $u^\perp$ in R^n contains $(n-1)$ linearly independent vectors among the $\{u_i\}_{i=1}^m$. Using this fact it can be deduced that Z has always at least $2m$ $(n-1)$-dimensional faces. The situation where Z has exactly $2m$ such faces can be analysed using results about arrangements of points and hyperplanes in projective space. This analysis was done by Weil [W]. It turns out that the number of $(n-1)$-dimensional faces of a zonotope Z is twice its number of summands if and only if Z is the Minkowski sum of $n/2$ 2-dimensional polygons (if n is even) or $(n-1)/2$ 2-dimensional polygons and another line (if n is odd). In other words, passing to the Banach space X such that $Z = B_X$ (the unit ball of X) we get that

$$(1.13)' \quad X = \Big(\sum_{j=1}^{n/2} \oplus Y_j\Big)_\infty\ ,\ \dim Y_j = 2\ ,\ 1 \le j \le n/2\ ,\ n \text{ even}$$

$$(1.13)'' \quad X = \Big(\sum_{j=0}^{(n-1)/2} \oplus Y_j\Big)_\infty\ ,\ \dim Y_0 = 1\ ,\ \dim Y_j = 2\ ,\ 1 \le j \le (n-1)/2\ ,\ n \text{ odd}\ .$$

It follows from this fact and (iii) and (iv) above that a symmetric polytope K satisfies that $Z(Z(K))$ is equal to ρK for some $\rho > 0$ if and only if K is the unit ball of a space of the form (1.13). As shown in [W] another equivalent condition on a polytope K to be a unit ball of a space of the form (1.13) is that K is in the range of the iterated Minkowski map $\overbrace{Z \circ Z \circ Z \circ \cdots \circ Z}^{k-\text{times}}$ for every finite k. As already pointed out above the argument in [W] is based on counting the number of faces of a polytope (as well as combinatorial statements involving arrangements in projective spaces). This method of proof breaks down if we pass to general convex bodies. This leads to some interesting open questions in the isometric theory of Banach spaces (which we did not study). For which Banach space X is the space isometric to Y where $B_Y = Z(B_X)$? For which Banach space X is $B_X = \overbrace{Z(Z(\cdots(Z}^{k}(K_k))$ for some body K_k and every $k = 1, 2, \ldots$?

(vi) Let X be a Banach space with a 1-symmetric basis $\{e_i\}_{i=1}^n$, i.e., satisfying $\|\sum_{i=1}^n a_i e_i\| = \|\sum_{i=1}^n \pm a_i e_{\pi(i)}\|$ for every choice of scalars, signs and permutation π. By using an averaging procedure it was proved by C. Schütt [Schü] that for every $u = \sum_{i=1}^{n} a_i e_i$ with $|||u||| = \left(\sum_{i=1}^{n} a_i^2\right)^{1/2} = 1$ we have

$$2^{-1/2}\,\mathrm{Vol}_{n-1}(B_{X_{n-1}}) \le h_{Z(B_X)}(u) \le cn^{-1/2}\,\mathrm{Vol}_{n-1}(\partial B_X) \tag{1.14}$$

where c is an absolute constant, $\mathrm{Vol}_{n-1}(\partial B_X)$ denotes the surface area of the unit sphere of X and $B_{X_{n-1}}$ is the unit ball in $\mathrm{span}\{e_i\}_{i=1}^{n-1}$. Since, for example, for every fixed $1 \le p < \infty$

$$\mathrm{Vol}_{n-1}\big(\partial B(\ell_p^n)\big)/n^{1/2}\,\mathrm{Vol}_{n-1}\big(B(\ell_p^{n-1})\big)$$

is bounded and bounded away from 0 as $n \to \infty$ it follows from (1.14) that $Z\big(B(\ell_p^n)\big)$ is up to a constant (depending only on p but not on n) an Euclidean ball.

It would be of interest to study the Minkowski map for more examples of convex bodies.

We turn now to the main subject of the this note namely the degree of continuity of the Minkowski correspondence $K \leftrightarrow Z(K)$ when restricted to convex bodies having the origin as center of symmetry. It is not hard to see (cf [G]) that even if we consider only rectangular boxes and if we allow them to converge to degenerate convex sets (i.e., become unbounded or tend to sets with empty interior), taking as distance the usual Hausdorff distance, then the Minkowski map may fail to be uniformly continuous in both directions. Therefore in order to study the continuity problem in a meaningful way we have to normalize the situation by controlling from above the diameter of the sets and controlling from below the radius of the ball around 0 which they contain. One way to do this is to measure distances in an affine invariant way (provided both sets undergo the same affine transformation) by putting

$$\delta(K,H) = \min\{\log\lambda\ ;\ \lambda^{-1}H \subset K \subset \lambda H\}\ . \tag{1.15}$$

In situations in which we have control on the diameter and in-sphere of K and H, $\delta(K,H)$ is clearly equivalent to the Hausdorff metric. In view of the affine invariance of δ and the result of F. John [J] we may always assume when we consider δ that

$$B^n \subset K \subset \sqrt{n}B^n \tag{1.16}$$

where B^n denotes the Euclidean ball.

Another distance which is of interest, especially in the Banach space context is the Banach Mazur distance. It measures distances between equivalence classes of $\widetilde{H}$ and $\widetilde{K}$ of symmetric convex bodies modulo affine equivalence (i.e., $H \in \widetilde{H} \Leftrightarrow TH \in \widetilde{H}$, $T \in \mathrm{GL}(R^n)$)

$$\widetilde{\delta}(\widetilde{H},\widetilde{K}) = \min\left\{\delta(H,K)\ ;\ H \in \widetilde{H}\ ,\ K \in \widetilde{K}\right\}\ . \tag{1.17}$$

(In Banach space theory $\tilde{\delta}$ is usually defined without using log. In our context, especially since we want to consider Lipschitz or Hölder maps with respect to $\tilde{\delta}$, it becomes too unnatural to omit the log.)

The map $K \to Z(K)$ behaves nicely with respect to δ. If $\lambda^{-1}H \subset K \subset \lambda H$ then, because of the obvious monotonicity of the Minkowski map, $\lambda^{-n+1}Z(H) \subset Z(K) \subset \lambda^{n-1}Z(H)$ and thus the map is a Lipschitz map (with respect to δ) for small distances with a Lipschitz constant depending only on the dimension. If we consider this map in the setting of equivalence classes of convex bodies (which can be done in view of Proposition 2) we get a map which is globally Lipschitz with respect to $\tilde{\delta}$.

The inverse map $Z(K) \to K$ is much harder to analyse and does not behave as well (of course, only for $n \geq 3$, in view of example (iv) above). For example it is not monotone. The relation $Z(K) \supset Z(H)$ does not even imply that $\mathrm{Vol}_n(K) \geq \mathrm{Vol}_n(H)$ (see [Schn1] for a detailed discussion of this point). In the smooth case the map $Z(K) \to K$ was the subject of several papers in differential geometry. Assume that K is a C^2 convex body with non-vanishing Gauss curvature. Then by classic elementary differential geometry the measure σ_K on S^{n-1} defined in (1.3) is given by

$$d\sigma_K(v) = \frac{1}{\kappa(v)} d\sigma_n(v) \tag{1.18}$$

where $\kappa(v)$ denotes the Gauss curvature of K at the point on ∂K corresponding to v via the Gauss map and σ_n is the normalized rotation invariant measure on S^{n-1}. In the smooth case the problem of finding K given $Z(K)$ thus becomes the problem of reconstructing a convex set from its Gauss curvature. Analytically, it can be expressed as a solution of the following non-linear differential equation (called Monge Ampère equation):

$$\det\left(\frac{\partial^2 h}{\partial x_i \partial x_j} + h\delta_{i,j}\right) = \frac{1}{\kappa(x)}, \qquad |||x||| = 1 \tag{1.19}$$

where h is assumed to be homogeneous of degree 1, i.e., $h(x) = |||x|||h(x/|||x|||)$, $x \neq 0$. The solution h will be the support function of the set K. The interest in differential geometry is to prove that the set K is smooth provided $\kappa(x)$ is given as smooth. One formulation of the result obtained in this direction is the following.

Theorem 3. *Assume that $\kappa(u)$ is positive function on S^{n-1} which is of class C^m for $m \geq 3$, and satisfies $\int_{S^{n-1}} u\kappa(u)^{-1} d\sigma_n(u) = 0$ (this is (1.6) in the present setting). Then the unique (up to translation) convex body K which has $\kappa(u)$ as its Gauss curvature (via the Gauss map) has a support function of class C^{m+1}. If κ is analytic so is the support function of K.*

Theorem 3 was proved first in the case $n = 3$ by H. Lewy in the analytic setting and by Pogorelov and Nirenberg (independently) in the C^m setting. The case $n > 3$ was solved by Pogorelov (see the book [Po] which has a general discussion of the Minkowski map). There were some questions about the validity of the original argument of Pogorelov and another proof was given by Cheng and Yau [C.Y]. Both of these proofs use geometric concepts and methods. Recently there appeared purely analytic proofs of Theorem 3 which directly solve the Monge Ampère equation (1.19) without using geometric tools (see e.g., [C.N.S]).

As is well known, it can be shown by using spherical harmonic functions (cf. [Schn1]) that if $h_{Z(K)}(u)$ is sufficiently smooth so is also the Radon Nikodym derivative $d\sigma_K/d\sigma_n$ where σ_K is given by (1.4) (the arguments used in the beginning of section 2 implicitly prove this point also). Hence it follows from Theorem 3 that if $h_{Z(K)}$ is sufficiently smooth or analytic so is h_K (provided we restrict ourselves to symmetric bodies K).

Theorem 3 does not, however, give information on the degree of continuity of the map $Z(K) \to K$ in the δ or $\tilde{\delta}$ metric. Our interest in this question arose in connection with the study done in [B.L.M] on the approximation of zonoids (and especially the Euclidean ball B^n) by zonotopes. It is quite easy to see and well known that if we are interested in approximating a convex body $K \subset R^n$ by a polytope, up to ε say, in the δ metric we can do this by a polytope having at most $a(K)\varepsilon^{-(n-1)/2}$ extreme points. It is also known that for sufficiently smooth convex bodies we need at least that many extreme points. In case K is of class C^3 with a strictly positive Gauss curvature κ there is available even a precise asymptotic formula for the right constant $a(K)$ (cf. [Schn2]). By duality we have the same estimate for the number of $(n-1)$-dimensional faces needed for approximating a convex body up to ε by a polytope. Thus, for example, there are polytopes $P(n,\varepsilon)$ in R^n so that $\delta(P(n,\varepsilon), B^n) \le \varepsilon$ which have at most $a(n)\varepsilon^{-(n-1)/2}$ faces. Hence (as remarked in [Be.Mc]) it follows by passing to the projection body of $P(n,\varepsilon)$ and using example (iii) above that we may approximate B^n up to ε by a zonotope having at most $c(n)\varepsilon^{-(n-1)/2}$ summands. While for $n = 2$ this result is trivially optimal it turns out that for large n we can do much better. It was proved in [F.L.M] that one can approximate B^n up to ε by a zonotope having $cn\varepsilon^{-2}|\log\varepsilon|$ summands, and in [Gor] it was shown that one can drop the factor $|\log\varepsilon|$. In [Sche] it was shown that an arbitrary zonoid in R^n can be approximated by a zonotope having $cn^2\varepsilon^{-2}|\log\varepsilon|$ summands. In [B.L.M] sharper results were obtained (but the improvements concern the dependence on n and not that on ε which is of interest in the present context) while on the other hand it was proved that if Z is a zonotope with $\delta(Z, B^n) \le \varepsilon$ then Z has at least $c(n)\varepsilon^{-2(n-1)/(n+2)}$ summands.

From the above we get immediately information on the possible degree of continuity of the inverse Minkowski map. If $Z = Z(n,\varepsilon)$ is a polytope with $\delta(Z,B^n) \le \varepsilon$ having $c(n)\varepsilon^{-2}$ summands then $Z = Z(P)$ for some symmetric polytope $P = P(n,\varepsilon)$ having $c(n)\varepsilon^{-2}$ $(n-1)$-dimensional faces. Thus, if ρ_n is such that $Z(\rho_n B^n) = B^n$ we get that

$$\delta(P,\rho_n B^n) \ge c_1(n)\varepsilon^{4/(n-1)} \ge c_1(n)\big(\delta(Z(P),Z(\rho_n B^n))\big)^{4/(n-1)}\,. \tag{1.20}$$

An estimate similar to (1.20) can also be obtained by taking a Banach space theory point of view. Let $\mathcal{F}$ be a family of n-dimensional Banach spaces so that $\widetilde{\delta}(B_X,B_Y) \ge \varepsilon$ for every $X \ne Y$ in $\mathcal{F}$. It is easily verified that there is such a family with m elements where $\log m \ge c_2(n)\varepsilon^{-(n-1)/2}$. Let π be any map (not necessarily the Minkowski map) which assigns to every n dimensional Banach space an n-dimensional subspace of L_1 (or alternatively, consider π as a map between equivalence classes of convex bodies modulo affine equivalence). Assume that for some $\tau > 0$ we have that $\widetilde{\delta}(\pi B_X,\pi B_Y) \ge \tau$ for all $X \ne Y$ in $\mathcal{F}$. By the result of Schechtman quoted above there is a map φ which assigns to very n-dimensional zonoid the unit ball of an n dimensional subspace of ℓ_1^k with $k = c_3(n)\tau^{-2}|\log\tau|$ so that $\widetilde{\delta}(\varphi K,K) \le \tau/4$. Hence $\widetilde{\delta}(\varphi\pi B_X,\varphi\pi B_Y) \ge \tau/2$ for all $X \ne Y$ in $\mathcal{F}$. An easy computation (using an Auerbach basis) shows that the number of $\tau/2$ separated n-dimensional subspaces of a fixed k-dimensional space is at most $(32n/\tau)^{nk}$. Combining these facts we get that $m \le (32n/\tau)^{nk}$ or that $c_4(n)\tau^{-2}(\log\tau)^2 \ge \varepsilon^{-(n-1)/2}$. Hence the estimate we get on the Hölder exponent of π^{-1} is essentially $4/(n-1)$, as in (1.20).

A very easy direct geometric example gives however, a sharper result than (1.20). Let B^n be as usual the n-dimensional Euclidean ball and for $\varepsilon > 0$ let $K_\varepsilon = \{u = (u_1,\dots,u_n) \in B^n\ ,\ |u_1| < 1-\varepsilon\}$. Then clearly $\delta(K_\varepsilon,B^n) \ge \varepsilon/2$ while $\delta(Z(K_\varepsilon),Z(B^n)) \le c(n)\varepsilon^{n/2}$, i.e.,

$$\delta(K_\varepsilon,B^n) \ge c_2(n)\big(\delta(Z(K_\varepsilon),Z(B^n))\big)^{2/n} \tag{1.21}$$

The next theorem gives an estimate from below on the Hölder exponent of the inverse Minkowski map.

Theorem 4. *For every $n \ge 3$ and every $\gamma < 2/n(4+n)$ there are constants $c_1(n,\gamma)$ and $c_2(n,\gamma)$ depending only on n and γ so that*

$$\delta(H,K) \le c_1(n,\gamma)\delta(Z(H),Z(K))^\gamma \tag{1.22}$$

for every two convex symmetric bodies H and K in R^n with $\delta(Z(H),Z(K)) \le 1$ and

$$\widetilde{\delta}(\widetilde{H},\widetilde{K}) \le c_2(n,\gamma)\widetilde{\delta}(Z(\widetilde{H}),Z(\widetilde{K}))^\gamma \tag{1.23}$$

for any two equivalence classes of convex symmetric bodies in R^n modulo affine equivalence.

Clearly the second assertion of Theorem 4 is a consequence of the first. It is very likely that the exponent appearing in Theorem 4 can be improved. The discussion above as well as other facts indicate that the right exponent should be of the order of n^{-1} and not n^{-2} as in the statement of the theorem. The proof of Theorem 4, to be presented in the next section, is divided into two steps which are parallel to the steps indicated above for the proof of Theorem 1. The first step involves quantitative estimates related to the inversion of the Radon transform (1.4). This is done by using spherical harmonics. It is close to the argument used in [B.L.M] to estimate from below the degree of approximation of B^n by zonotopes (since only a small detour from the main arguments is needed we shall in fact reproduce this approximation result from [B.L.M] as Proposition 6 in section 2). The second step is based on a result of Diskant [D2] which investigates the question under which condition there is almost an equality in (1.9). Each of the two steps of the proof contributes a factor of about n^{-1} to the Hölder exponent. It may be that both steps can be combined to a single argument which will yield the right order of magnitude for the exponent. It would also be of interest to obtain good estimates for the exponent for small values of n. For example, in view of the example mentioned before the statement of Theorem 4 it is natural to ask whether the right exponent for $n = 3$ is $2/3$.

2. The Proof of Theorem 4

This section is devoted mainly to the proof of Theorem 4. After the end of the proof we shall present a proof of (a weakened version) of the result of Diskant which is used for proving Theorem 4. The argument we use for deriving Diskant's result uses the Knothe map [K] between convex sets.

We shall first collect some facts concerning spherical harmonics as a preparation for the proof of Theorem 4.

Let μ be a symmetric positive measure on S^{n-1} and put

$$h(x) = \int_{S^{n-1}} |\langle x, y\rangle| d\mu(y) \qquad x \in S^{n-1}. \tag{2.1}$$

For each $k \geq 1$ let $\{Y_{k,j}\ ,\ 1 \leq j \leq M(n,k)\}$ be an orthonormal basis of the spherical harmonics of degree k on S^{n-1}. By the addition theorem (see [Mü], p. 10)

$$\sum_j Y_{k,j}(\xi) Y_{k,j}(\eta) = M(n,k) P_k(\langle \xi, \eta\rangle) \qquad \xi, \eta \in S^{n-1} \tag{2.2}$$

where $P_k(t)$ is the Legendre polynomial of degree k and dimension n, given by Rodrigues' formula (see [Mü], p. 17)

$$(2.3)\qquad P_k(t) = (-\tfrac{1}{2})^k \frac{\Gamma((n-1)/2)}{\Gamma(k+(n-1)/2)}(1-t^2)^{\frac{3-n}{2}} \left(\frac{d}{dt}\right)^k \{(1-t^2)^{k+(n-3)/2}\} .$$

By the Funk-Hecke formula (see [Mü], p. 20)

$$(2.4)\qquad [h, Y_{k,j}] = \int_{S^{n-1}} h(x) Y_{k,j}(x) d\sigma_n(x) =$$

$$= \int_{S^{n-1}} \left[\int_{S^{n-1}} |\langle x, y\rangle| Y_{k,j}(x) d\sigma_n(x)\right] d\mu(y) = \lambda_k \int_{S^{n-1}} Y_{k,j}(y) d\mu(y)$$

where σ_n denotes the normalized rotation invariant measure on S^{n-1}, $[\cdot,\cdot]$ denotes the inner product in $L_2(\sigma_n)$, $\lambda_k = 0$ for odd k and for even k (by (2.3))

$$(2.5)\qquad \frac{\omega_n}{\omega_{n-1}}\lambda_k = \int_{-1}^{1} |t|(1-t^2)^{(n-3)/2} P_k(t) dt =$$

$$= \frac{(-1)^{(k-2)/2}\Gamma((n-1)/2)\Gamma(k-1)}{2^{k-1}\Gamma(k/2)\Gamma((k+n+1)/2)}$$

where ω_n is the usual surface area of S^{n-1}, and in particular

$$(2.6)\qquad |\lambda_k| \approx k^{-\frac{n}{2}-1} \quad \text{as} \quad k \to \infty\ ,\ k \text{ even} .$$

Put

$$(2.7)\quad Y_0 = \int_{S^{n-1}} h(x) d\sigma_n(x) = \mu(S^{n-1}) \int_{S^{n-1}} |x_1| d\sigma_n(x)\ , \qquad Y_k = \sum_j [h, Y_{k,j}] Y_{k,j}\ , \quad k \geq 1 .$$

Then formally

$$(2.8)\qquad h = \sum_{k \text{ even}} Y_k \quad , \quad \mu = \sum_{k \text{ even}} \lambda_k^{-1} Y_k .$$

Recall also the following identity ([Mü], p.30)

$$(2.9)\qquad \frac{1-r^2}{(1+r^2-2rt)^{n/2}} = \sum_{k=0}^{\infty} M(n,k) r^k P_k(t)\ , \qquad |t| \leq 1\ ,\ 0 \leq r < 1 .$$

By (2.2) we deduce that for $0 \leq r < 1$ and $\xi, \eta \in S^{n-1}$

$$(2.10)\qquad \frac{1-r^2}{(1+r^2-2r\langle\xi,\eta\rangle)^{n/2}} = \sum_{k=0}^{\infty} r^k \sum_j Y_{k,j}(\xi) Y_{k,j}(\eta) .$$

By integrating with respect to μ we get that

$$\mu_r(\xi) \stackrel{\text{def}}{=} \int\limits_{S^{n-1}} \frac{1-r^2}{(1+r^2-2r\langle\xi,\eta\rangle)^{n/2}} \cdot d\mu(\eta) = \sum_{k \text{ even}} \lambda_k{}^{1} r^k Y_k . \tag{2.11}$$

Let $F : S^{n-1} \to R$ be a function satisfying a Lipshitz condition and put $\|F\|_{\text{Lip}} = \sup|F(\xi) - F(\eta)|/\|\xi - \eta\|$ and

$$F_r(\xi) = \int\limits_{S^{n-1}} \frac{1-r^2}{(1+r^2-2r\langle\xi,\eta\rangle)^{n/2}} F(\eta)d\sigma_n(\eta) \qquad \xi \in S^{n-1}, \ 0 \le r < 1 . \tag{2.12}$$

We have

$$\|F_r - F\|_\infty \le \sup_\xi \int\limits_{S^{n-1}} \frac{(1-r^2)|F(\xi) - F(\eta)|}{(1-r^2-2r\langle\xi,\eta\rangle)^{n/2}} d\sigma_n(\eta) \le \tag{2.13}$$

$$\le c_1(n)(1-r)\|F\|_{\text{Lip}} \int_0^1 \frac{t^{n-1}}{(1-r)^n + t^n} dt \le c_2(n)(1-r)\log|1-r|\|F\|_{\text{Lip}} .$$

Let now μ' be another symmetric measure on S^{n-1} and let h', Y'_k and μ'_r be defined in alaogy to (2.1), (2.7) and (2.11). By Parsevals theorem, (2.6) and (2.8) we have for $0 \le r < 1$

$$\|\mu_r - \mu'_r\|_{L_2(\sigma_n)} = \Big(\sum_{k \text{ even}} \lambda_k^{-2} r^{2k} |Y_k - Y'_k|^2_{L_2(\sigma_n)}\Big)^{1/2} \le \tag{2.14}$$

$$\le \big(\max_{k \text{ even}} \lambda_k^{-1} r^k\big) \|h - h'\|_{L_2(\sigma_n)} \le c_3(n)(1-r)^{-\frac{n}{2}-1}\|h - h'\|_{L_2(\sigma_n)} .$$

Hence with F as above

$$\Big|\int\limits_{S^{n-1}} F d\mu - \int\limits_{S^{n-1}} F d\mu'\Big| \le \Big|\int\limits_{S^{n-1}} F_r d(\mu - \mu')\Big| + \Big|\int\limits_{S^{n-1}} (F - F_r) d\mu\Big| + \Big|\int\limits_{S^{n-1}} (F - F_r) d\mu'\Big| =$$

$$= \Big|\int\limits_{S^{n-1}} F(\mu_r - \mu'_r) d\sigma_n\Big| + \Big|\int\limits_{S^{n-1}} (F - F_r) d\mu\Big| + \Big|\int\limits_{S^{n-1}} (F - F_r) d\mu'\Big| \le \tag{2.15}$$

$$\le \|F\|_{L_2(\sigma_n)} \|\mu_r - \mu'_r\|_{L_2(\sigma_n)} + \|F - F_r\|_\infty (\mu(S^{n-1}) + \mu'(S^{n-1})) .$$

Hence by (2.13) and (2.14)

$$\Big|\int\limits_{S^{n-1}} F d\mu - \int\limits_{S^{n-1}} F d\mu'\Big| \le \tag{2.16}$$

$$\le c_3(n)(1-r)^{-\frac{n}{2}-1}\|F\|_2\|h - h'\|_2 + c_2(n)(1-r)\log|1-r|\|F\|_{\text{Lip}}(\mu(S^{n-1}) + \mu'(S^{n-1})) .$$

Finally by taking supremum over r we get that for every $\alpha < 2/(4+n)$ there is a $C(\alpha, n)$ so that

$$\Big|\int\limits_{S^{n-1}} F d\mu - \int\limits_{S^{n-1}} F d\mu'\Big| \le C(\alpha, n)(\|F\|_2\|h - h'\|_2)^\alpha (\|F\|_{\text{Lip}}(\mu(S^{n-1}) + \mu'(S^{n-1})))^{1-\alpha} \tag{2.17}$$

provided $\|F\|_2\|h - h'\|_2 \le \|F\|_{\mathrm{Lip}}(\mu(S^{n-1}) + \mu'(S^{n-1}))$. Inequality 2.17 will be one of the basic tools for proving Theorem 4. The second tool for proving the theorem is the following result.

Proposition 5 (Diskant). *Let H and K be symmetric convex bodies in R^n. Assume that for some $0 < \tau < 1/2$*

$$|\operatorname{Vol}_n H - V_1(K,H)| < \tau \operatorname{Vol}_n H \tag{2.18}$$

$$|\operatorname{Vol}_n K - V_1(H,K)| < \tau \operatorname{Vol}_n K \tag{2.19}$$

where $V_1(H,K)$ and $V_1(K,H)$ denote mixed volumes (see (1.8)). Then

$$\delta(K,H) \le c(n)\tau^{1/n} \tag{2.20}$$

Diskant has several papers dealing with Proposition 5 and related questions. In [D1] he proves this proposition for $H = B^n$ the Euclidean ball using an interesting symmetrization argument. In [D2] he gives another argument which proves the proposition as stated. We shall not reproduce here the proof from [D2]. Instead we shall present below yet another way to prove Proposition 5 (in a weaker form).

Proof of Theorem 4. As mentioned already after the statement of the theorem it suffices to prove (1.22). Let H and K be symmetric convex bodies in R^n and put

$$h_{Z(H)}(x) = \tfrac{1}{2}\int_{S^{n-1}} |\langle x,y\rangle| d\sigma_H(y)$$

$$h_{Z(K)}(x) = \tfrac{1}{2}\int_{S^{n-1}} |\langle x,y\rangle| d\sigma_K(y)\,.$$

We assume as we may (in view of (1.16)) that

$$|||x||| \le h_K(x) \le \sqrt{n}|||x||| \tag{2.21}$$

and also that for some $0 < \varepsilon < 1/2$

$$(1-\varepsilon)h_{Z(K)} \le h_{Z(H)} \le (1+\varepsilon)h_{Z(K)} \tag{2.22}$$

It follows from the monotonicity of the Minkowski map,(2.21) and (2.22) that for some constant $c_1(n)$

$$c_1(n)^{-1}|||x||| \le h_{Z(H)}(x)\,,\quad h_{Z(K)}(x) \le c_1(n)|||x|||\,. \tag{2.23}$$

Since $\int_{S^{n-1}} h_{Z(x)}(x)d\sigma_n(x) = \frac{1}{2}(\int_{S^{n-1}} |x_1|d\sigma_n(x))\sigma_K(S^{n-1})$ it follows from (2.23) that the surface areas and thus the volumes of H and K are bounded by a constant depending only on n. A simple geometric argument, using the lower bound on $h_{Z(H)}$ in (2.23) and the upper bound on the volume of H, yields that the diameter of H is bounded by a constant depending only on n, i.e.,

$$h_H(x) \le c_2(n)|||x||| \ . \tag{2.24}$$

Since h_K and h_H are norms we deduce from (2.21) and (2.24) that

$$\|h_K\|_{\text{Lip}} \le \sqrt{n} \ , \qquad \|h_H\|_{\text{Lip}} \le c_2(n) \ . \tag{2.25}$$

We may now apply (2.17) with $F = h_H$ (or h_K) and $\mu = \sigma_K$, $\mu' = \sigma_H$ and get that for $0 < \varepsilon < \varepsilon_0(n)$ and every $\alpha < 2/(4+n)$ there are constants $c_1(\alpha, n)$ and $c_2(\alpha, n)$ so that

$$\begin{aligned} |\operatorname{Vol}_n H - V_1(K,H)| &= n^{-1}\Big|\int_{S^{n-1}} h_H(x)d(\sigma_H(x) - \sigma_K(x))\Big| \\ &\le c_1(\alpha,n)\varepsilon^\alpha \le c_2(\alpha,n)\varepsilon^\alpha \operatorname{Vol}_n H \end{aligned} \tag{2.26}$$

and similarly

$$|\operatorname{Vol}_n K - V_1(H,K)| \le c_2(\alpha,n)\varepsilon^\alpha \operatorname{Vol}_n K \ . \tag{2.27}$$

In the proof of (2.27) we also use the fact that $\operatorname{Vol}_n(K)$ is bounded from below by a constant depending only on n. This follows from the fact that ∂K cannot have points too close to the origin. Otherwise K would be between two close parallel hyperplanes and this jointly with the bound on the diameter of K would contradict (2.23). Thus, we verified that (2.18) and (2.19) hold with $\tau = c_2(\alpha, n)\varepsilon^\alpha$ and hence (2.20) is also valid for this τ. This gives exactly (1.22). □

We recall now the definition of the Knothe map f (cf. [K]) between two open and bounded convex sets H and K in R^n. It is a map which has the form $f(x) = (f_1(x), f_2(x), \dots, f_n(x))$, $x \in H$ where

$$\begin{cases} f_1(x) = f_1(x_1) \\ f_2(x) = f_2(x_1, x_2) \\ \quad \cdot \quad \cdot \quad \cdot \quad \cdot \\ f_n(x) = f_n(x_1, x_2, \dots, x_n) \end{cases} \tag{2.28}$$

$$\frac{\partial f_j}{\partial x_j}(x) \ge 0 \quad 1 \le j \le n \ , \ x \in H \ , \tag{2.29}$$

$$\text{(2.30)} \qquad \operatorname{Jac}(f) = \prod_{j=1}^{n} \frac{\partial f_j}{\partial x_j} \equiv \operatorname{Vol}_n(K)/\operatorname{Vol}_n(H)\,.$$

In order to define f we put for an open set A

$$A_{x_1,\dots,x_j} = \{u \in A\,,\; u_i = x_i\,,\; 1 \le i \le j\}$$

We shall use such sets only whenever they are non empty and consider them as subsets in R^{n-j}. The function f_1 is defined by

$$\int_{-\infty}^{x_1} \operatorname{Vol}_{n-1}(H_t)dt = \left[\int_{-\infty}^{f_1(x_1)} \operatorname{Vol}_{n-1}(K_t)dt\right] \frac{\operatorname{Vol}_n(H)}{\operatorname{Vol}_n(K)}$$

and generally by induction on j

$$\text{(2.31)} \quad \int_{-\infty}^{x_j} \operatorname{Vol}_{n-j}(H_{x_1,x_2,\dots,x_{j-1},t})dt =$$

$$= \left[\int_{-\infty}^{f_j(x_1,\dots,x_j)} \operatorname{Vol}_{n-j}(K_{f_1(x),f_2(x),\dots,f_{j-1}(x),t})dt\right] \frac{\operatorname{Vol}_{n-j+1}(H_{x_1,\dots,x_{j-1}})}{\operatorname{Vol}_{n-j+1}(K_{f_1(x),\dots,f_{j-1}(x)})}$$

It is clear from (2.31) that for every j

$$\frac{\partial f_j(x)}{\partial x_j} = \frac{\operatorname{Vol}_{n-j}(H_{x_1,\dots,x_j}) \cdot \operatorname{Vol}_{n-j+1}(K_{f_1(x),\dots,f_{j-1}(x)})}{\operatorname{Vol}_{n-j}(K_{f_1(x),\dots,f_j(x)}) \cdot \operatorname{Vol}_{n-j+1}(H_{x_1,\dots,x_{j-1}})}$$

and this implies (2.29) and (2.30). It will be convenient to use in the sequel the notation $\frac{\partial f_j(x)}{\partial x_j} = \partial_j f_j(x)$. We shall now use this map to prove a version of Proposition 5 in which $1/n$ is replaced by $2/n$ in (2.20). Such an argument may be useful also in realted contexts.

Proof of a (weakened version of) Proposition 5. We remark first that there is no loss of generality to assume that $\operatorname{Vol}_n(H) = \operatorname{Vol}_n(K) = 1$ in the assumption of Proposition 5. Indeed, (2.18) and (2.19), if combined with (1.9), yield that

$$|\operatorname{Vol}_n(H)/\operatorname{Vol}_n(K) - 1| \le c_1(n)\tau$$

and thus by replacing τ by $c_2(n)\tau$ we may normalize both sets to have volume 1.

Thus, let H and K be two symmetric bodies in R^n of volume 1 satisfying (2.18) and (2.19). Let f be the Knothe map between their interiors. Clearly $f(0) = 0$. We have (use (1.7)) for

$0 \leq t \leq 1$ and with μ denoting the usual Lebesgue measure in R^n

$$\int_H \prod_{j=1}^{n} ((1-t) + t\partial_j f_j(x))\, d\mu(x) = \mathrm{Vol}_n([(1-t)I + tf]H)$$

$$\leq \mathrm{Vol}_n((1-t)H + tK) = \sum_{k=0}^{n} \binom{n}{k} V_k(H,K)(1-t)^{n-k} t^k .$$

Hence, by (2.18),

$$(2.32) \quad \int_H \Big(\sum_{j=1}^{n} (\partial_j f_j(x) - 1)\Big) d\mu(x) = \lim_{t\downarrow 0} \Big(\int_H \prod_{j=1}^{n} ((1-t) + t\partial_j f_j(x))\, d\mu(x) - 1\Big)/t \leq$$

$$\leq \left[\sum_{k=0}^{n} \binom{n}{k} V_k(H,K)(1-t)^{n-k} t^k\right]'_{t=0} = n(V_1(H,K) - V_0(H,K)) =$$

$$= n(V_1(H,K) - 1) \leq n\tau .$$

Note that since $\prod_{j=1}^{n} \partial_j f_j(x) = 1$ on H we have that

$$(2.33) \qquad \sum_{j=1}^{n} (\partial_j f_j(x) - 1) \geq 0 \qquad x \in H .$$

Observe next that there is a positive constant $c_2(n)$ so that

$$\log(1+t) \leq t - c_2(n)\min(t^2, n^2) , \qquad -1 < t < \infty .$$

Hence, by (2.32)

$$(2.34) \sum_{j=1}^{n} \Big(\int_H (n \wedge |\partial_j f_j(x) - 1|)\, d\mu(x)\Big)^2 \leq \sum_{j=1}^{n} \int_H (n \wedge |\partial_j f_j(x) - 1|)^2 d\mu(x)$$

$$\leq c_2(n)^{-1}\Big(\int_H \sum_{j=1}^{n} (\partial_j f_j(x) - 1)\, d\mu(x) + \int_H \log\Big(\prod_{j=1}^{n} \partial_j f_j(x)\Big) d\mu(x)\Big) \leq c_2(n)^{-1} n\tau .$$

Now, let $1 \leq i \leq n$ and put $A^i = H \cap \{x, \partial_i f_i(x) \geq n+1\}$. Then by (2.32) and (2.33)

$$(2.35) \qquad \int_{A^i} (\partial_i f_i(x) - 1)\, d\mu(x) \leq \int_{A^i} n(\partial_i f_i(x) - n)\, d\mu(x) \leq$$

$$\leq n \int_{A^i} \Big(\sum_{j=1}^{n} \partial_j f_j(x) - n\Big) d\mu(x) \leq n \int_H \Big(\sum_{j=1}^{n} \partial_j f_j(x) - n\Big) d\mu(x) \leq n^2\tau .$$

It follows from (2.34) and (2.35) that for $1 \leq i \leq n$

$$(2.36) \qquad \int_H |\partial_i f_i(x) - 1|\, d\mu(x) \leq c_3(n)\tau^{1/2} .$$

We shall use (2.36) for $i = 1$ and denote the unit vector in the direction of the first coordinate by e_1. Let $\lambda = h_H(e_1)$ (the support function of H at e_1) and assume that for some $\delta > 0$, $\lambda \geq (1+\delta)h_K(e_1)$. We shall derive an estimate on δ. We first rewrite (2.36) for $i = 1$ as

(2.37)
$$\int_0^\lambda |1 - f_1'(t)| \operatorname{Vol}_{n-1}(H_t)dt \leq c_3(n)\tau^{1/2} .$$

By the convexity and symmetry of H we have

(2.38)
$$\lambda \operatorname{Vol}_{n-1}(H_0) \geq \operatorname{Vol}_n(H)/2 = 1/2$$

(2.39)
$$\operatorname{Vol}_{n-1}(H_t) \geq \left(\frac{\lambda - t}{t}\right)^{n-1} \operatorname{Vol}_{n-1}(H_0) .$$

Hence, by (2.37), (2.38), (2.39) and the definition of δ we get for every $0 < \sigma < 1$

$$c_3(n)\tau^{1/2} \geq \sigma^{n-1} \operatorname{Vol}_{n-1}(H_0) \int_0^{(1-\sigma)\lambda} |f_1'(t) - 1| dt \geq$$
$$\geq \sigma^{n-1} \operatorname{Vol}_{n-1}(H_0)((1-\sigma)\lambda - f_1((1-\sigma)\lambda)) \geq$$
$$\geq \sigma^{n-1} \operatorname{Vol}_{n-1}(H_0)((1-\sigma)\lambda - f_1(\lambda)) \geq \sigma^{n-1}((1-\sigma) - (1+\delta)^{-1})/2 .$$

By taking $\sigma = \tau^{1/2n}$ above we deduce that $\delta \leq c_4(n)\tau^{1/2n}$. Since we may replace H by K and since we were free to choose the coordinate system in any direction, it follows that

$$(1 + c_4(n)\tau^{1/2n})^{-1} h_K(u) \leq h_H(u) \leq (1 + c_4(n)\tau^{1/2n}) h_K(u) \qquad u \in R^n$$

and this proves (2.20) with $\tau^{1/2n}$ instead of $\tau^{1/n}$. □

To conclude this paper we reproduce the proof of the following

Proposition 6 [B.L.M]. *If Z is a zonotope $\sum_{j=1}^N I_j$ for which $\delta(Z, B^n) \leq \varepsilon$ then*

(2.40)
$$N \geq c(n)\varepsilon^{-2(n-1)/(n+2)} .$$

Proof: By the assumption there is a symmetric measure $\mu = \sum_{j=1}^N a_j \delta_{\eta_j}$, $\eta_j \in S^{n-1}$ with $a_j > 0$ so that

(2.41)
$$\left| 1 - \sum_{j=1}^N a_j |\langle \xi, \eta_j \rangle| \right| \leq \varepsilon , \qquad \xi \in S^{n-1} .$$

There is clearly no loss of generality to assume that

$$(2.42) \qquad \beta_n \sum_{j=1}^{N} a_j = 1 \text{ , where } \beta_n = \int_{S^{n-1}} |\langle x, y\rangle| d\sigma_n(x) \text{ .}$$

We now apply (2.14) to this μ taking as $\mu' = \beta_n^{-1}\sigma_n$ we get for $0 < r < 1$

$$(2.43) \qquad \left\| \beta_n^{-1} - \sum_{j=1}^{N} \frac{a_j(1-r^2)}{(1+r^2-2r\langle \xi, \eta_j\rangle)^{n/2}} \right\|_{L_2(\sigma_n)}^2 \leq c_3^2(n)(1-r)^{-n-2}\varepsilon^2 \text{ .}$$

In view of (2.10) the left hand side of (2.43) equals

$$\beta_n^{-2} - 2\beta_n^{-1} \sum_{j=1}^{N} a_j + \sum_{i,j=1}^{N} \frac{a_i a_j (1-r^4)}{(1+r^4-2r^2\langle \eta_i \eta_j\rangle)^{n/2}} \text{ .}$$

Hence, using (2.42), we get

$$(2.44) \qquad \begin{aligned} c_3^2(n)\varepsilon^2(1-r)^{-n-2} &\geq \sum_{j=1}^{N} \frac{a_j^2(1-r^4)}{(1+r^4-2r^2)^{n/2}} - \beta_n^{-2} \\ &\geq \beta_n^{-2}(N^{-1}(1-r^2)^{-n+1} - 1) \text{ .} \end{aligned}$$

By taking r in (2.44) so that $(1-r^2)^{-n+1} = 2N$ we get (2.40). □

References

[B.L] J. Bourgain and J. Lindenstrauss, Nouveaux résultats sur les zonoides et les corps de projection, C.R. Acad. Sci. Paris (to appear).

[B.L.M] J. Bourgain, J. Lindenstrauss and V. Milman, Approximation of zonoids by zonotopes, Preprint IHES 1987.

[Be.Mc] U. Betke and P. McMullen, Estimating the sizes of convex bodies from projections, J. Lon. Math. Soc. 27 (1983), 525-538.

[Bol] E.D. Bolker, A class of convex bodies, Trans. Amer. Math. Soc. 145 (1969), 323-346.

[Bo.F] T. Bonnesen and W. Fenchel, Theorie der konvexen Körper, Ergebnisse der Mathematik 3, Springer-Verlag 1934.

[Ca] S. Campi, Recovering a centered convex body from the areas of its shadows, a stability estimate, Annali Di Math. (to appear).

[C.N.S] L. Caffarelli, L. Nirenberg and J. Spruck, The Dirichlet problem for non linear second order elliptic equations 1, Monge Ampère equations, Com. Pure Appl. Math. 37 (1984).

[C.Y] S.Y. Cheng and S.T. Yau, On the regularity of the solution of the n-dimensional Minkowski problem, Com. Pure Appl. Math. 29 (1976), 495-516.

[Ch] G. Choquet, Lectures on analysis, Vol. III, W.A. Benjamin, Reading, Mass. 1969.

[D1] V.I. Diskant, Bounds for convex surfaces with bounded curvature functions, Siberian Math. J. 12 (1971), 78-89.

[D2] V.I. Diskant, Bounds for the discrepancy between convex bodies in terms of the isoperimetric difference, Siberian Math. J. 13 (1972), 529-532.

[E] H.G. Eggleston, Convexity, Cambridge Tracts No. 47, 1958.

[F.L.M] T. Figiel, J. Lindenstrauss and V. Milman, The dimension of almost spherical sections of convex bodies, Acta Math. 129 (1977), 53-94.

[G] P.R. Goodey, Instability of projection bodies, Geom. Dedicata 20 (1986), 295-305.

[Gor] Y. Gordon, Some inequalities for Gaussian processes and applications, Isr. J. Math. 50 (1985), 265-289.

[J] F. John, Extremum problems with inequalities as subsidiary conditions, Courant anniversary volume, Interscience N.Y. 1949, 187-204.

[K] H. Knothe, Contributions to the theory of convex bodies, Mich. Math. J. 4 (1957), 39-52.

[Mü] C. Müller, Spherical harmonics, Springer Lecture Notes 17, 1966.

[P] C.M. Petty, Projection bodies, Proc. Colloq. on Convexity, Copenhagen, 1967, 234-241.

[Po] A.V. Pogorelov, The Minkowski multidimensional problem, John Wiley Publ. Washington D.C. 1978.

[S.W] R. Schneider and W.Weil, Zonoids and related topics, in Convexity and its Applications, Birkhauser Verlag 1983, 296-317.

[Sche] G. Schechtman, More on embedding subspaces of L_p in ℓ_p^n, Compositio Math. 61 (1987), 159-170.

[Schn1] R. Schneider, Zur einem Problem von Shephard über die Projektionen konvexer Körper, Math. Z. 101 (1967), 71-82.

[Schn2] R. Schneider, Zur optimalen Approximation konvexer Hyperflächen durch Polyeder, Math, Ann. 256 (1981), 289-301.

[Schü] C. Schütt, personal communication.

[W] W. Weil, Über die Projektionenkörper konvexer Polytope, Arch. Math. 22 (1971), 664-672.

ON A GEOMETRIC INEQUALITY

J. Bourgain — IHES
M. Meyer — University Paris VI
V. Milman — Tel Aviv University
A. Pajor — University Paris VII

Let $X = (\mathbb{R}^n, \|\cdot\|)$ be a normed space and $K = K(X) = \{x \in \mathbb{R}^n : \|x\| \leq 1\}$ its unit ball. We denote $|K| = \mathrm{Vol}\, K$, where Vol means the standard Lebesgue volume on $\mathbb{R}^n$ equipped with the standard Euclidean structure. Let $\lambda_i \in \mathbb{R}$ and $\bar{\lambda} = (\lambda_i)_{i=1}^m$. Consider the following symmetric norm $|||\cdot|||$ on $\mathbb{R}^m$,

$$|||\bar{\lambda}|||_K = \int_{x_1 \in K} \cdots \int_{x_m \in K} \Big\|\sum_1^m \lambda_i x_i\Big\| \frac{dx_1 \cdots dx_m}{|K|^m} . \tag{1.1}$$

At the Denmark Conference on Probability in Banach spaces (June, 1986), the following question was asked (by V.M.):

Is it true that, for any X (or, in other terms, for any centrally symmetric compact body $K \subset \mathbb{R}^n$ and the norm generated by this body), the symmetric norm $|||\cdot|||_K$ is, up to $\log n$, close to the standard Euclidean norm?

Moreover, is it true that under cotype condition on X this norm is equivalent to the Euclidean norm?

Precisely:

Problem. Does there exist a constant $C = C(q, C_q(X))$ depending only on $2 \leq q < \infty$ and on the cotype q constant $C_q(X)$, such that

$$\frac{1}{C}\Big(\sum_1^m \lambda_i^2\Big)^{1/2} \leq |||\bar{\lambda}|||_K \leq C\Big(\sum_1^m \lambda_i^2\Big)^{1/2} ?$$

Note that the question appears first as some kind of a generalization of the Khinchine inequality $\operatorname*{Ave}_{\varepsilon_i=\pm 1}\left|\sum_1^m \varepsilon_i \lambda_i\right| \approx \left(\sum_1^m \lambda_i^2\right)^{1/2}$, complementary to the Kahane-type generalization: in the Kahane inequality the numbers λ_i are replaced by vectors $x_i \in X$ and, as in the original Khinchine inequality, an equivalence between different L_p-norms is established:

$$\Big\|\sum_1^m \varepsilon_i x_i\Big\|_{L_p(X)} \stackrel{\text{def}}{=} \Big(\operatorname*{Ave}_{\varepsilon_i=\pm 1}\Big\|\sum_1^m \varepsilon_i x_i\Big\|^p\Big)^{1/p} < C(p)\Big\|\sum_1^m \varepsilon_i x_i\Big\|_{L_1(X)} .$$

(Kahane [Ka] originally proved $C(p) \leq C \cdot p$ and later Kwapien [Kw] improved this to be $C(p) \leq C\sqrt{p}$; see [Ta] for a modern approach.)

In our generalization, we still let λ_i be numbers but we replace the independent random variables ε_i by vector valued random variables $x_i \in (K, \mu_{\text{Vol}})$, where the probability measure $\mu_{\text{Vol}}(A) = \frac{\text{Vol}\, A}{\text{Vol}\, K}$ for any Borel subset $A \subset K$.

Note also that using a Borell lemma [Bo] in a standard way – see [MSch], Appendix III – we easily establish an equivalence for the different L_p-norms when $1 \leq p < \infty$:

$$\left(\int_{\bar{x} \in \prod_1^m K} \left|\sum_1^m \lambda_i x_i \right|^p \frac{d\bar{x}}{|K|^m} \right)^{1/p} < C \cdot p \cdot \int_{\bar{x} \in \prod_1^m K} \left|\sum_1^m \lambda_i x_i \right| \frac{d\bar{x}}{|K|^m}$$

(we abbreviate $(x_i \in K)_{i=1}^m = \bar{x}$ for $\bar{x} \in \prod_1^m K$; likewise $dx_1 \cdot \ldots \cdot dx_m = d\bar{x}$).

It is clear at the present time, that the positive solution of the problem will imply numerous consequences in Local Theory.

In this note, we show a low bound in the most interesting case of $m \geq n$ and $\lambda_i = 1$.

Observe that in the special case of X with an unconditional basis, the problem was solved by G. Pisier (unpublished). This case looks easy. Unfortunately, all consequences of the positive solution of the problem were well known in this case.

In what follows, we also consider a more general expression. Our norm $\|\cdot\|$ is defined by a fixed body K (we emphasize this sometimes writing $\|\cdot\|_K$) but we integrate over different bodies (central symmetric) $\{T_i \subset \mathbb{R}^n\}_{i=1}^m$:

$$J(K; T_1, \ldots, T_m) = \int_{x_1 \in T_1} \cdots \int_{x_m \in T_m} \left\|\sum_1^m x_i\right\|_K \frac{dx_1 \cdots dx_m}{\prod |T_i|} .$$

In the case $T_i = T$ for all $i = 1, \ldots, m$, we write simply $J(K; T, m)$.

Theorem 1.1. *There exists a universal constant $c > 0$ such that for $m \geq n$*

$$J(K; T, m) \geq c\sqrt{m}|T|^{1/n}/|K|^{1/n} .$$

In a general case (for $m = n$)

$$J(K; T_1, \ldots, T_n) \geq c\sqrt{n}\left(\prod_1^n |T_i|\right)^{1/n^2}/|K|^{1/n} .$$

Corollary 1.2.

$$|||\bar{\lambda}|||_{K;T} = \int_{\bar{x} \in \prod T} \left\|\sum_1^n \lambda_i x_i\right\|_K \frac{d\bar{x}}{|T|^n} \geq c\sqrt{n}\left(\prod |\lambda_i|\right)^{1/n} \left(\frac{|T|}{|K|}\right)^{1/n} .$$

(Define $T_i = |\lambda_i| T$ and use Theorem 1.1.)

We prove Theorem 1.1 in Section 2 using Steiner symmetrization in a situation similar to that used by Blaschke [Bl].

Define for $p \geq 0$ and $m \geq n$

$$I_p(T_1, \dots, T_m) = \int_{T_1} \cdots \int_{T_m} \Big(\mathrm{Vol}\Big(\sum_{k=1}^{k=m} [0, x_k]\Big)\Big)^p dx_1, \dots, dx_m ,$$

where $\sum$ means here the Minkowski sum in $\mathbb{R}^n$ of the segments $[0, x_k] = \{\lambda x_k , 0 \leq \lambda \leq 1\}$.

Theorem 1.3. *Let T_i be convex compact bodies in $\mathbb{R}^n$ and let D_i be Euclidean balls with, respectively, the same volume as T_i. Then for $m \geq n$ and $p \geq 0$*

$$I_p(T_1, \dots, T_m) \geq I_p(D_1, \dots, D_m) .$$

Note, that for $p \geq 1$ and $m = n$, this is a known result proved by Buseman [Bu]; in that case

$$\left|\sum_{k=1}^{k=n} [0, x_k]\right| = |\det(x_1, \dots, x_n)| \,.\,.$$

However, we have to use Theorem 1.3 for $p = 1/n$ and $m \geq n$.

To describe another approach to Theorem 1.1, developed in section 3, we need additional notations. We may assume that our body T stays in the isotropic position with respect to the standard Euclidean ball D of $\mathbb{R}^n$ (see survey [MP] in the next volume of GAFA Seminar). It means that for every $y \in S^{n-1} = \partial D$ (i.e., the euclidean norm $|y| = 1$):

$$\int_{x \in T} |(x, y)|^2 \frac{dx}{|T|} = \text{Const.} \qquad \text{(independent of } y) .$$

Then, considering $t > 0$ such that $|tT| = 1$, we denote

$$L_T^2 = \int_{x \in tT} |(x, y)|^2 dx \qquad \text{for some (any) } y \in S^{n-1} .$$

It is known that $L_T \geq c > 0$ for some universal constant c (independent of body T or dimension n). However, whether there exists a universal constant C such that $L_T \leq C$ is an open problem. (For a complete discussion of this subject see [MP].) We also use the following standard notation (see [MSch] or any other survey on Local Theory for a few facts on Local Theory which we use below)

$$M_K = \int_{S^{n-1}} \|x\|_K \, d\mu(x)$$

where μ is the probabilistic rotation invariant measure on S^{n-1}. Note *that we consider the average M_K with respect to the Euclidean structure which is the isotropic one for the body T.*

Theorem 1.4. *There exist universal constants $c_i > 0$, $i = 1,2$, such that for any $m \geq n$*

$$\int\limits_{\bar{x}\in\Gamma T} \int_\Omega \Big\|\sum_1^m g_i(w) x_i\Big\|_K dw \frac{d\bar{x}}{|T|^m} \geq c_1\sqrt{m}\cdot n L_T \cdot M_K |T|^{1/n}$$

$$\geq c_2\sqrt{m}\frac{|T|^{1/n}}{|K|^{1/n}},$$

where $\{g_i(w)\}_{i=1}^m$ are independent normalized Gaussian random variables.

The inequality $\int\limits_{\bar{x}\subset^- T} \int_\Omega \|\sum_1^m g_i(w)x_i\|_K dw \frac{d\bar{x}}{|T|^m} \geq c_2\sqrt{m}\frac{|T|^{1/n}}{|K|^{1/n}}$, is also a consequence of Theorem 1.1, since the Gaussian averages dominate the Rademacher average.

To pass from the above inequality to Theorem 1.1, we have (in this approach) to use a cotype condition on space $X = (\mathbb{R}^n, \|\cdot\|_K)$. Then Gaussian and Rademacher averages become equivalent up to $C(q, C_q(X))$ depending on $q \geq 2$ and on the cotype q constant $C_q(X)$ of space X (see [MaPi])

$$C(q, C_q(X)) \cdot \operatorname*{Ave}_{\varepsilon_i=\pm 1} \Big\|\sum \varepsilon_i x_i\Big\|_K \geq \int_\Omega \Big\|\sum g_i(w)\Big\|_K dw .$$

By symmetry of T, we now transform the integral from the left side in Theorem 1.4 to $J(K;T;m)$.

We use below the standard notation $K^\circ = \{x \in \mathbb{R}^n : (x,y) \leq 1 \text{ for } \forall y \in K\}$ for the polar of K.

2. Approach through symmetrization

We start by recalling the following inequalities

$$\frac{1}{M_K} \leq \text{v.r.}\, K = \left(\frac{\operatorname{Vol} K}{\operatorname{Vol} D}\right)^{1/n} \leq M_{K^\circ}. \tag{2.1}$$

The right side is the Urysohn inequality and the left side is an immediate consequence of the Hölder inequality applied to

$$\text{v.r.}\, K = \left(\int\limits_{S^{n-1}} \frac{d\mu(x)}{\|x\|_K^n}\right)^{1/n} \geq \int\limits_{S^{n-1}} \frac{d\mu(x)}{\|x\|_K} \geq \left(\int\limits_{S^{n-1}} \|x\| d\mu(x)\right)^{-1}.$$

For our purpose we have to strengthen (2.1), replacing $M(K)$, which corresponds to a gaussian average, by a suitable Rademacher average.

Let $e = \{e_i\}_1^n$ be an orthonormal basis. Define

$$r_e(K) = \operatorname*{Ave}_{\varepsilon_i = \pm 1} \left\| \sum_1^n \varepsilon_i e_i \right\|_K .$$

Then there exist universal constants c_1 and $c_2 > 0$ such that

$$c_2 \frac{\sqrt{n}}{r_e(K)} \leq \text{v.r.}\, K \leq c_1 \frac{r_e(K^\circ)}{\sqrt{n}} . \tag{2.2}$$

The right side inequality was used in [P] and the left one follows from the right side (applied to K°) and the inverse Santalo inequality (see[BM]). However, this is an unnecessarily difficult path and we will give a direct argument. By the way, we prove a slightly stronger version.

Note that the left side of (2.2) is equivalent to the following inequality: for any $x_1, \ldots, x_n$ in $\mathbb{R}^n$

$$|K|^{1/n} \operatorname*{Ave}_{\varepsilon_i = \pm 1} \left\| \sum_1^n \varepsilon_i x_i \right\|_K \geq c |\det(x_1, \ldots, x_n)|^{1/n} \tag{2.3}$$

for some universal constant $c > 0$. (We replace $e = (e_i)$ by any $\bar{x} = (x_i \in \mathbb{R}^n)_{i=1}^n$, emphasizing that (2.2) is satisfied for any Euclidean structure: any linearly independent set $\{x_1, \ldots, x_n\}$ can be considered an orthonormal basis for some euclidean norm.)

Proposition 2.1. *There exists a number $c > 0$ such that for any $\{x_1, \ldots, x_m\} \subset \mathbb{R}^n$, $m \geq n$,*

$$\operatorname*{Ave}_{\varepsilon_i = \pm 1} \left\| \sum_1^m \varepsilon_i x_i \right\|_K \geq c \frac{\sqrt{n}}{\sqrt{m}} \left(\frac{\left| \sum_1^m [-x_i, x_i] \right|}{|K|} \right)^{1/n} . \tag{2.4}$$

Remark 2.2. Clearly, $\left| \sum_1^m [-x_i, x_i] \right| = 2^n \sum_{\substack{I \subset \{1, \ldots, m\} \\ |I| = n}} |\det\{x_i\}_{i \in I}|$. So, Proposition 2.1, for $m = n$ implies (2.3).

To prove Proposition 2.1, we use a result of [CP] on entropy numbers. Recall that if E and F are finite dimensional normed spaces with the unit balls B_E and B_F and if $v : E \to F$ is a linear operator, the kth entropy number $e_k(v)$ of v is defined by: $e_k(v) = \inf \{ r > 0 ,\ \exists y_1, \ldots, y_k$ in F such that $u(B_E) \subset \bigcup_{j=1}^{r=k} (y_j + r B_F) \}$.

Proof of Proposition 2.1.

Let $f_1, \dots, f_m$ be the natural basis of ℓ_∞^m and define a linear operator $u : \ell_\infty^m \to X$ by $uf_i = x_i$, $1 \le i \le m$. Let $i : \ell_\infty^m \to \ell_2^m$ be the identity operator and $\tilde{u} : \ell_2^m \to X$, $\tilde{u} = ui^{-1}$. By an estimate due to [CP], we have

$$e_n(\tilde{u}) \le \frac{c}{\sqrt{n}} \operatorname*{Ave}_{\varepsilon_i = \pm 1} \left\| \sum_1^m \varepsilon_i x_i \right\| .$$

Denoting by B_∞^m the unit ball of ℓ_∞^m, we also have

$$\left(\frac{|u(B_\infty^m)|}{|K|}\right)^{1/n} = \left(\frac{|u(B_\infty^m)|}{|B_x|}\right)^{1/n} \le 2e_n(u)$$

$$e_n(u) \le \|i\| e_n(\tilde{u}) = \sqrt{m} e_n(\tilde{u})$$

$$|u(B_\infty^m)| = \left|\sum_{i=1}^{i=m} [-x_i, x_i]\right| = 2^n \left|\sum_{i=1}^{i=m} [0, x_i]\right|$$

The result follows from these relations. □

We will also give an alternative proof of Proposition 2.1 in a slightly weaker form with n/m instead of $\sqrt{n/m}$. However, the next proof is independent of the result on entropy numbers of operators and can be more convenient for geometers.

Lemma 2.3. *Let* $\| \;\|$ *be a norm on* $\mathbb{R}^n$ *and* $f : \mathbb{R}^n \to \mathbb{R}_+$ *be a bounded function. Then* $F_p(f) = \left(\int_{\mathbb{R}^n} \|x\|^p f(x) dx / \|f\|_\infty \int_{\|x\| \le 1} \|x\|^p dx \right)^{1/n+p}$ *is increasing on* $\mathbb{R}_+$.

Proof: Let $s \ge 0$, $p \ge q \ge 0$; then:

$$\int_{\mathbb{R}^n} \|x\|^p f(x) dx \ge \int_{\|x\| \le s} (\|x\|^p - s^{p-q} \|x\|^q) f(x) dx + s^{p-q} \int \|x\|^q f(x) dx$$

$$\ge \|f\|_\infty s^{n+p} \int_{\|x\| \le 1} (\|x\|^p - \|x\|^q) dx + s^{p-q} \int \|x\|^q f(x) dx .$$

Choosing $s = F_q(f)$, we get the lemma.

Lemma 2.4. *There exists $c > 0$ such that, if X, Y are respectively m and n dimensional normed spaces with the unit balls B_X and B_Y, $m \ge n$, and if $u : X \to Y$ is a rank-n linear operator, we have*

$$V(u) = \frac{1}{|B_X|} \cdot \int_{B_X} \|u(x)\|_Y dx \ge c \frac{n}{m} \left(\frac{|u(B_X)|}{|B_Y|}\right)^{1/n}$$

where $|\quad|$ denotes volume on $\mathbb{R}^n$ or $\mathbb{R}^m$.

Proof: Let $E = \operatorname{Ker} u$ and $F = \{x, \langle x,y\rangle = 0 \text{ for any } y \in E\}$; then F is n-dimensional and $\tilde{u} = u|_F : F \to Y$ is an isomorphism. Let $C = \tilde{u}^{-1}(B_Y)$; C is the unit ball in F, for the associated norm $\|\quad\|_C$. For $x \in F$ let $f(x) = |B_X \cap (x+E)|$; we have $\|f\|_\infty = f(o)$ and $\int_F f(x)dx = |B_X|$. Moreover, if P_F denotes the orthogonal projection onto F, by [RSh]

$$\binom{m}{n}|B_X| \geq f(o)|P_F B_X| = f(o)\tilde{u}^{-1}(u(B_X)) . \tag{2.4}$$

By Fubini theorem and Lemma 2.3 applied to $p = 1$ and $q = 0$,

$$V(u) = \frac{1}{|B_X|}\int_F \|x\|_C f(x)dx \geq \frac{n}{n+1}\left(\frac{|B_X|}{\|f\|_\infty |C|}\right)^{1/n} .$$

From (2.4), we get:

$$V(u) \geq \frac{n}{n+1}\binom{m}{n}^{-1/n}\left(\frac{|\tilde{u}^{-1}(u(B_X))|}{|\tilde{u}^{-1}(B_Y)|}\right)^{1/n} \geq c\frac{n}{m}\left(\frac{|u(B_X)|}{|B_Y|}\right)^{1/n} . \qquad \square$$

Now the proof of Proposition 2.1 (with n/m instead of $\sqrt{n/m}$) follows easily. It sufficies to apply Lemma 2.4 to an operator $u : \ell_\infty^m \to X = (\mathbb{R}^n, \|\cdot\|_K)$ defined by $u(e_k) = x_k$, $1 \leq k \leq m$ ($\{e_k\}_1^m$ is, as usual, the natural basis of ℓ_∞^m) and to observe that $V(u) \leq \operatorname{Ave}_{\epsilon_i = \pm 1} \|\sum_i^m \epsilon_i x_i\|$.

Remark. In Lemma 2.4, one can give better estimates, depending on X. For instance, if $X = \ell_2^m$, one can replace $\frac{n}{m}$ by $\sqrt{\frac{n}{m}}$ because $\left(\frac{|B_X|}{f(o)|P_F B_X|}\right)^{1/n} = \left(\frac{v_m}{v_{m-n}v_n}\right)^{1/n} \geq c\left(\frac{n}{m}\right)^{1/2}$ where v_k is the volume of the Euclidean ball in $\mathbb{R}^k$.

We are now ready to prove Theorem 1.3.

Proof of Theorem 1.3. Note first that for every linear operator $A : \mathbb{R}^n \to \mathbb{R}^n$, $\det A = 1$, we have $I_p(A(T_1), \ldots, A(T_m)) = I_p(T_1, \ldots, T_m)$. It is well known that, for given bodies $\{T_i\}_{i=1}^m$, there exists a sequence S_p of Steiner symmetrizations, such that, for every i, $1 \leq i \leq m$, the sequence $(S_p \circ S_{p-1} \circ \ldots \circ S_1)(T_i)$ converge to the Euclidean ball D_i, $|D_i| = |T_i|$. Therefore, it suffices to prove that $I_p(T_1, \ldots, T_m) \geq I_p(\tilde{T}_1, \ldots, \tilde{T}_m)$, where $\tilde{T}_1, \ldots, \tilde{T}_m$ are obtained from $T_1, \ldots, T_m$ by Steiner symmetrization about $\{x_n = 0\}$. Let PT_k, $1 \leq k \leq m$ be the orthogonal projections of T_k onto $\{x_n = 0\}$; then

$$T_k = \{(Y_k, y_k) \ , \ Y_k \in PT_k \ , \ (u_k - r_k)(Y_k) \leq y_k < (u_k + r_k)(Y_k)\}$$

for some functions $u_k : PT_k \to \mathbb{R}$ and $r_k : PT_k \to \mathbb{R}_+$. Denote $I = \prod_{k=1}^{k=m} [-r_k, r_k](Y_k)$ and $u = \left(u_k(Y_k)\right)_{k=1}^{m} \in \mathbb{R}^m$. By Remark 2.2 we have

$$I_p(T_1,\ldots,T_n) = \int_{PT_1} \cdots \int_{PT_m} \left(\int_{I+u} \Big(\sum_{\substack{S\subset\{1,\ldots,m\}\\ |S|=n}} |\langle x_S, \varphi(Y_S)\rangle| \Big)^p dx \right) dY$$

where for $x = (x_1,\ldots,x_m) \in \mathbb{R}^m$ and $Y = (Y_1,\ldots,Y_m) \in (\mathbb{R}^{n-1})^m$, $x_S = (x_i)_{i\in S} \in \mathbb{R}^n$ and $\varphi(Y_S)$ is the exterior product (in $\mathbb{R}^n$) of the lines $(Y_i)_{i\in S}$ of the matrix of Y. Similarly

$$I_p(\widetilde{T}_1,\ldots,\widetilde{T}_m) = \int_{PT_1} \cdots \int_{PT_m} \left(\int_{I} \Big(\sum_{\substack{S\subset\{1,\ldots,m\}\\ |S|=n}} |\langle x_S, \varphi(Y_S)\rangle| \Big)^p dx \right) dY .$$

Using Lemma 2.5 below it is easy to conclude.

Lemma 2.5. *Let T a convex body in $\mathbb{R}^m$, symmetric with respect to 0, $\varphi : \mathbb{R}^m \to \mathbb{R}_+$ be convex and even and $g : \mathbb{R}_+ \to \mathbb{R}_+$ be non decreasing. Then for any $u \in \mathbb{R}^m$*

$$\int_{T+u} g(\varphi(x))\,dx \geq \int_{T} g(\varphi(x))\,dx .$$

Proof: Note that $\int_{T+u} g(\varphi(x))dx = \int \theta(u,t) g'(t)dt$ where $\theta(u,t) = |(T+u)\cap\{\varphi \geq t\}|$. Also $\theta(u,t) = |T| - |(T+u)\cap\{\varphi < t\}|$. By the Brunn-Minkowski theorem, a function $f_t(u) = |(T+u)\cap\{\varphi < t\}|^{1/m}$, for any fixed $t \geq 0$, is concave (since $\{\varphi < t\}$ is convex) and even (since φ is even and T is symmetric). It follows that $\max_u f_t(u) = f_t(0)$ for every $t \geq 0$ and therefore $\min_u \theta(u,t) = \theta(0,t)$ for every $t \geq 0$. □

Now, to derive Theorem 1.1, we integrate formula (2.4) and apply Theorem 1.3 for $p = 1/n$. It remains to compute an integral which we present in the following lemma:

Lemma 2.6. *Let D be the Euclidean ball in $\mathbb{R}^n$ and define for every $p \geq 0$ and $m \geq n$,*

$$J(p,n,m) = \frac{1}{|D|^m} \int_{D_m} \left| \sum_1^m [0, x_i] \right| dx_1 \ldots dx_m .$$

Then

$$J(p,n,n) = \frac{n!}{(n+p)\cdots(1+p)} \left(\frac{\Gamma\left(1+\frac{n}{2}\right)}{\Gamma\left(1+\frac{n+p}{2}\right)} \right)^n \prod_{k=1}^{k=n} \frac{\Gamma\left(\frac{k+p}{2}+1\right)}{\Gamma\left(\frac{k}{2}+1\right)},$$

and for $p = 1/n$, there exists some constants $d > c > 0$ such that, for any $m \geq n$,

$$c\frac{m}{n} \leq J\left(\frac{1}{n}, n, m\right) \leq d\frac{m}{n} .$$

Proof: For the first formula, use induction on n and observe that

$$J(p,n,n) = \frac{1}{|D|^n}\left(\int_D |x|^p dx\right)\int_{D^{n-1}} |\det(\tilde{x}_2,\ldots,\tilde{x}_n)|^p dx_2\ldots dx_n$$

where $\tilde{x}_B$ denotes the projection of x_k on $\mathbb{R}^{n-1}$. We get

$$J(p,n,n) = \frac{n}{n+p}\cdot\left(\frac{\Gamma\left(1+\frac{n}{2}\right)}{\Gamma\left(1+\frac{n-1}{2}\right)}\frac{\Gamma\left(1+\frac{n+p-1}{2}\right)}{\Gamma\left(1+\frac{n+p}{2}\right)}\right)^{n-1}\cdot J(p,n-1,n-1)$$

It is clear that for any $p>0$, $J(p,n,n)\leq 1$ and an easy computation with Stirling formula gives a constant $c>0$ such that

$$J(1/n,n,n)\geq c \quad \text{for any} \quad n\,.$$

Using Remark 2.2, we get, by Hadamard inequality for the upper bound and Hölder inequality for the lower bound

$$\binom{m}{n}^p J(p,n,n)\leq J(p,n,m)\leq\binom{m}{n}^p \qquad 0\leq p\leq 1\,,\ n\leq m\,.$$

Since $\binom{m}{n}^{1/n}\approx\frac{m}{n}$, this gives the inequalities for $p=\frac{1}{n}$.

3. Proof of Theorem 1.4

We use notations introduced in Section 1. We may also assume the normalization $\operatorname{Vol} T = 1$ and that the canonical Euclidean ball D of $\mathbb{R}^n$ is in the isotropic position with respect to T. Fix $\xi\in\mathbb{R}^n$ of the Euclidean norm $|\xi|=1$. Define $\chi_T(x)$ to be the characteristic function of body T and consider a function

$$f(x) = |(x,\xi)|\chi_T(x)$$

where (x,ξ) is the inner product of x and ξ in the canonical Euclidean structure of $\mathbb{R}^n$. It follows from [Bo] (see [MSch], App. III) or [GrM]) that for some universal C and $p\geq 1$

$$\|f(x)\|_{L_p(\mathbb{R}^n)}\leq C\cdot p\cdot\|f(x)\|_{L_1(\mathbb{R}^n)}\,.$$

This means that the Orlicz norm L_{ψ_1} (for $\psi_1(t)=e^t-1$) of $f(x)$ is equivalent to the L_1 norm (see [BLM], Section 6, for definitions and an account of the results which we use next). Moreover, $\|f(x)\|_{L_2}=L_T$ (see Sec 1) independent of ξ and $\|f\|_{L_1}\simeq\|f\|_{L_2}$. This information allows us to use Lemma 6.1 from [BLM] (we use this Lemma in the same spirit as it is used in Section 6 of [BLM]). As a consequence, we have

Lemma 3.1. *There exist universal constants c, c_1 and $c_2 > 0$ such that for any integer $m > cn$ there exists a set $\mathcal{A} \subset \prod_1^m T \stackrel{\text{def}}{\equiv} T^m \subset \mathbb{R}^{n \cdot m}$ such that for any $\bar{x} = (x_1 \in T, \dots, x_m \in T) \in \mathcal{A}$ and any $\xi \in \mathbb{R}^n$*

$$\frac{1}{c_1} L_T |\xi| \le \frac{1}{m} \sum_{i=1}^m |(x_i, \xi)| \le c_1 L_T |\xi| \tag{3.1}$$

(briefly, we write $\frac{1}{m}\sum_1^m |(x_i \xi)| \sim L_T |\xi|$). Also $\mathcal{A}$ has "almost" full measure: the $n \cdot m$- dimensional volume

$$\mathrm{Vol}(\mathcal{A}) > 1 - \exp(-c_2 m)$$

(note that $\mathrm{Vol}\, T^m = 1$; we briefly call $\mathcal{A}$ a "random" subset of T^m).

It follows from (3.1) that

$$\Big(\sum_1^m |(x_i, \xi)|^2\Big)^{1/2} \ge \frac{1}{c_1} \sqrt{m} L_T |\xi| = \frac{1}{c_1} \sqrt{m} L_T \sqrt{\sum_{j=1}^n |(\xi, e_j)|^2} \tag{3.2}$$

for $\bar{x} = (x_i \in T) \in \mathcal{A}$ and any $\xi \in \mathbb{R}^n$.

Note that (3.2) compares two ℓ_2-metric on $\mathbb{R}^n$ and therefore, it implies, by the Slepian comparison principle ([Sl]), the following inequality for the expectations of supremum of two Gaussian processes (defined by $\xi \in K^\circ$)

$$\int \Big\| \sum_1^m g_i(w) x_i \Big\|_K dw \ge \frac{1}{c_1} \sqrt{m} L_T \int \Big\| \sum_1^n g_i(w) e_i \Big\|_K dw1 , \tag{3.3}$$

where $\{g_i(w)\}$ are the canonical independent Gaussian variables. It is again well known in Local Theory (see, e.g., [MSch], Ch.9.5) that $\int \| \sum_1^n g_i(w) e_i \|_K dw \simeq \sqrt{n} M_K$ (as defined in sec. 1). Therefore, we have from (3.3) (for some $c > 0$)

$$\int \Big\| \sum_1^m g_i(w) x_i \Big\|_K dw \ge c\sqrt{mn} L_T \cdot M_K .$$

Recall now that the last inequality is satisfied for any $\bar{x} = (x_i) \in \mathcal{A}$ (i.e., for a "random" set of vectors from T). So, integrating it by $\bar{x} \in T^m$, we derive Theorem 1.4 in the case $|T| = 1$. By a homotetical renorming, we receive the first inequality in Theorem 1.4. To derive the second step, use the left side of (2.1) noting that $|\,\mathrm{Vol}\, D|^{1/n} \sim 1/\sqrt{n}$:

$$\frac{1}{M_K} \le \text{v.r.}\, K \simeq \sqrt{n} |K|^{1/n} .$$

References

[BLM] J. Bourgain, J. Lindenstrauss and V. Milman, Approximation of zonoids by zonotopes, Preprint I.H.E.S., September 1987, 62pp.

[BM] J. Bourgain and V.D. Milman, Sections euclidiennes et volume des corps symetriques convexes dans $\mathbb{R}^n$, C.R. Acad. Sc. Paris, t.300 Serie 1, N. 13 (1985) 435-438. (Also: New volume ratio properties for convex symmetric bodies in $\mathbb{R}^n$, Invent. Math. 88 (1987), 319-340.

[Bl] W. Blaschke, Vorlesungen über Differentialgeometrie II, Springer, Berlin, 1923.

[Bo] C. Borell, The Brunn-Minsowski inequality in Gauss spaces, Inventiones Math. 30 (1975), 207-216.

[Bu] H. Buseman, Volume in terms of concurrent cross sections, Pacific J. Math. 3 (1953), 1-12.

[CP] B. Carl and A. Pajor, Gelfand numbers of operators with values in Hilbert spaces (preprint).

[GrM] M. Gromov and V.D. Milman, Brunn theorem and a concentration of volume of convex bodies, GAFA Seminar Notes, Israel 1983-1984.

[Ka] J.P. Kahane, Series of Random Functions, Heath Math. Monographs, Lexington, Mass, Heath & Co., 1968.

[Kw] S. Kwapień, Isomorphic characterizations of inner product spaces by orthogonal series with vector valued coefficients, Studia Math. 44 (1972) 583-595.

[MaPi] B. Maurey and G. Pisier, Series de variables aléatoires vectoriėlles indépendantes et propriétés géométriques des espaces de Banach, Studia Math. 58 (1976) 45-90.

[MP] V.D. Milman and A. Pajor, Isotropic position, inertia ellipsoids and zonoid of the unit ball of a normed n-dimensional space, GAFA Israel Seminar, Springer-Verlag, Lecture Notes in Mathematics.

[MSch] V.D. Milman and G. Schechtman, Asymptotic theory of finite dimensional normed spaces, Springer Lecture Notes #1200 (1986).

[P] A. Pajor, Sous-espaces ℓ_1^n des espaces de Banach, Hermann, Editeurs des Sciences et des Arts, Paris (1985).

[RSh] C.A. Rogers, G.C. Shephard, The difference body of a convex body, Arch. Math. 8 (1957), 220-233.

[Sl] D. Slepian, The one-sided barrier problem for Gaussian noise, Bell System Tech. J. 41 (1962), 463-501.

[Ta] M. Talagrand, An isoperimetric theorem on the cube and the Khinchine-Kahane inequalities, Preprint.

A FEW OBSERVATIONS ON THE CONNECTIONS BETWEEN LOCAL THEORY AND SOME OTHER FIELDS

V.D. Milman

The Raymond and Beverley Sackler
Faculty of Exact Sciences
Tel Aviv University
Israel

1. Waring type theorems and isometrically euclidean sections of some convex bodies.

The remarks of this section were observed jointly by M. Gromov and myself. We will see that the classical approach of Hurwitz [Hu] – Hilbert [Hi] – Schmidt [S] to the Waring problem is connected to an isometric version of Dvoretzky's Theorem for ℓ_p^n, p is an even integer; and, moreover, it suggests a problem which we would call the arithmetic version of Dvoretzky's Theorem.

Let $N = N(k; d)$ be the dimension of the space of all homogeneous forms of degree $2k$ of d variables (i.e., N is equal to the number of representation of $2k$ as a sum of at most d positive integers).

Consider a $2k$-form of x_i, $i = 1, \dots, d$, and let $\bar{\lambda} = (\lambda_1, \dots, \lambda_d) \in S^{d-1}$, i.e., $\sum \lambda_i^2 = 1$,

$$g(x_1, \dots, x_d) = \int_{\bar{\lambda} \in S^{d-1}} \Big(\sum_{i=1}^{d} \lambda_i x_i\Big)^{2k} d\mu(\bar{\lambda})$$

where μ is the probability rotation invariant measure on S^{d-1}. Clearly g depends only on $|x| = \big(\sum_1^d x_i^2\big)^{1/2}$, i.e., $g(\bar{x}) = f(|\bar{x}|)$, and g is a homogeneous $2k$-form. Therefore

$$g(x_1, \dots, x_d) = c\Big(\sum_1^d x_i^2\Big)^k .$$

By the Caratheodory principle, the center of gravity g in an N-dimensional space can be represented as an average of $N+1$ forms, i.e.,

$$\Big(\sum_1^d x_i^2\Big)^k = \sum_{j=0}^{N} \Big(\sum_{i=1}^{d} \lambda_{ij} x_i\Big)^{2k} . \tag{1.1}$$

This equality defines, of course, an isometric embedding of ℓ_2^d in ℓ_p^{N+1} for $p = 2k$. Indeed, our embedding is the following (linear) map $\varphi : \ell_2^d \to \ell_{2k}^{N+1}$; for $\bar{x} = (x_i)_1^d \in \ell_2^d$, $\varphi(\bar{x}) = \bar{t} = (t_j = \sum_1^d \lambda_{ij} x_i)_{j=0}^N \subset \ell_{2k}^{N+1}$. The isometry property of the embedding φ is the equality (1.1).

Because $N \simeq d^p$ (for a fixed $p = 2k$ and $d \to \infty$), we see that an isometric embedding of ℓ_2^d into ℓ_p^N $(p = 2k)$ can be realized for $d \sim N^{1/p}$.

Theorem. *The space ℓ_p^n for $p = 2k$ being an even integer contains an isometric copy of ℓ_2^d for $d \sim n^{1/p}$.*

Note that in [FLM] an ε-embedding of ℓ_2^d in ℓ_p^N was realized for $d \sim f(\varepsilon)N^{2/p}$ (for any $p > 2$) and it is known to be the best possible [BDGJN]. Also T. Figiel informed me that he observed that ℓ_2^d can be embedded isometrically in ℓ_p^N for $p = 2k$ and N large enough, but without any estimate on $d(N;p)$.

In connection with the Waring problem, we note in passing that the presentation (1.1) (for the case $d = 5$) of a power of a quadratic form, as some sum of powers of linear forms, is essential for the problem of Waring, as was realized by Hurwitz [Hu]. He has proved it for $k = 2$ and 4. It was proved by Hilbert [Hi] in the general case and was drastically simplified by Schmidt [S]. We have demonstrated Schmidt's approach. I would also like to present the E. Lucas [L]-Hurwitz identity which proves (1.1) for $d = 4$ and $k = 2$:

$$6\left(\sum_{i=1}^{4} x_i^2\right)^2 = \sum_{1\le i<j\le 4} (x_i + x_j)^4 + \sum_{1\le i<j\le 4} (x_i - x_j)^4 .$$

It gives the concrete embedding of 4-dimensional euclidean space $\ell_2^4 \hookrightarrow \ell_4^{12}$ (in 12-dimensional ℓ_p, $p = 4$). Above discussion raises the following question which I would describe as an arithmetic version of Dvoretzky's theorem.

Problem. For any integers d and p, does there exist $N = N(d;p)$ such that for any homogeneous form $F(x_1,\dots,x_N)$ of N variables and degree p, we may find a d-dimensional subspace E of $\mathbb{R}^N$ such that F, restricted on E, is the $p/2$ - power of a quadratic form?

Note that, for p odd, such a subspace E does exist, as was shown by Birch [B] and it means, of course, that the restriction of F on subspace E is zero. So, the problem only remains for even forms.

In the next three observations we will show applications of Topology to the study of convex bodies in $\mathbb{R}^n$.

2. Rattray's Theorem

This theorem [Ra] states that for any continuous antipodal map $\varphi : S^{n-1} \to S^{n-1}$ (i.e., $\varphi(-x) = -\varphi(x)$) there exists an orthonormal basis $\bar{e} = (e_1,\dots,e_n)$ such that $\varphi(\bar{e}) = (\varphi(e_1),\dots,\varphi(e_n))$ is also an orthonormal basis.

An interpretation of the theorem is nice and obvious:

Corollary 2.1. *Let $K \subset I\!R^n$ be a symmetric convex compact body. Then there exists an n-dimensional rectangular cylinder C_n circumscribed around K (i.e., $K \subset C_n$) and which touches K at least at some orthogonal basis $(x_1, \dots, x_n) \subset K$.*

Proof: Let K_ε be a symmetric compact body which is an ε-approximation of K in the Hausdorff sense such that K_ε is uniformly smooth and convex. Indeed, we need only that the support map $\psi_\varepsilon(x) : \partial K_\varepsilon \to I\!R^n$ defined on the boundary ∂K_ε by $(\psi_\varepsilon(x), x) = \max\{(y,x) \mid y \in K_\varepsilon\}$ will be uniquely defined and continuous. Let $\|\cdot\|_\varepsilon$ be the norm defined by the unit ball K_ε. Then projecting this map onto the unit euclidean sphere, i.e., constructing the map

$$\varphi_\varepsilon(y) = \psi_\varepsilon\left(\frac{y}{\|y\|_\varepsilon}\right) \Big/ |\psi_\varepsilon(y/\|y\|_\varepsilon)| : S^{n-1} \longrightarrow S^{n-1} ,$$

we are in a position to use Rattray's theorem. So, we find orthogonal vectors $x_1^\varepsilon, \dots, x_n^\varepsilon$, $\|x_i^\varepsilon\|_\varepsilon = 1$, such that their images $\{y_i^\varepsilon = \psi_\varepsilon(x_i^\varepsilon)\}_1^n$ under the map ψ_ε are also orthogonal. Send $\varepsilon \to 0$ and use an obvious compactness argument to receive $\{x_i\}_1^n \subset \partial K \cap E$ such that there are support functionals f_i to K at x_i defined by $f_i(x) = (y_i, x)$ and $\{y_i\}_1^n$ are orthogonal. Clearly

$$C_n = \{x \in I\!R^n \mid |(y_i, x)| < |y_i| \ , \ i = 1, \dots, n\}$$

is a circumscribed rectangular cylinder which touches K (at least) at the orthogonal points $\{x_i\}_1^n \subset K$.

3. The best dependence on ε in the general form of Dvoretzky's theorem and the Knaster hypothesis.

The following version of *Dvoretzky's Theorem* [Dv] is well known (see [MSch]): For every $\varepsilon > 0$ there exists an $f(\varepsilon) > 0$ such that any symmetric convex compact set $K \subset I\!R^n$ has a centrally symmetric k-dimensional section $K \cap E$ (i.e., there exists a k-dimensional subspace E) which is ε-close (in the Hausdorff metric induced by the euclidean distance on $I\!R^n$) to a k-dimensional euclidean ball if

$$k \simeq f(\varepsilon) \log n . \tag{3.1}$$

(Indeed, we found it more convenient to use later the multiplicative distance.) The logarithmic dependence on n cannot be improved for a general body K. However, the right dependence on ε is not clear. It was shown in [M] that if the (multiplicative) distance $d(K, D) = \inf\{a \cdot b \mid$

$K \subset aD \subset abK\} < C$ (D is the euclidean ball) then $k \simeq \frac{\varepsilon^2}{\log 1/\varepsilon} n/C^2$ and Y. Gordon [Go] (very recently) took out the logarithmic term: $k \sim \varepsilon^2 n/C^2$ (using a different approach, by the way). This dependence is already the best possible (Figiel [F]). However, the dependence on n in this case is, as we see, much better than in a general case. So, although $f(\varepsilon) \gtrsim \varepsilon^2$ follows from the above remarks, it is not clear that $\log n$ in (3.1) does not allow us to improve dependence on ε.

As we will see next, some old topological hypothesis of Knaster (1947) which extends the previously known hypothesis of Rademacher solved by Kakutani [Ka] in 1942, would imply a much better estimate on $f(\varepsilon)$ in (3.1).

Problem. (Knaster [Kn], Problem 4). Let $f : S^n \to I\!R$ be any continuous function and let $(x_1, \dots, x_{n+1}) \subset S^n$ be any subset of $(n+1)$ points on the euclidean sphere S^n. Does there exist an orthogonal transformation $T \in O(n+1)$ such that $f(Tx_i) = f(Tx_j)$ for any $i, j = 1, \dots, n+1$ (i.e., the function f is a constant on the set $\{Tx_i\}_1^{n+1}$)?

It is still an open question in full generality. However, the answer is known to be positive for many special cases. For example, if $\{x_i\}_1^{n+1}$ is an orthonormal basis then the answer is positive (Yamabe-Yujòbô [YY], who extended the similar result of Kakutani [Ka] for $n = 2$). Also if f is an odd function (i.e., $f(-x) = -f(x)$) then the answer is positive for *any* set $\{x_i\}_1^{[n/2]}$ (Geraghty [Ge]).

Statement (conditional). Assuming Knaster's hypothesis to be true, we would have the following estimate in the Dvoretzky's theorem

$$k > \alpha(n, \varepsilon) \log n / \log 1/\varepsilon \ ,$$

where $\alpha(n, \varepsilon) \to 1$ if $n \to \infty$ and $\varepsilon \to 0$.

Proof: Let $N(K; 2\varepsilon)$ be the cardinality of the best (minimal) 2ε-net $\mathcal{N}$ of the unit k-dimensional euclidean sphere S^k. A trivial computation shows that $N(k, 2\varepsilon) \le M(K, \varepsilon)$ where $M(k, \varepsilon)$ is the maximal number of disjoint ε-balls with centers on S^k. Let $(*)_\varepsilon$ be an ε-cap on S^k. Then (see, e.g., [MSch])

$$n = N(k, 2\varepsilon) \le M(k, \varepsilon) \le \frac{\mathrm{Vol}_k S^k}{\mathrm{Vol}_k (*)_\varepsilon} \lesssim \frac{2}{\sqrt{k}} \frac{1}{(\varepsilon/2)^{k-1}} \ .$$

Therefore

$$k \gtrsim \frac{\log n}{\log 2/\varepsilon} \ .$$

To arrive at a better estimate, as stated above, we have to use an advance on a problem of the best packings of ε-caps on a sphere. Using, e.g., a result of [KL], we will get

$$k \ge \alpha(n, \varepsilon) \log n / \log 1/2\varepsilon$$

for some $\alpha(n,\varepsilon) \to 1$ for $n \to \infty$ and $\varepsilon \to 0$.

Consider now an n-dimensional normed space $X = (\mathbb{R}^n, \|\cdot\|)$ and let $\mathbb{R}^n$ be equipped with the standard euclidean structure $(\mathbb{R}^n, |\cdot|)$. Then $f(x) = \|x\|$ is the function on S^{n-1} - the unit euclidean sphere. We assume $n = N(k, 2\varepsilon)$ and fix 2ε-net $\mathcal{N}$, $|\mathcal{N}| = n$, on the unit sphere S^{k-1} of some (fixed) k-dimensional subspace E of $\mathbb{R}^{n-1}$. Assuming the Knaster hypothesis being true, we would find an orthogonal rotation $A \in SO(n)$ such that our function $f(x)$ is constant on a set $A\mathcal{N}$: $f(x) = \text{const}$ for any $x_i \in A\mathcal{N}$. Clearly $A\mathcal{N}$ is a 2ε-net of a k-dimensional section AS^{k-1}. So $\|x_i\| = \text{const}$, $x_i \in A\mathcal{N}$, (and $|x_i| = 1$). By homothetic normalization, we may assume

$$\|x_i\| = 1 = |x_i| .$$

Then $\text{Conv}\{\pm x_i\}_{i=1}^n$ is the largest norm with this property which means that (for an easily computed $c > 0$)

$$\|x\| \le \frac{1}{1 - c\varepsilon^2} |x| .$$

For any $x \in AS^{k-1}$ there exists $x_i \in A\mathcal{N}$ such that $|x - x_i| \le 2\varepsilon$. Then

$$\|x\| \ge \|x_i\| - \|x - x_i\| \ge 1 - \frac{1}{1 - c\varepsilon^2}|x - x_i| \ge 1 - \frac{2\varepsilon}{1 - c\varepsilon^2} .$$

Therefore $d(K \cap AE, D \cap AE) < 1 + 2\varepsilon + O(\varepsilon^2)$.

4. Borsuk-Ulam Theorem and 2-dimensional Dvoretzky's Theorem

The next observation was explained to me by M. Gromov (around 10 years ago). We start with a well known generalization (see, e.g., [N]) of the classical Borsuk Theorem. Consider S^n for n-odd. Then the group $\mathbb{Z}_p$ acts freely on S^n: introduce a complex structure in $\mathbb{R}^{n+1} = \mathbb{C}^k$, $k = (n+1)/2$; then the product by the primitive roots $\sqrt[p]{1}$ of one will define a free action of $\mathbb{Z}_p$. Then the mentioned generalization of the Borsuk Theorem says that for any continuous function $f : S^n \to \mathbb{R}$ and $n = p - 1$ there exists an orbit $\{x_1, \ldots, x_p\}$ for the action such that $f(x_i) = \text{const}$. Clearly $\{x_i\}_1^p$ belong to a 2-dimensional subspace $E \hookrightarrow \mathbb{R}^{n+1}$ and are ε-net on the sphere $S(E) = S^1$ for an $\varepsilon = 2\pi/2p$. An easy 2-dimensional geometry shows now that *any* norm on E such that $\|x_i\| = |x_i| = 1$, $i = 1, \ldots, p$, must satisfy

$$\frac{1}{1 + \frac{\varepsilon^2}{6}} |x| < \|x\| < \frac{1}{1 - \frac{\varepsilon^2}{2}} |x| .$$

Therefore,

Statement. Any $X = (\mathbb{R}^n, \|\cdot\|)$ contains a 2-dimensional subspace E_2 such that

$$d(K \cap E_2, D_2) \leq 1 + \frac{2\pi^2}{3(n+1)^2},$$

where K is the unit ball of X and D_2 is a 2-dimensional euclidean ball.

References

[B] B.J. Birch, Homogeneous forms of odd degree in a large number of variables, Math. 4(1957), 102-105.

[BDGJN] G. Bennett, L.E. Dor, V. Goodman, W.B. Johnson and C.M. Newman, On uncomplemented subspaces of L_p, $1 < p < 2$, Israel J. Math. 26 (1977), 178-187.

[Dv] A. Dvoretzky, Some results on convex bodies and Banach spaces, Proc. Symp. on Linear spaces, Jerusalem 1961, 123-160.

[F] T. Figiel, Local theory of Banach spaces and some operator ideals, Proc. ICM, Warsaw (1983), 961-976.

[FLM] T. Figiel, J. Lindenstrauss and V.D. Milman, The dimension of almost spherical sections of convex bodies, Acta Math., 139 (1977), 53-94.

[Ge] M. Geraghty, Applications of Smith index to some covering and frame theorems, Nederl. Akad. Wetensch. Indag. Math. 23 (1961), 219-228.

[Go] Y. Gordon, Some inequalities for gaussian processes and applications, Israel J. Math. 50 (1985), 265-289.

[Hi] D. Hilbert, Beweis für die Darstellbarkeit der ganzen Zahlen durch eine feate Anzahl n-ten Potenzen, Math. Ann. 67 (1909).

[Hu] A. Hurwitz, Über die Darstellung der ganzen Zahlen als Summen von n-ten Potenzen ganter Zahlen, Math. Ann. 65 (1908), 424-427.

[Ka] S. Katutani, A proof that there exists a circumscribing cube around any bounded closed convex set in $\mathbb{R}^3$, Ann. of Math. 43 (1942), 739-741.

[Kn] B. Knaster, Colloq. Math. 1 (1947), 30-31.

[KL] G.A. Kabatyanskii and V.D. Levenshtein, Bounds for packings on a sphere and in space, Problemy Peredachi Inform. 14, No. 1 (1978), 3-25.

[L] E. Lucas, Nouv. Corresp. Math., 2 (1876), 101.

[M] V.D. Milman, A new proof of the theorem of A. Dvoretzky on sections of convex bodies, Funkcional. Anal i Proložen. 5 (1971), 28-37 (Russian).

[MSch] V.D. Milman and G. Schechtman, Asympototic theory of finite dimensional normed spaces, Springer Lecture Notes #1200 (1986).

[N] M.H.A. Newman, Fixed point and coincidence theorems, J. London Math. Soc. 27 (1952), 135-140.

[Ra] B. Rattray, An antipodal-point, orthogonal point theorem, Ann. of Math. (2) 60 (1954), 502-512.

[S] E. Schmidt, Zum Hilbertschen Baveise des Waringschen Theorems, Math. Ann. 77 (1913), 271-274.

[YY] Yamabe and Yujobô, On the continuous function defined on a sphere. Osaka Math. J. 2 (1950), 19-22.